CALCULUS

A Modern Approach

ABOUT THE AUTHOR

C. O. Oakley received the degrees of B.S. in engineering at the University of Texas, M.S. in mathematics at Brown University, and Ph.D. in mathematics at the University of Illinois. He has taught at all these institutions, as well as at Bryn Mawr College, the University of Delaware, and the University of Washington. Recently retired as head of the Mathematics Department at Haverford College, he is now visiting professor at Villanova University. He has lectured at the universities in Australia, where he spent two years on a Fulbright Fellowship. Professor Oakley has served as a Governor of the Mathematical Association of America and is currently a member of several national mathematical committees. His publications include more than twenty-five mathematical papers in American and foreign learned journals, *Analytic Geometry* (a companion College Outline), and *Sets, Relations and Functions*. He has also published three books (with C. B. Allendoerfer): *Principles of Mathematics, Fundamentals of Freshman Mathematics,* and *Fundamentals of College Algebra.* The first two of these have been translated into Japanese, Serbo-Croatian, and Spanish.

CALCULUS

A Modern Approach

Cletus O. Oakley

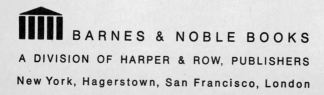
BARNES & NOBLE BOOKS

A DIVISION OF HARPER & ROW, PUBLISHERS

New York, Hagerstown, San Francisco, London

L. C. Catalogue Card Number: 78-146258
ISBN: 0-06-460134-X

85 86 87 88 89 90   20 19 18 17 16 15 14

Published in the United States of America
by Harper & Row, Publishers, Inc.

Preface

An *outline* must of necessity be a condensed version of a subject, but we have tried to cover as much as possible of what is ordinarily called *calculus* in the allotted space. (Some books on calculus are published in two volumes of about 700 pages each.)

For the most part, the topics are what might be covered in a two- or three-semester course. Special features of this book not usually found in calculus texts are the chapters on vectors and elementary differential equations.

The Outline also serves as a self-study guide, and the problems are graded in order of difficulty with the easier ones presented first. There are many problems with complete solutions, and all the answers are given. Some additional theory is provided in the solutions of many of the problems.

I am indebted to James McEnerney and Justine Baker, who were helpful in many capacities, especially with the typing, checking problems and solutions, and proofreading.

Contents

CALCULUS

A Modern Approach

CHAPTER 1

MATHEMATICAL METHODS AND REFERENCE FORMULAS

1.1 Introduction

As a student you may wonder, at times, how to attack a problem. There are some definite mathematical procedures which, if followed, should be of help. These are outlined below in sufficient detail to give you some insight into "how to solve it."

I. *Read the problem—carefully.* Unless you understand the problem, the chances are slight that you will come up with the answer.

II. *What is given?* That is, what are the data that constitute the **hypothesis?** Mathematical theorems are *"if, then"* propositions. *If* something is assumed true, *then* something follows, namely, the *conclusion* you are seeking.

III. *Choose a notation.* Both numerical and abstract mathematical questions are usually formulated in terms of *letters*, the *knowns* and the *unknowns*. Often the proper choice of notation will help to carry the ideas. A complicated and clumsy notation may be a hindrance.

IV. *Determine the connection* between the hypotheses and the unknowns, or variables. Usually this will give rise to *equations*. If possible, draw *diagrams;* they often suggest a method of attack.

V. *Devise a plan for solving* the equations. Make free use of known theorems or related problems. If you cannot solve the main problem directly, it may be possible to solve, first, an auxiliary problem which will then make the original problem readily solvable.

VI. *Restate the problem*, if possible, when you are unable to proceed. It is often possible to prove a *substitute, but equivalent,* problem.

VII. *Generalize the problem.* Frequently, it is very much easier to prove a more general theorem, from which the particular theorem follows, than it is to prove the special case.

VIII. *Carry out the solution.*

IX. *Check your solution.* It may not be the solution! You may have made a mistake, or the problem may not have a solution. This brings up the question of the original data, which may be either insufficient, just exactly sufficient, redundant, or contradictory.

1

X. *Force order into your writing.* Write each line as if it were going to be set in print. Scribble all you like on scratch paper, but maintain complete order on your problem sheet. This is important; it will help you measurably in the total formulation of your ideas.

Certain reference formulas from algebra, plane and solid geometry, trigonometry, and analytic geometry of two and three dimensions will be of great help in the study of the calculus. Some of these formulas are listed here, along with the equations and graphs of many of the standard curves for quick review and reference.

1.2 Algebra

(1) *Quadratic equation.* The *roots* (solutions) of the quadratic equation $ax^2 + bx + c = 0$ are x_1 and x_2, where a, b, and c are real numbers, if

$$x_1 = \frac{-b + \sqrt{b^2 - 4ac}}{2a} \quad \text{and} \quad x_2 = \frac{-b - \sqrt{b^2 - 4ac}}{2a}$$

The expression $\Delta = b^2 - 4ac$ is called the **discriminant**.

(a) If $\Delta > 0$, the roots are *real* and distinct.

(b) If $\Delta = 0$, the roots are *real* and equal, that is, $x_1 = x_2$.

(c) If $\Delta < 0$, the roots are *complex;* i.e., of the form $a \pm ib$, where $i = \sqrt{-1}$. The roots x_1 and x_2 are obtained by *completing the square.* To do this, we first divide $ax^2 + bx + c = 0$ by a so as to reduce the coefficient of x^2 to 1; thus,

$$x^2 + \frac{b}{a}x + \frac{c}{a} = 0, \quad \text{or} \quad x^2 + \frac{b}{a}x = -\frac{c}{a}$$

Next, add to each side of the last equation *the square of half the coefficient of x;* thus

$$x^2 + \frac{b}{a}x + \left(\frac{b}{2a}\right)^2 = \left(\frac{b}{2a}\right)^2 - \frac{c}{a}$$

This completes the square, since now the left-hand side is a *perfect square;* namely, $\left(x + \frac{b}{2a}\right)^2$; hence,

$$\left(x + \frac{b}{2a}\right) = \pm \frac{\sqrt{b^2 - 4ac}}{2a}$$

and
$$x = \frac{-b \pm \sqrt{b^2 - 4ac}}{2a}$$

The expressions for x_1 and x_2 are obtained by taking the positive sign before the radical for x_1 and the negative sign for x_2.

(2) *Factorial notation.* The symbol $n!$, called *n factorial,* stands for the product of the first n (positive) integers; thus,

$$n! = 1 \cdot 2 \cdot 3 \cdots n, \quad \text{but } 0! = 1, \text{ by definition}$$

(3) *Binomial theorem.* The expression of $(a + b)^n$ is,

$$(a + b)^n = a^n + na^{n-1}b + \frac{n(n-1)}{2!}a^{n-2}b^2$$

$$+ \frac{n(n-1)(n-2)}{3!}a^{n-3}b^3 + \cdots$$

$$+ \frac{n(n-1)(n-2)\cdots(n-r+2)}{(r-1)!}a^{n-r+1}b^{r-1}$$

$$+ \cdots + b^n$$

The rth term in this expansion is,

$$\frac{n(n-1)(n-2)\cdots(n-r+2)}{(r-1)!}a^{n-r+1}b^{r-1}$$

(4) *Logarithms*

If $a^b = x$, then, by definition of logarithm, $\log_a x = b$; also,

$$\log_b a = \frac{1}{\log_a b}$$

To any base:

(a) $\log MN = \log M + \log N$

(b) $\log M^n = n \log M$

(c) $\log \dfrac{M}{N} = \log MN^{-1} = \log M - \log N$

(d) $\log \sqrt[n]{M} = M^{1/n} = \dfrac{1}{n} \log M$

(5) *Inequalities*

For each pair of real numbers a and b, exactly one of the following relations is true.

$$a < b, \quad a = b, \quad \text{or} \quad a > b$$

(a) If $a < b$ and $b < c$, then $a < c$.

(b) If $a < b$, then $a + c < b + c$.

(c) If $a < b$ and $0 < c$, then $ac < bc$.

(6) *Determinants.* The left-hand member of the identity,

(a)
$$\begin{vmatrix} a_1 & b_1 \\ a_2 & b_2 \end{vmatrix} = a_1 b_2 - b_1 a_2$$

is called *a determinant of the second order.* It is another, and useful, way of writing the algebraic quantity on the right. Similarly, for determinants of the *third* and *fourth* orders,

(b)
$$\begin{vmatrix} a_1 & b_1 & c_1 \\ a_2 & b_2 & c_2 \\ a_3 & b_3 & c_3 \end{vmatrix} = a_1 b_2 c_3 + b_1 c_2 a_3 + c_1 a_2 b_3 - c_1 b_2 a_3 - b_1 a_2 c_3 - a_1 c_2 b_3$$

(c)
$$\begin{vmatrix} a_1 & b_1 & c_1 & d_1 \\ a_2 & b_2 & c_2 & d_2 \\ a_3 & b_3 & c_3 & d_3 \\ a_4 & b_4 & c_4 & d_4 \end{vmatrix} = a_1 A_1 - a_2 A_2 + a_3 A_3 - a_4 A_4$$

where $A_1 = \begin{vmatrix} b_2 & c_2 & d_2 \\ b_3 & c_3 & d_3 \\ b_4 & c_4 & d_4 \end{vmatrix}$; namely, A_1 is that determinant of the third order which remains when that *row* and that *column* containing a_1 are disregarded. The other A's are determined similarly. Determinants (a) and (b) are said to be expanded into their equivalent algebraic forms on the right-hand side.

(7) *Simultaneous equations*

(a) Two *linear* equations in *two* unknowns

$$a_1 x + b_1 y = c_1$$
$$a_2 x + b_2 y = c_2$$

The solution of these equations in determinant form is

$$x = \frac{\begin{vmatrix} c_1 & b_1 \\ c_2 & b_2 \end{vmatrix}}{D} \qquad y = \frac{\begin{vmatrix} a_1 & c_1 \\ a_2 & c_2 \end{vmatrix}}{D} \qquad D = \begin{vmatrix} a_1 & b_1 \\ a_2 & b_2 \end{vmatrix} \neq 0$$

(b) Three linear equations in *three* unknowns

$$\begin{aligned} a_1 x + b_1 y + c_1 z &= d_1 \\ a_2 x + b_2 y + c_2 z &= d_2 \\ a_3 x + b_3 y + c_3 z &= d_3 \end{aligned} \qquad D = \begin{vmatrix} a_1 & b_1 & c_1 \\ a_2 & b_2 & c_2 \\ a_3 & b_3 & c_3 \end{vmatrix} \neq 0$$

$$x = \frac{\begin{vmatrix} d_1 & b_1 & c_1 \\ d_2 & b_2 & c_2 \\ d_3 & b_3 & c_3 \end{vmatrix}}{D} \quad y = \frac{\begin{vmatrix} a_1 & d_1 & c_1 \\ a_2 & d_2 & c_2 \\ a_3 & d_3 & c_3 \end{vmatrix}}{D} \quad z = \frac{\begin{vmatrix} a_1 & b_1 & d_1 \\ a_2 & b_2 & d_2 \\ a_3 & b_3 & d_3 \end{vmatrix}}{D}$$

(c) One linear and one quadratic equation, each in two unknowns, x and y,

$$ax + by + c = 0$$

$$Ax^2 + Bxy + Cy^2 + Dx + Ey + F = 0$$

Solve the linear equation for one of the variables, say x, in terms of the other and substitute this value for x into the quadratic equation. This will result in one quadratic equation in one letter y, which can be solved by the quadratic equation formula, yielding two possible values of y. Substituting these values back into the linear equation will give two corresponding values of x.

(8) *Three special relations.* Three special ratios that the student should be thoroughly familiar with are: $0/A$, $A/0$, and $0/0$, where $A \neq 0$.

(a) Let $0/A = x$, i.e., $0 = Ax$. There is only one value of x that will yield zero when multiplied by A; namely, zero itself; hence,

$$\frac{0}{A} = 0$$

(b) Let $A/0 = x$, i.e., $A = 0x$. There is no value of x which, when multiplied by zero, will yield A. Division by zero is impossible.

(c) Let $0/0 = x$, i.e., $0 = 0x$. Any number x will satisfy this equation; hence,

$$\frac{0}{0} \text{ is indeterminate}$$

1.3 Geometry

(1) *Radian measure*

(a) A central angle subtended by an arc equal in length to the radius of the circle is called a *radian*.

(b) If r is the radius of a circle and if θ, measured in radians, is the central angle subtended by an arc S, then

$$S = r\theta$$

(c) Relations between degree measure and radian measure:

$$360° = 2\pi \text{ radians} = 1 \text{ revolution (or circumference)}$$
$$1° \cong 0.01745 \text{ radian}; 1 \text{ radian} \cong 57.3°$$

In general,

$$\text{degrees} = \text{radians} \times 180/\pi$$
$$\text{radians} = \text{degrees} \times \pi/180$$

(2) *Mensuration formulas.* Let r denote the radius, θ the central angle in radians, S the arc, h the altitude, b the length of base, s the slant height, and A the area of base.

FIGURE	CIRCUMFERENCE	AREA	VOLUME
(a) Circle	$2\pi r$	πr^2	
(b) Circular sector		$\frac{1}{2} r^2 \theta$	
(c) Triangle		$\frac{1}{2} bh$	
(d) Trapezoid		$\frac{1}{2}(b_1 + b_2)h$	
(e) Right prism			Ah
(f) Right circular cylinder (limiting case of a right prism)	$2\pi rh$		$Ah = \pi r^2 h$
(g) Pyramid			$\frac{1}{3} Ah$
(h) Right circular cone (limiting case of a pyramid)	$\pi rs = \pi r \sqrt{r^2 + h^2}$		$\frac{1}{3} \pi r^2 h$
(i) Sphere	$4\pi r^2$		$\frac{4}{3} \pi r^3$

(j) Similar areas are to each other as the squares of corresponding dimensions.

(k) Similar volumes are to each other as the cubes of corresponding dimensions.

(3) *Pythagorean theorem.* The square on the hypotenuse of any right triangle is equal to the sum of the squares on the two sides: $c^2 = a^2 + b^2$.

(4) *Definitions*

(a) A **median** of a triangle is a line joining the midpoint of a side and the opposite vertex.

(b) A **rhombus** is an equilateral parallelogram.

(c) An **isosceles trapezoid** is a trapezoid in which the nonparallel sides are equal in length and make equal interior angles with the base.

(d) A polygon is *convex* if each interior angle is less than or equal to 180°.

1.4 Trigonometry

(1) *Definitions.* In Fig. 1.1, the circle is of radius 1, θ is the arc length measured (counterclockwise positive, clockwise negative) from $P(1, 0)$, ϕ is an angle in standard position (counterclockwise positive, clockwise negative), usually measured in degrees. Since, in the calculus,

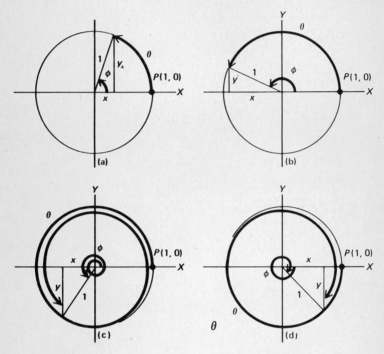

FIG. 1.1

radian measure, which is equivalent to arc length, is much more useful than degree measure, it will be used, for the most part, throughout this book.

(a) $\sin \theta = y$ (d) $\csc \theta = 1/y$

(b) $\cos \theta = x$ (e) $\sec \theta = 1/x$

(c) $\tan \theta = y/x$ (f) $\cot \theta = x/y$

(If the circle is of radius r, we define $\sin \theta = y/r$, $\cos \theta = x/r$, etc.)

(2) *Signs of the trigonometric functions*

QUADRANT	SIN	COS	TAN	CSC	SEC	COT
I	+	+	+	+	+	+
II	+	−	−	+	−	−
III	−	−	+	−	−	+
IV	−	+	−	−	+	−

(3) *Functions of special angles*

QUAD-RANT	DE-GREES	RADI-ANS	SIN	COS	TAN	CSC	SEC	COT
I	0, 360	$0, 2\pi$	0	1	0	−	1	−
	30	$\frac{1}{6}\pi$	$\frac{1}{2}$	$\frac{1}{2}\sqrt{3}$	$\frac{1}{3}\sqrt{3}$	2	$\frac{2}{3}\sqrt{3}$	$\sqrt{3}$
	45	$\frac{1}{4}\pi$	$\frac{1}{2}\sqrt{2}$	$\frac{1}{2}\sqrt{2}$	1	$\sqrt{2}$	$\sqrt{2}$	1
	60	$\frac{1}{3}\pi$	$\frac{1}{2}\sqrt{3}$	$\frac{1}{2}$	$\sqrt{3}$	$\frac{2}{3}\sqrt{3}$	2	$\frac{1}{3}\sqrt{3}$
II	90	$\frac{1}{2}\pi$	1	0	−	1	−	0
	120	$\frac{2}{3}\pi$	$\frac{1}{2}\sqrt{3}$	$-\frac{1}{2}$	$-\sqrt{3}$	$\frac{2}{3}\sqrt{3}$	−2	$-\frac{1}{3}\sqrt{3}$
	135	$\frac{3}{4}\pi$	$\frac{1}{2}\sqrt{2}$	$-\frac{1}{2}\sqrt{2}$	−1	$\sqrt{2}$	$-\sqrt{2}$	−1
	150	$\frac{5}{6}\pi$	$\frac{1}{2}$	$-\frac{1}{2}\sqrt{3}$	$-\frac{1}{3}\sqrt{3}$	2	$-\frac{2}{3}\sqrt{3}$	$-\sqrt{3}$
III	180	π	0	−1	0	−	−1	−
	210	$\frac{7}{6}\pi$	$-\frac{1}{2}$	$-\frac{1}{2}\sqrt{3}$	$\frac{1}{3}\sqrt{3}$	−2	$-\frac{2}{3}\sqrt{3}$	$\sqrt{3}$
	225	$\frac{5}{4}\pi$	$-\frac{1}{2}\sqrt{2}$	$-\frac{1}{2}\sqrt{2}$	1	$-\sqrt{2}$	$-\sqrt{2}$	1
	240	$\frac{4}{3}\pi$	$-\frac{1}{2}\sqrt{3}$	$-\frac{1}{2}$	$\sqrt{3}$	$-\frac{2}{3}\sqrt{3}$	−2	$\frac{1}{3}\sqrt{3}$
IV	270	$\frac{3}{2}\pi$	−1	0	−	−1	−	0
	300	$\frac{5}{3}\pi$	$-\frac{1}{2}\sqrt{3}$	$\frac{1}{2}$	$-\sqrt{3}$	$-\frac{2}{3}\sqrt{3}$	2	$-\frac{1}{3}\sqrt{3}$
	315	$\frac{7}{4}\pi$	$-\frac{1}{2}\sqrt{2}$	$\frac{1}{2}\sqrt{2}$	−1	$-\sqrt{2}$	$\sqrt{2}$	−1
	330	$\frac{11}{6}\pi$	$-\frac{1}{2}$	$\frac{1}{2}\sqrt{3}$	$-\frac{1}{3}\sqrt{3}$	−2	$\frac{2}{3}\sqrt{3}$	$-\sqrt{3}$

$\sqrt{2} = 1.414, \frac{1}{2}\sqrt{2} = 0.707, \sqrt{3} = 1.732, \frac{1}{2}\sqrt{3} = 0.866, \frac{1}{3}\sqrt{3} = 0.577$

(4) *Fundamental identities*

(a) $\sin x = \dfrac{1}{\csc x}$

(b) $\cos x = \dfrac{1}{\sec x}$

(c) $\tan x = \dfrac{1}{\cot x}$

(d) $\tan x = \dfrac{\sin x}{\cos x}$

(e) $\sin^2 x + \cos^2 x = 1$ (f) $1 + \tan^2 x = \sec^2 x$

(g) $1 + \cot^2 x = \csc^2 x$

(5) *Reduction formula rules*

(a) Any trigonometric function of the angle $\left(k\dfrac{\pi}{2} \pm \alpha\right)$ is equal to $\pm$ the same function of α, if k is even, and to $\pm$ the cofunction of α, if k is odd.

(b) The $+$ sign is used if the original function of the original angle $\left(k\dfrac{\pi}{2} \pm \alpha\right)$ is plus; the $-$ sign is used if the original function is negative.

The sign of the original function of $\left(k\dfrac{\pi}{2} \pm \alpha\right)$ is determined by the usual quadrantal conventions. To summarize:

$$\text{Any function of } \left(k\frac{\pi}{2} \pm \alpha\right) = \pm \begin{cases} \text{Same function of } \alpha, \text{ if } k \text{ is even.} \\ \text{Cofunction of } \alpha, \text{ if } k \text{ is odd.} \\ \text{Use sign of original function of} \\ \qquad \left(k\dfrac{\pi}{2} \pm \alpha\right) \end{cases}$$

(6) *Functions of the sum and difference of two angles*

(a) $\sin(x \pm y) = \sin x \cos y \pm \cos x \sin y$

(b) $\cos(x \pm y) = \cos x \cos y \mp \sin x \sin y$

(c) $\tan(x \pm y) = \dfrac{\tan x \pm \tan y}{1 \mp \tan x \tan y}$

(7) *Multiple angle formulas*

(a) $\sin 2x = 2 \sin x \cos x$

(b) $\cos 2x = \cos^2 x - \sin^2 x = 2\cos^2 x - 1 = 1 - 2\sin^2 x$

(c) $\tan 2x = \dfrac{2 \tan x}{1 - \tan^2 x}$

(d) $\sin \dfrac{x}{2} = \sqrt{\dfrac{1 - \cos x}{2}}$

(e) $\cos \dfrac{x}{2} = \sqrt{\dfrac{1 + \cos x}{2}}$

(f) $\tan \dfrac{x}{2} = \sqrt{\dfrac{1 - \cos x}{1 + \cos x}} = \dfrac{1 - \cos x}{\sin x} = \dfrac{\sin x}{1 + \cos x}$

(8) *Sum and product formulas*

(a) $\sin x + \sin y = 2 \sin \frac{1}{2}(x+y) \cos \frac{1}{2}(x-y)$

(b) $\sin x - \sin y = 2 \cos \frac{1}{2}(x+y) \sin \frac{1}{2}(x-y)$

(c) $\cos x + \cos y = 2 \cos \frac{1}{2}(x+y) \cos \frac{1}{2}(x-y)$

(d) $\cos x - \cos y = -2 \sin \frac{1}{2}(x+y) \sin \frac{1}{2}(x-y)$

(e) $\sin x \sin y = \frac{1}{2} \cos(x-y) - \frac{1}{2} \cos(x+y)$

(f) $\sin x \cos y = \frac{1}{2} \sin(x-y) + \frac{1}{2} \sin(x+y)$

(g) $\cos x \cos y = \frac{1}{2} \cos(x-y) + \frac{1}{2} \cos(x+y)$

(9) *Formulas for plane triangles.* Let a, b, c be sides; A, B, C, opposite angles; $s = \dfrac{a+b+c}{2}$, the semiperimeter;

$$r = \sqrt{\frac{(s-a)(s-b)(s-c)}{s}}$$

the radius of the inscribed circle; R, the radius of the circumscribed circle; and K, the area.

(a) Law of sines: $\dfrac{a}{\sin A} = \dfrac{b}{\sin B} = \dfrac{c}{\sin C} = 2R$

(b) Law of cosines: $a^2 = b^2 + c^2 - 2bc \cos A$

(c) Law of tangents: $\dfrac{a+b}{a-b} = \dfrac{\tan \frac{1}{2}(A+B)}{\tan \frac{1}{2}(A-B)}$

(d) Tangent of half angle: $\tan \frac{1}{2}A = \dfrac{r}{s-a}$

(e) Area:

$$K = \frac{1}{2}ab \sin C = \sqrt{s(s-a)(s-b)(s-c)} = rs = \frac{abc}{4R}$$

(10) *Relation between trigonometric and exponential functions*

(a) $$\sin x = \frac{e^{ix} - e^{-ix}}{2i}$$

(b) $$\cos x = \frac{e^{ix} + e^{-ix}}{2}$$

Note that $i = \sqrt{-1}$.

(11) *Hyperbolic functions*

(a) $$\sinh x = \frac{e^x - e^{-x}}{2}$$

(b) $$\cosh x = \frac{e^x + e^{-x}}{2}$$

1.5 Plane Analytic Geometry

(1) *Distance between two points* $d = \sqrt{(x_2 - x_1)^2 + (y_2 - y_1)^2}$

(2) *Slope of line through two points* $m = \dfrac{y_2 - y_1}{x_2 - x_1}$

(3) *Angle between two lines*

(a) $\tan \theta_{12} = \dfrac{m_2 - m_1}{1 + m_1 m_2}$, where θ_{12} designates the angle from line 1 to line 2 (counterclockwise), and m_1 and m_2 are the slopes of lines 1 and 2, respectively.

(b) $\cos \theta = \lambda_1 \lambda_2 + \mu_1 \mu_2$, where λ_1 is the cosine of the angle between the first line and the X-axis and μ_1 is the cosine of the angle between the first line and the Y-axis, and similarly for λ_2 and μ_2.

(4) *Two lines are parallel* if $m_1 = m_2$ (slopes) or if $\lambda_1 \lambda_2 + \mu_1 \mu_2 = 1$, where the λ's and μ's are called **direction cosines.**

(5) *Two lines are perpendicular* if $m_1 = -\dfrac{1}{m_2}$ (slopes) or if $\lambda_1 \lambda_2 + \mu_1 \mu_2 = 0$ (direction cosines).

(6) *Forms of the equation of a straight line and their graphs* (Figs. 1.2-1.5).

(a) General $ax + by + c = 0$

(b) Two-point $\dfrac{y - y_1}{x - x_1} = \dfrac{y_2 - y_1}{x_2 - x_1}$ (Fig. 1.2)

(c) Point-slope $y - y_1 = m(x - x_1)$

(d) Slope-intercept $y = mx + b$ (Fig. 1.3)

(e) Two-intercept $\dfrac{x}{a} + \dfrac{y}{b} = 1$ (Fig. 1.4)

(f) Parallel to Y-axis $x = k$

(g) Parallel to X-axis $y = k$

(h) Normal form $x \cos \theta + y \sin \theta - p = 0$ (Fig. 1.5)

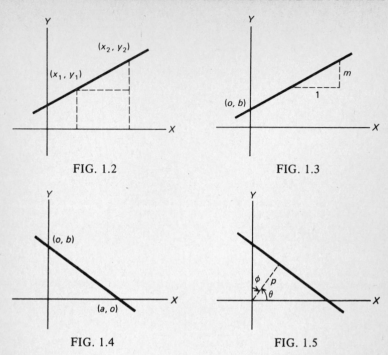

FIG. 1.2

FIG. 1.3

FIG. 1.4

FIG. 1.5

Note that (h) could also be written $x \cos\theta + y \cos\phi - p = 0$, where ϕ is the angle between the normal and the Y-axis. Or, writing $\lambda = \cos\theta$ and $\mu = \cos\phi$, we have $\lambda x + \mu y - p = 0$, λ and μ being **direction cosines** of the normal to the line. Note also that $\lambda^2 + \mu^2 = 1$.

(7) *To reduce the general equation $ax + by + c = 0$ to normal form,* divide through by $\pm\sqrt{a^2 + b^2}$ and choose the sign opposite to that of c.

(8) *The distance from a line $ax + by + c = 0$ to a point $P(x_1, y_1)$ is* given by

$$d = \frac{ax_1 + by_1 + c}{\pm\sqrt{a^2 + b^2}}$$

(9) *Standard forms of the conic sections* (Figs. 1.6–1.13)

(a) PARABOLA

Equation	$y^2 = 4px$	$(y - k)^2 = 4p(x - h)$
Coordinates of vertex	$V(0, 0)$	$V(h, k)$
Coordinates of focus	$F(p, 0)$	$F(h + p, k)$

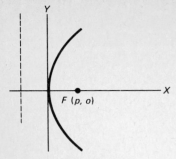

FIG. 1.6

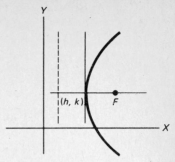

FIG. 1.7

Equation of directrix D	$x = -p$	$x = h - p$
Length of latus rectum	$LL' = 4p$	$LL' = 4p$

(b) CIRCLE

Equation	$x^2 + y^2 = r^2$	$(x - h)^2 + (y - k)^2 = r^2$
Coordinates of center	$C(0, 0)$	$C(h, k)$
Radius	r	r

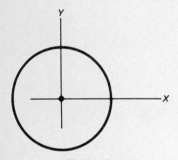

FIG. 1.8

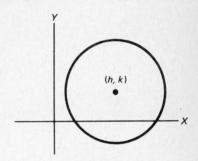

FIG. 1.9

(c) ELLIPSE

Equation	$\dfrac{x^2}{a^2} + \dfrac{y^2}{b^2} = 1$	$\dfrac{(x - h)^2}{a^2} + \dfrac{(y - k)^2}{b^2} = 1$
Coordinates of vertices	$V(a, 0), V'(-a, 0)$	$V(h + a, k), V'(h - a, k)$
Coordinates of foci	$F(ae, 0), F'(-ae, 0)$	$F(h + ae, k), F'(h - ae, k)$
Coordinates of center	$C(0, 0)$	$C(h, k)$

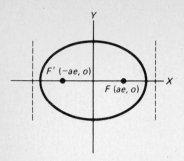

FIG. 1.10   FIG. 1.11

Equations of directrices D $\qquad x = \pm \dfrac{a}{e}$ $\qquad\qquad x = h \pm \dfrac{a}{e}$

Semimajor axis $\qquad a$

Semiminor axis $\qquad b$

Eccentricity $\qquad e = \dfrac{\sqrt{a^2 - b^2}}{a} < 1$

Length of latus rectum $\qquad LL' = \dfrac{2b^2}{a}$

(d) HYPERBOLA

Equation $\qquad \dfrac{x^2}{a^2} - \dfrac{y^2}{b^2} = 1 \qquad \dfrac{(x - h)^2}{a^2} - \dfrac{(y - k)^2}{b^2} = 1$

Coordinates of vertices $\qquad V(a,0), V'(-a,0) \qquad V(h + a, k), V'(h - a, k)$

Coordinates of foci $\qquad F(ae, 0), F'(-ae, 0) \quad F(h + ae, k), F'(h - ae, k)$

Coordinates of center $\qquad C(0, 0) \qquad\qquad C(h, k)$

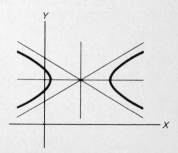

FIG. 1.12   FIG. 1.13

Equations of directrices D $x = \pm \dfrac{a}{e}$ $x = h \pm \dfrac{a}{e}$

Equations of asymptotes $y = \pm \dfrac{b}{a} x$ $y - k = \pm \dfrac{b}{a}(x - h)$

Semitransverse axis a

Semiconjugate axis b

Eccentricity $e = \dfrac{\sqrt{a^2 + b^2}}{a} > 1$

Length of latus rectum $LL' = \dfrac{2b^2}{a}$

(e) Polar equation of a conic with focus at the pole and directrix perpendicular to the polar axis,

$$\rho = \frac{\pm ep}{1 \mp e \cos \theta}$$

1.6 Solid Analytic Geometry

(1) *Distance between two points:* $P_1(x_1, y_1, z_1)$ and $P_2(x_2, y_2, z_2)$ is given by

$$d = \sqrt{(x_2 - x_1)^2 + (y_2 - y_1)^2 + (z_2 - z_1)^2}$$

(2) *Direction cosines* $\lambda, \mu,$ *and* ν *of a line*

$$\lambda = \cos \alpha, \mu = \cos \beta, \nu = \cos \gamma$$

where α, β, and γ are the angles made by the line and the positive directions of the coordinate axes X, Y, Z, respectively, and $\lambda^2 + \mu^2 + \nu^2 = 1$. For some purposes, especially in physics, it is often useful to use a right-handed system of coordinates in which the X- and Y-axes are interchanged.

(3) *Direction numbers* a, b, c are proportional to the direction cosines; thus, $a = k\lambda$, $b = k\mu$, and $c = k\nu$.

(4) *Equation of a plane* (Figs. 1.14 and 1.15)

(a) General $ax + by + cz + d = 0$, where a, b, c are direction numbers of a line perpendicular to the plane.

(b) Intercept $\dfrac{x}{a} + \dfrac{y}{b} + \dfrac{z}{c} = 1$

(c) Normal form $\lambda x + \mu y + \nu z - p = 0$

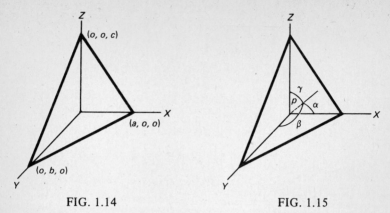

FIG. 1.14   FIG. 1.15

To reduce the general equation $ax + by + cz + d = 0$ to normal form, divide through by $\pm\sqrt{a^2 + b^2 + c^2}$, and choose the sign opposite to that of d; hence,

$$\frac{ax + by + cz + d}{\pm\sqrt{a^2 + b^2 + c^2}} = 0$$

(5) *The distance from a plane* $ax + by + cz + d = 0$ *to a point* $P(x_1\,y_1\,z_1)$ is given by

$$D = \frac{ax_1 + by_1 + cz_1 + d}{\pm\sqrt{a^2 + b^2 + c^2}}$$

(6) *Equations of a line*

(a) Intersection of two planes:

$$a_1x + b_1y + c_1z + d_1 = 0$$
$$a_2x + b_2y + c_2z + d_2 = 0$$

For this line

$$\lambda : \mu : \nu = \begin{vmatrix} b_1 & c_1 \\ b_2 & c_2 \end{vmatrix} : - \begin{vmatrix} a_1 & c_1 \\ a_2 & c_2 \end{vmatrix} : \begin{vmatrix} a_1 & b_1 \\ a_2 & b_2 \end{vmatrix}$$

(b) Through (x_1, y_1, z_1) with direction numbers a, b, c

$$\frac{x - x_1}{a} = \frac{y - y_1}{b} = \frac{z - z_1}{c}$$

This is sometimes called the symmetric form.

(c) Through two points

$$\frac{x - x_1}{x_2 - x_1} = \frac{y - y_1}{y_2 - y_1} = \frac{z - z_1}{z_2 - z_1}$$

Note that here

$$\lambda : \mu : \nu = (x_2 - x_1) : (y_2 - y_1) : (z_2 - z_1)$$

(7) *Angle between two lines*

$$\cos \theta = \lambda_1 \lambda_2 + \mu_1 \mu_2 + \nu_1 \nu_2$$

The lines are parallel if $\lambda_1 \lambda_2 + \mu_1 \mu_2 + \nu_1 \nu_2 = 1$

The lines are perpendicular if $\lambda_1 \lambda_2 + \mu_1 \mu_2 + \nu_1 \nu_2 = 0$

(8) *The angle between two planes* is given by the angle between normals to the planes.

(9) *Standard forms of the quadric surfaces* (Figs. 1.16-1.23)

(a) SPHERE $x^2 + y^2 + z^2 = r^2$
(Fig. 1.16)

(b) ELLIPSOID $\dfrac{x^2}{a^2} + \dfrac{y^2}{b^2} + \dfrac{z^2}{c^2} = 1$
(Fig. 1.17)

(c) HYPERBOLOID OF ONE SHEET (Fig. 1.18)

$$\frac{x^2}{a^2} + \frac{y^2}{b^2} - \frac{z^2}{c^2} = 1$$

(d) HYPERBOLOID OF TWO SHEETS (Fig. 1.19)

$$\frac{x^2}{a^2} - \frac{y^2}{b^2} - \frac{z^2}{c^2} = 1$$

(e) ELLIPTIC PARABOLOID (Fig. 1.20)

$$\frac{x^2}{a^2} + \frac{y^2}{b^2} = cz$$

(f) HYPERBOLIC PARABOLOID (Fig. 1.21)

$$\frac{x^2}{a^2} - \frac{y^2}{b^2} = cz$$

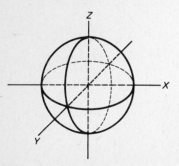

FIG. 1.16

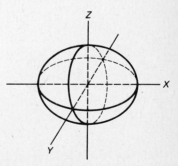

FIG. 1.17

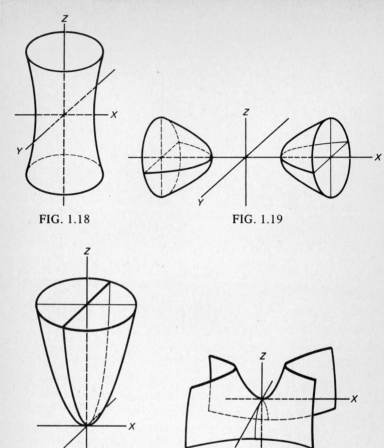

FIG. 1.18

FIG. 1.19

FIG. 1.20

FIG. 1.21

(g) ELLIPTIC CONE
(Fig. 1.22)

$$\frac{x^2}{a^2} + \frac{y^2}{b^2} - \frac{z^2}{c^2} = 0$$

(h) ELLIPTIC CYLINDER
(Fig. 1.23)

$$\frac{x^2}{a^2} + \frac{y^2}{b^2} = 1$$

Note that the equation of the elliptic cylinder contains only two variables and that the cylinder extends in the direction of the axis of the missing variable. This is typical: the equation of any cylindrical surface whose elements are parallel to one of the coordinate axes will not contain the variable of that axis.

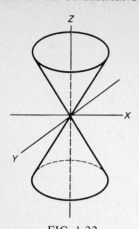

FIG. 1.22

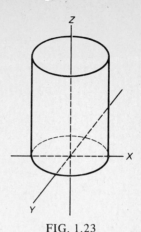

FIG. 1.23

1.7 Principles of Graphing

The graph of the equation $y = f(x)$, or of the equation $F(x,y) = 0$, is the set of points whose **rectangular coordinates** (x_i, y_i) satisfy the equation. Similarly, the graph of $\rho = f(\theta)$, or of $F(\rho, \theta) = 0$, is the set of points whose **polar coordinates** (ρ_i, θ_i) satisfy the equation (Figs. 1.24 and 1.25). In either rectangular or polar coordinates parametric equations might be used: $x = f(t), y = g(t)$; or $\rho = f(t), \theta = g(t)$. In either case, corresponding values of x and y, and ρ and θ are determined by assigning arbitrary values to the parameter t; these number pairs (x,y) and (ρ, θ) are then plotted as before. Often a graph is indicated by a curve drawn through a few of the plotted points. The main items of consideration in the tracing of a curve (plotting a graph) are:

(1) *Intercepts.* Plot the points $(x_i, 0)$ and $(0, y_j)$; that is, the points of the graph lying on the coordinate axes.

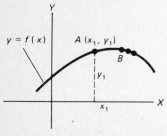

FIG. 1.24

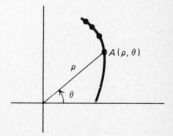

FIG. 1.25

(2) *Symmetry*

(a) The graph is symmetric with respect to the X-axis, if the equation remains unchanged when $-y$ is substituted for $+y$.

(b) The graph is symmetric with respect to the Y-axis, if the equation remains unchanged when $-x$ is substituted for $+x$.

(c) The graph is symmetric with respect to the origin, if the equation remains unchanged when $-x$ and $-y$ are substituted for $+x$ and $+y$, respectively.

(3) *Extent*

(a) Determine whether the graph lies wholly in a finite portion of space.

(b) Indicate the points of discontinuity.

(c) Determine and sketch the asymptotes.

(d) To plot the graph of the equation $z = f(x, y)$ we may take (rectangular coordinates) a right-handed system of axes as in Fig. 1.26, or a left-handed system as in Fig. 1.27. Cylindrical coordinates may be indicated (Fig. 1.28), where the given equation is of the form $z = f(\rho, \theta)$. Again, a spherical system of coordinates is used (Fig. 1.29), where the

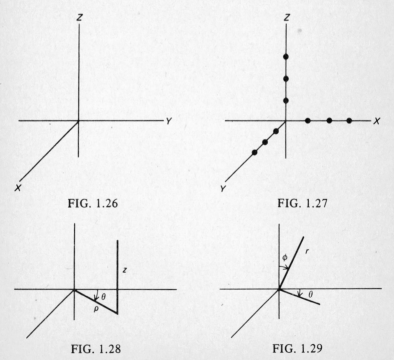

FIG. 1.26   FIG. 1.27

FIG. 1.28   FIG. 1.29

equation is of the form $r = f(\theta, \phi)$. No matter which system is used, care must be exercised in the choice of units. For example, note that in Fig. 1.27 the x and the z units are geometrically equal in length, since these axes lie in the plane of the paper, but the y unit is indicated by a smaller segment, since there is foreshortening in the y direction. The graph is the set of points whose coordinates satisfy the equation. When connected appropriately, a few points suggest that the graph is a surface. It is worth mentioning that the surfaces drawn will not be in true perspective, but nevertheless will look something like perspective figures. The main items of consideration in the sketching of surfaces are:

(4) *Intercepts*

(a) Plot the points lying on the coordinate axes.

(b) Plot the points lying in the coordinate planes. The curves joining these points are called the **traces**.

(c) Determine the character of the traces in planes parallel to the coordinate planes, even though these curves may not be included in the graph as finally sketched.

(5) *Symmetry*

The graph is symmetric with respect to the XY-plane if the equation remains unchanged when $-z$ is substituted for $+z$. Similarly, for the other coordinate planes.

(6) *Extent* (See (3) (a) above.)

1.8 Graphs for Reference

Students should have at least an acquaintance with certain other equations and their graphs, which are reproduced in Figs. 1.30 through 1.61.

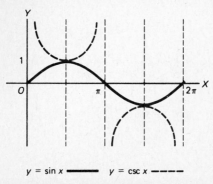

$y = \sin x$ ——— $y = \csc x$ – – – –

FIG. 1.30

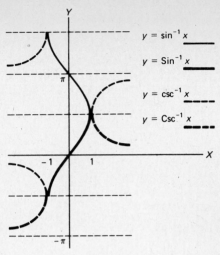

FIG. 1.31

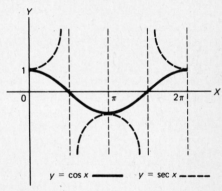

FIG. 1.32

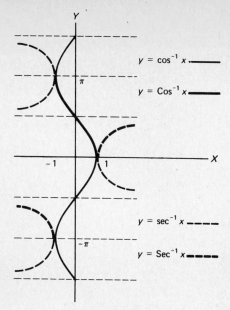

FIG. 1.33

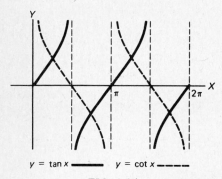

FIG. 1.34

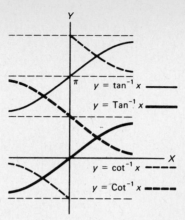

Fig. 1.35

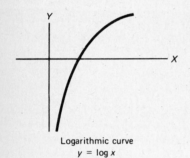

Logarithmic curve
$y = \log x$
Fig. 1.36

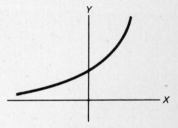

Exponential curve
$y = e^x$
Fig. 1.37

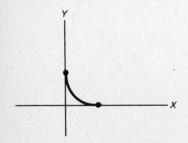

Parabola tangent to axes
$x^{\frac{1}{2}} + y^{\frac{1}{2}} = a^{\frac{1}{2}}$
Fig. 1.38

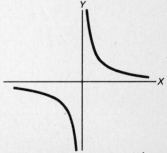

Rectangular hyperbola $\quad y = \dfrac{1}{x}$
Fig. 1.39

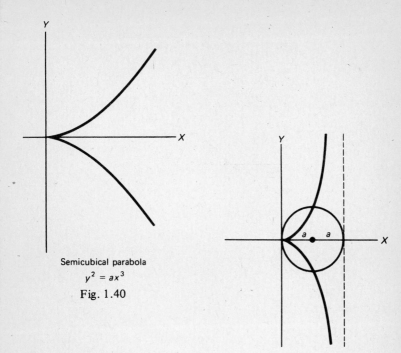

Semicubical parabola
$$y^2 = ax^3$$
Fig. 1.40

Cissoid of Diocles
$$y^2 = \frac{x^3}{2a - x}$$
Fig. 1.41

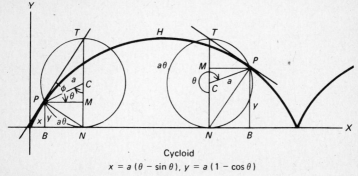

Cycloid
$$x = a\,(\theta - \sin\theta),\ y = a\,(1 - \cos\theta)$$
Fig. 1.42

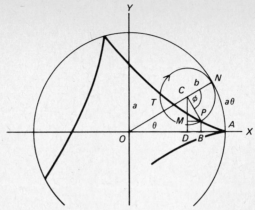

Hypocycloid

$$x = (a - b) \cos \theta + b \cos \frac{a - b}{b} \theta$$

$$y = (a - b) \sin \theta - b \sin \frac{a - b}{b} \theta$$

Fig. 1.43

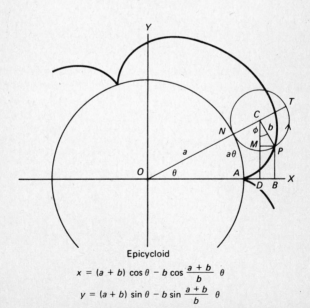

Epicycloid

$$x = (a + b) \cos \theta - b \cos \frac{a + b}{b} \theta$$

$$y = (a + b) \sin \theta - b \sin \frac{a + b}{b} \theta$$

Fig. 1.44

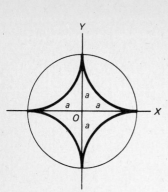

Astroid
(Hypocycloid of four cusps)
$$x^{\frac{2}{3}} + y^{\frac{2}{3}} = a^{\frac{2}{3}}$$
Fig. 1.45

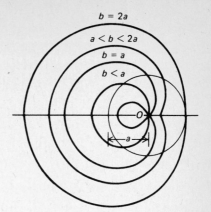

Limaçon of Pascal
$$\rho = b - a \cos \theta$$
(cardioid if $b = a$. See Fig. 1.47.)
Fig. 1.46

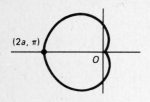

Cardioid
$$\rho = a (1 - \cos \theta)$$
Fig. 1.47

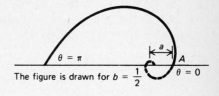

Logarithmic, or equiangular, spiral
$$\rho = ae^{b\theta}$$
Fig. 1.48

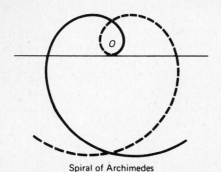

Spiral of Archimedes
$$\rho = a\theta$$
Fig. 1.49

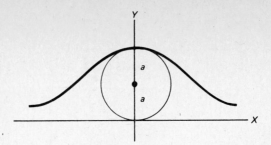

Witch of Agnesi

$$y = \frac{8a^3}{4a^2 + x^2}$$

Fig. 1.50

Folium of Descartes

$$x^3 + y^3 - 3axy = 0$$

Fig. 1.51

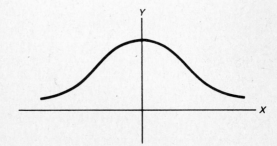

Probability curve

$$y = e^{-x^2}$$

Fig. 1.52

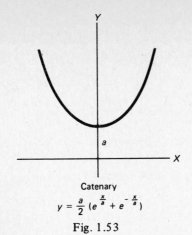

Catenary

$$y = \frac{a}{2}(e^{\frac{x}{a}} + e^{-\frac{x}{a}})$$

Fig. 1.53

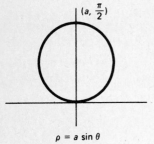

$\rho = a \sin \theta$

Fig. 1.54

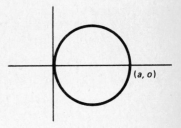

$\rho = a \cos \theta$

Fig. 1.55

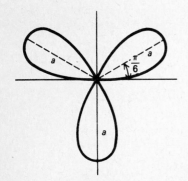

Three-leaved rose

$p = a \sin 3\theta$

Fig. 1.56

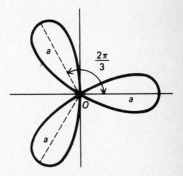

Three-leaved rose

$\rho = \cos 3\theta$

Fig. 1.57

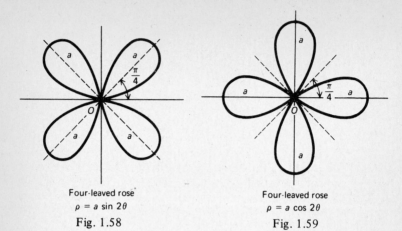

Four-leaved rose
$\rho = a \sin 2\theta$
Fig. 1.58

Four-leaved rose
$\rho = a \cos 2\theta$
Fig. 1.59

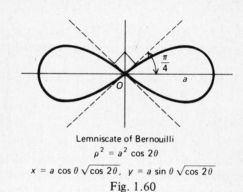

Lemniscate of Bernouilli
$\rho^2 = a^2 \cos 2\theta$
$x = a \cos \theta \sqrt{\cos 2\theta}, \quad y = a \sin \theta \sqrt{\cos 2\theta}$
Fig. 1.60

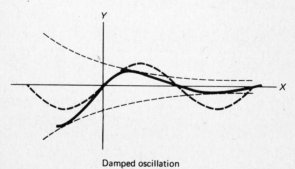

Damped oscillation
$y = e^{-x} \sin x$
Fig. 1.61

CHAPTER 2
SETS, RELATIONS, AND FUNCTIONS

2.1 Sets

The term **set** is usually undefined in mathematics, but can be described as a well-defined collection of things. Thus, we may speak of the set of letters of the English alphabet, written $\{a, b, c, \ldots, z\}$. The letters themselves are **elements** of the set. The set of prime numbers not exceeding 13 is written $\{2, 3, 5, 7, 11, 13\}$. The set of numbers x such that $x^2 = 1$ is, when written in set-builder notation, $\{x | x^2 = 1\}$; the vertical bar | means *such that*. Another way of describing this set is, of course, $\{-1, 1\}$.

Set A is a **subset** of set B if and only if each element of A is an element of B. The notation for this is $A \subset B$, which means A *is contained in B*, or A *is a subset of B*. To indicate that x is an element of A, we write $x \in A$. The **null set** ϕ contains no element; it is also called the **void**, or **empty set**. For "if and only if," we often write "iff."

2.2 Cartesian Product

If X is the set of real numbers and if Y is the set of real numbers, then the set $X \times Y$, called the **Cartesian product**, consists of the set of all ordered pairs of real numbers (x, y), where $x \in X$ and $y \in Y$. Or, stated geometrically, $X \times Y$, the Cartesian product of the real line with itself, is *the set of all points in the plane.*

2.3 Relations

A **relation** is a *subset* of the set $X \times Y$ of ordered pairs (x, y) of real numbers, and the graph is simply the plot of the ordered pairs. The set of first members x of the ordered pairs (x, y) is called the **domain** of the relation. The set of second members y is called the **range**, and each y is the **image** of an x. Conversely, each x is said to be the *inverse* image of a y. Some rule (equation, inequality, graph, table, etc.) determines which y's correspond to which x's in the ordered pairs.

2.4 Functions

A real-valued *function* is a special kind of relation; namely, a set of ordered pairs of real numbers such that no two of the ordered pairs

31

have the same first member. Some notations for a function f are

$f : (x,y)$ f is the set of ordered pairs (x,y).

$f : X \longrightarrow Y$, or $X \xrightarrow{f} Y$ f is a mapping of X onto Y; that is, f assigns to each $x \in X$ a unique $y \in Y$.

$f = \{(x,y) | y = f(x)\}$ f is the set of ordered pairs (x,y) such that y is determined by the equation (rule) $y = f(x)$.

$f : [x, f(x)]$ f is the set of ordered pairs $[x, f(x)]$

$y = f(x)$ y is the *value* of f at x.

Sometimes x is called the **independent variable** and y is called the **dependent variable**, or the *value* of f at x, which is the way $f(x)$ is read. The notations for domain and range of f are, respectively, D_f and R_f. If a is an element of D_f, then $f(a)$ is an element of R_f. If function f, determined by $z = f(x,y)$, is the set of ordered triplets (x,y,z), then no two ordered triplets have the same first two members.

(1) *Operations with functions.* If f and g are functions on the set X of real numbers, then for any $x \in X$, $f(x) + g(x) = g(x) + f(x)$; $f(x)g(x) = g(x)f(x)$; $f(x)/g(x) = f(x)[1/g(x)] = [1/g(x)]f(x)$, if $g(x) \neq 0$; and $cf(x) = f(cx)$, where $c =$ constant.

(2) *Inverse functions.* If a function $f(x,y)$ is such that no two ordered pairs (x,y) have the same second element y, then there is an associated *inverse* function, $f^{-1}(y,x)$, where the domain and range of f^{-1} are the range and domain, respectively, of f. Thus, if f is given by $y = f(x)$, then f^{-1} is obtained by solving $y = f(x)$ for x in terms of y; hence, $f^{-1}y = f^{-1}[f(x)] = x$. To express f^{-1} in the usual notation, using x as independent variable, we must then interchange letters. For example, if f is given by $y = 3x - 5$, then f^{-1} is given by $x = (y + 5)/3$ with y the independent variable. Therefore, if $y = 3x - 5$, then $f^{-1}(y) = f^{-1}(3x - 5) = x$; that is, if $y = f(x)$, then $f^{-1}(y) = f^{-1}[f(x)] = x$. The variables, of course, may be interchanged for convenience; hence, $x = (y + 5)/3$ may be rewritten as $y = (x + 5)/3$, which is now in terms of a function $y = f(x)$ instead of an inverse function $x = f^{-1}(y)$.

(3) *A special function*, which is extremely important in the calculus, is the **exponential function** $f(x) = e^x$. The *inverse* exponential function is called the **natural logarithm** which is written $\log_e$ or $\ln$; it is customary to drop the subscript e and just to write **log**. Thus, if $y = e^x$, then $f^{-1}(y) = \log y = \log e^x = x$, or $x = \log y$; if we think of the log as a function instead of an inverse function, we may write $y = \log x$.

(4) *Inverse trigonometric functions* are of great importance in the calculus. But, since the trigonometric functions are periodic, their in-

verses exist only as relations and not as functions. To overcome this difficulty, we define new trigonometric functions over a restricted domain. For example, we define Sin x (read "Cap-sin x") to be the same as sin x with the domain restricted to the interval $\left[-\dfrac{\pi}{2}, \dfrac{\pi}{2}\right]$. Thus,

$$\text{Sin } x = \sin x \text{ for } -\frac{\pi}{2} \leqslant x \leqslant \frac{\pi}{2}$$

But Sin x is not defined outside of this interval, which is the **domain of definition**. Since the inverse sine exists, we may write $y =$ Sin $x(-\pi/2 \leqslant x \leqslant \pi/2, -1 \leqslant y \leqslant 1)$, thus $f^{-1}(y) = \text{Sin}^{-1} y = x$, or $x =$ Sin$^{-1} y$. The inverse sine function is sometimes referred to as the **arc sine**; hence, Sin$^{-1} y = $ Arcsin y. The inverse sine function may also be written as a function; thus, $f(x) = \text{Sin}^{-1} x = $ Arcsin x. The inverse functions of the cosine, tangent, cotangent, secant, and cosecant are similarly constructed. The table below gives the *principal values* for trigonometric functions and their inverses.

FUNCTION	DOMAIN	RANGE
Sin x	$\left[-\dfrac{\pi}{2}, \dfrac{\pi}{2}\right]$	$[-1, 1]$
Sin^{-1}x	$[-1, 1]$	$\left[-\dfrac{\pi}{2}, \dfrac{\pi}{2}\right]$
Cos x	$[0, \pi]$	$[-1, 1]$
Cos^{-1}x	$[-1, 1]$	$[0, \pi]$
Tan x	$\left(-\dfrac{\pi}{2}, \dfrac{\pi}{2}\right)$	$(-\infty, \infty)$
Tan^{-1}x	$(-\infty, \infty)$	$\left(-\dfrac{\pi}{2}, \dfrac{\pi}{2}\right)$
Cot x	$(0, \pi)$	$(-\infty, \infty)$
Cot^{-1}x	$(-\infty, \infty)$	$(0, \pi)$
Sec x	$\left[-\pi, -\dfrac{\pi}{2}\right) ; \left(-\dfrac{\pi}{2}, 0\right]$	$(-\infty, -1] ; [1, \infty)$
Sec^{-1}x	$(-\infty, -1] ; [1, \infty)$	$\left[-\pi, -\dfrac{\pi}{2}\right) ; \left(-\dfrac{\pi}{2}, 0\right]$
Csc x	$\left[-\dfrac{\pi}{2}, 0\right) ; \left(0, \dfrac{\pi}{2}\right]$	$(-\infty, -1] ; [1, \infty)$
Csc^{-1}x	$(-\infty, -1] ; [1, \infty)$	$\left[-\dfrac{\pi}{2}, 0\right) ; \left(0, \dfrac{\pi}{2}\right]$

(5) *A composite function* is a function of a function; hence, if F is a real-valued composite function, the domain D_F of F is the set Y of real numbers y determined by a function, say g, such that $y = g(x)$, and the range is determined by another function f, so that $F = f[g(x)]$. For example, if $y = g(x) = \sqrt{x+1}$ and $f(y) = f[g(x)] = y^2 + 3$, then

$$F = f(y) = y^2 + 3 = f[g(x)] = [g(x)]^2 + 3 = (\sqrt{x+1})^2 + 3 = x + 4$$

Similarly, if $G = g[f(y)]$ and $x = f(y)$, then

$$G = g(x) = \sqrt{x+1} = g[f(y)] = \sqrt{f(y)+1} = \sqrt{(y^2+3)+1} = \sqrt{y^2+4}$$

The domain D_F of F, therefore, consists of all $x \geqslant -1$, since $(x + 1)$ under the radical must be either zero or a positive number; i.e., a real number, and the domain D_G of G consists of all values of $y : -\infty < y < \infty$. The expression $y^2 + 4$ under the radical is always $\geqslant 4$ and, therefore, satisfies the requirement of being a positive, real number. The notion of the composite function is of special importance in the calculus.

2.5 Special Functions

The calculus makes constant use of several special functions.

(1) *The absolute value function* $(x, |x|)$, where

$$|x| = \begin{cases} x, 0 \leqslant x \\ -x, x < 0 \end{cases}$$

(2) *The constant function* (x, k), where k is a constant.

(3) *The identity function* (x, x)

(4) *The polynomial function* f given by

$$f(x) = a_0 x^n + a_1 x^{n-1} + \cdots + a_n$$

This function is generated by using the constant function, the identity function, and the operations of addition and multiplication.

(5) *The rational function* f given by

$$f(x) = \frac{P(x)}{Q(x)} = \frac{a_0 x^n + a_1 x^{n-1} + \cdots + a_n}{b_0 x^m + b_1 x^{m-1} + \cdots + b_m}$$

where P and Q are polynomial functions, and $Q(x) \neq 0$.

PROBLEMS WITH SOLUTIONS

1. If $f(x) = x^3$, find $f(1), f(-1), f(2)$, and $f(a)$.
   Solution. $f(1) = 1, f(-1) = -1, f(2) = 8$, and $f(a) = a^3$.

2. If $f(x) = x - \dfrac{1}{x}$, find $f(1), f(0), f(-2), f(100)$, and $f(1 + a)$.
   Solution. $f(1) = 0$, $f(0)$ does not exist (0 is not in the domain of
   f), $f(-2) = -3/2, f(100) = 99.99$, and $f(1 + a) = (1 + a) - \dfrac{1}{1 + a}$.

3. Find the domain, the range, and the function f given by $y = x^2$.
   Solution. The domain is the set of real numbers, the range is the
   set of nonnegative numbers, and the function f is the set of ordered pairs (x, x^2).

4. The equation $y = \sqrt{1 - x}$ defines a function $f(x, \sqrt{1 - x})$. Find the
   domain and range of f.
   Solution. The domain of f is $x \leqslant 1$, and the range of f is the set of
   nonnegative, real numbers.

5. For $f(x, \dfrac{1}{2} \cos x)$, find the domain and the range of f.
   Solution. The domain consists of all real numbers, and the range
   is $-\dfrac{1}{2} \leqslant x \leqslant \dfrac{1}{2}$.

6. If $g(x) = 3^x$, find $g(3), g(2), g(-4)$, and the domain and range of g.
   Solution. $g(3) = 27, g(2) = 9, g(-4) = 1/81$, the domain of g consists of all the reals, and the range of g consists of the positive
   reals.

7. The equation $y = (1 + x)/(1 - x)$ defines a function f. Find the
   domain of f.
   Solution. The domain of f consists of the set of reals except for 1
   (the denominator $1 - x$ cannot be zero).

8. Find several functions defined by the equation $x^2 + y^2 = 4$.
   Solution. One function is given by $y = +\sqrt{4 - x^2}$, another by $y = -\sqrt{4 - x^2}$, and still another is defined by the several equations
   and conditions:

$$y = \begin{cases} \sqrt{4 - x^2}, & -2 \leqslant x \leqslant 0 \\ -\sqrt{4 - x^2}, & 0 \leqslant x \leqslant 2 \end{cases}$$

(The case is similar for x as a function of y.)

9. Some of the ordered pairs of the function $h(t, \log_{10} t)$ are $(1,0)$, $(10,1)$, $(100,2)$, $(0.1,-1)$, $(0.01,-2)$, and $(10^{-5},-5)$. Find the domain and range of h.

   Solution. The domain of h is $0 < t$, and the range of h consists of all the reals.

10. If $z = x + 2y$, find $z(1,3)$, $z(1,-2)$, and $z(1/t,t)$.

    Solution. $z(1,3) = 7$, $z(1,-2) = -3$, and $z(1/t,t) = \dfrac{1}{t} + 2t$.

11. If $f(x) = \dfrac{2x}{1 + 3x}$ and $g(x) = x^2 + 2x + 3$, find the composite functions $g \circ f$ and $f \circ g$.

    Solution. The composite function $f \circ g$ is given by $y = \dfrac{2(x^2 + 2x + 3)}{1 + 3(x^2 + 2x + 3)}$, and $g \circ f$ is $y = \left(\dfrac{2x}{1 + 3x}\right)^2 + 2\left(\dfrac{2x}{1 + 3x}\right) + 3$.

12. If $f(x) = 1 - 2x^3 + \dfrac{4}{x}$, $x \neq 0$, find $f[f(x)]$.

    Solution. $f[f(x)] = 1 - 2\left(1 - 2x^3 + \dfrac{4}{x}\right)^3 + \dfrac{4}{1 - 2x^3 + \dfrac{4}{x}}$. Excluded

    from the domain of $f \circ f$ are 0 and those values of x for which $1 - 2x^3 + \dfrac{4}{x} = 0$; these are 1.10 and 1.27 (approximately).

13. If f is defined by $y = 1 + \dfrac{1}{x}$, $x \neq 0$, find the inverse function f^{-1} of f.

    Solution. $y = \dfrac{x + 1}{x}$, $xy = x + 1$, hence $x = \dfrac{1}{y - 1} = f^{-1}(y)$, $y \neq 0$.

    However, f^{-1} may be defined in terms of y by $f^{-1}(x) = \dfrac{1}{x - 1}$, $x \neq 1$.

14. For f defined by $y = 6x - 7$, find the inverse function f^{-1}.

Solution. $x = \dfrac{y+7}{6}$ so that f^{-1} is defined in terms of y by $y = \dfrac{x+7}{6}$.

15. If f is defined by $y = \sqrt{x}$, find the inverse function f^{-1}.
Solution. $x = y^2$ so that f^{-1} is defined in terms of y by $y = x^2$, $x \geqslant 0$.

PROBLEMS WITH ANSWERS

16. If $f(x) = x^3 - x + 6$, find $f(2), f(-1)$, and $f(\sqrt{2})$.
Ans. $12, 6, 2^{3/2} - 2^{1/2} + 6 = 2^{1/2} + 6$

17. If $f(x) = \sqrt{1 + \sqrt{1-x}}$, find $f(0), f(1), f(-1)$.
Ans. $\sqrt{2}, 1, \sqrt{1+\sqrt{2}}$

18. Which of the following tables define a function $f(x,y)$?

a.
x	3	3
y	4	2
 b.
x	2	3
y	-1	-1
 c.
x	1	1
y	2	3

Ans.   b.

19. In the following tables, which letter represents the independent variable?

a
u	0	0
v	5	6
 b.
s	1	2
t	2	2
 c.
z	2	3
w	3	2

Ans.   a. u is the independent variable. b. s is the independent variable. c. Either z or w is the independent variable.

20. If $f(x) = kx^2$, find $f(k), f(1/k)$, and $f(1-k)$.
Ans. $k^3, 1/k, k(1-k)^2$

21. If $f(x) = \dfrac{x-a}{x+a}$, $a \neq 0$, find $f(0), f(x-a), f(x+a), f(x/a)$, and $f(a/x)$.
Ans. $-1, (x-2a)/x, x/(x+2a), (x-a^2)/(x+a^2), (1-x)/(1+x)$

22. If $f(x) = \begin{cases} |x|, x > 1 \\ 1, x \leqslant 1 \end{cases}$, find $f(0), f(1), f(-1), f(2), f(-2), f(6), f(-6)$, and $f\left(\dfrac{1}{2}\right)$.
Ans. $1, 1, 1, 2, 2, 6, 6, 1$

23. If $g(x) = 2 \sin \frac{1}{2}x$, find $g(0), g(\pi/3), g(\pi/2), g(\pi), g(2\pi), g(1)$, and $g(-2)$.

Ans. $0, 1, \sqrt{2}, 2, 0, 2 \sin \frac{1}{2} = 0.95886, 2 \sin(-1) = -1.68294$

24. If $\phi(t) = 2^{1/t}$, find $\phi(1), \phi(-1), \phi(2), \phi(1/2)$, and $\phi(-1/4)$.

Ans. $2, \frac{1}{2}, \sqrt{2}, 4, \frac{1}{16}$

25. If $h(y) = \frac{1}{y} \log_{10} y$, find $h(1), h(2), h(10), h(100)$, and $h(1000)$.

Ans. $0, 0.15052, 0.1, 0.02, 0.003$

26. If $F(\theta) = \sec^2\theta - \tan^2\theta$, find $F(0)$, $F(1)$, $F(-1)$, $F(\pi/4)$, and $F(4\pi/7)$.

Ans. $1, 1, 1, 1, 1$

27. If $z = \sqrt{1 - x^2 - y^2}$, $x^2 + y^2 \leqslant 1$, find $z(1, 0)$, $z(0, -1)$, $z(0, 0)$, and $z(-1/\sqrt{2}, 1/\sqrt{2})$.

Ans. $0, 0, 1, 0$

28. If $f(x) = 1 + \frac{1}{x}$ and $g(x) = 1 - \frac{1}{x}$, $x \neq 0$, find (a) $f[g(x)]$, (b) $g[f(x)]$, (c) $f[g(1-x)]$, and (d) $g\left[f\left(1 - \frac{1}{x}\right)\right]$.

Ans. (a) $f[g(x)] = 1 + \dfrac{1}{1 - \dfrac{1}{x}} = \dfrac{2x - 1}{x - 1}, x \neq 0$ or 1

(b) $g[f(x)] = 1 - \dfrac{1}{1 + \dfrac{1}{x}} = \dfrac{1}{x + 1}, \; x \neq 0$ or -1

(c) $g(1 - x) = 1 - \dfrac{1}{1 - x} = \dfrac{x}{x - 1}, \; x \neq 0$ or 1, therefore,

$f[g(1 - x)] = 1 + \dfrac{1}{x/(x - 1)} = \dfrac{2x - 1}{x}, x \neq 0$ or 1

(d) $g\left[f\left(1 - \dfrac{1}{x}\right)\right] = \dfrac{x}{2x - 1}, x \neq 0, \dfrac{1}{2}$, or 1

29. Write down the ordered pairs of the function given by the table,

x	1	1	2	2
y	2	3	4	5

Ans. (2, 1), (3, 1), (4, 2), (5, 2)

30. Does $y > x$ define a function or a relation?

Ans. a relation

31. Which of the following relations possesses an inverse that is a function?

a.

t	1	1
w	1	2

b.

u	1	2
v	1	1

c.

θ	1	1	2
ψ	2	3	3

Ans. a. t is the independent variable, and the inverse function is $f(w,t)$, namely, (1, 1) and (2, 1).

b. u is the independent variable, and the inverse function f is (1, 1), (2, 1).

c. no inverse function

32. What is the composite function $f \circ g$, or $f[g(x)]$, where $f(x) = x^4 - x^2$ and $g(x) = x^2 + x - 1$? What is the domain of $f \circ g$?
Ans. $f[g(x)] = (x^2 + x - 1)^4 - (x^2 + x - 1)^2$; the domain of $f \circ g$ is all reals.

33. Given f defined by $f(x) = \sqrt{1 + x}$ with domain $x \geqslant -1$ and range the nonnegative reals, and g defined by $g(x) = -\sqrt{2 + x}$ with domain $x \geqslant -2$ and range the nonpositive reals, what is $f \circ g$ and what is its domain?
Ans. $f \circ g$ is defined by $f[g(x)] = \sqrt{1 - \sqrt{2 + x}}$ and the domain of $f \circ g$ is $-2 \leqslant x \leqslant -1$.

34. If $f(x) = \sqrt{1 - x}$ and $g(x) = \sqrt{x - 1}$, what is the domain of $f \circ g$?
Ans. $1 \leqslant x \leqslant 2$

CHAPTER 3

SEQUENCES AND LIMITS

3.1 Limit of a Sequence

The principal idea in the calculus is the concept of limits of sequences and of functions. A **sequence** is a set of numbers that is *ordered* by the positive integers $1, 2, 3, \ldots, n, \ldots$. Thus, if a positive integer is *assigned* to members of a set, these members are ordered sequentially. Some examples of sequences are

$$1, \frac{1}{2}, \frac{1}{3}, \ldots, \frac{1}{n}, \ldots \tag{1}$$

$$\frac{1}{2}, \frac{2}{3}, \frac{3}{4}, \ldots, \frac{n}{n+1}, \ldots \tag{2}$$

$$\log_{10} 1, \log_{10} 2, \log_{10} 3, \ldots, \log_{10} n, \ldots \tag{3}$$

And in general $a_1, a_2, a_3, \ldots, a_n, \ldots$

A sequence may also be thought of as a **function** whose domain is the set of positive integers and whose range consists of the elements of the sequence; for example, in sequence (1) above, $f(1) = 1$, $f(2) = 1/2$, $f(3) = 1/3, \ldots, f(n) = 1/n$. Set notation is often used to describe sequences; thus, for a *finite* sequence, we write, $\{a_i\}_{i=1}^{n} = a_1, a_2, a_3, \ldots, a_n$, where $i = 1, 2, 3, \ldots, n$. Similarly, for an *infinite* sequence, $\{a_i\}_{i=1}^{\infty} = a_1, a_2, a_3, \ldots, a_n, \ldots$ (∞ = infinity). Unless otherwise indicated, it will be assumed that $\{a_i\}$ is an infinite sequence.

Evidently, a *finite* sequence has a *last* term, but an *infinite* sequence has no final term, as in the examples shown above. Notice the general, or *n*th, terms of sequences (1), (2), and (3) which are respectively $1/n$, $n/(n+1)$, and $\log_{10} n$. In sequence (1), as n becomes very large, $1/n$ approaches the value zero; hence, we say that *as n approaches infinity*, $1/n$ *approaches zero*, or *the limit of $1/n$ is zero as n approaches infinity*. This is written symbolically

$$\lim_{n \to \infty} \frac{1}{n} = 0 \quad \text{or} \quad \frac{1}{n} \longrightarrow 0 \text{ as } n \longrightarrow \infty$$

Clearly, for sequence (2)

$$\lim_{n \to \infty} \frac{n}{n+1} = 1$$

In sequence (3), the general term $\log_{10} n$ becomes larger and larger *without bound*; hence, the sequence itself approaches infinity. In this case, we write

$$\lim_{n \to \infty} \log_{10} n = \infty$$

If the limit of an infinite sequence is a **real number,*** we say that *the sequence converges*; thus, sequences (1) and (2) converge. If a sequence does not converge to a real number, it is said to *diverge*. Since infinity is not a real number, sequence (3) diverges. If a sequence $\{a_i\}$ converges to a number L, we write

$$\lim_{n \to \infty} a_n = L$$

DEFINITION I.   *A sequence $a_1, a_2, \ldots, a_n, \ldots$ converges to a limit L, if for each arbitrarily small positive number ϵ there exists a positive integer N such that $|a_n - L| < \epsilon$ for $n \geqslant N$, where N depends on the choice of ϵ.*

This means essentially that from some term a_N on, i.e., a_{N+1}, a_{N+2}, $a_{N+3}, \ldots$, the difference (distance) between the values of each of these terms and the value of the limit L is arbitrarily small, close to zero; hence, all of these values are within ϵ-distance of L; ϵ can, therefore, be made as small as one would like to make it. This is why we say that *the choice of the value of $n > N$ depends on ϵ.*

The following is called **Cauchy's criterion** for convergence, and it is an even stronger test than Definition I.

DEFINITION II.   *A sequence converges to a limit, if for every positive number ϵ there exists a positive integer N such that $|a_m - a_p| < \epsilon$ for all integers $m, p \geqslant N$.*

This definition states that beyond some term a_N in the sequence, the difference between any two terms is arbitrarily small (less than ϵ); thus, certainly, $|a_m - L|$ and $|a_p - L|$ must both be less than ϵ, which satisfies the conditions of Definition I.

---

*Since the real numbers are a subset of the **complex numbers** $a + bi$ in which $b = 0$ and $i = \sqrt{-1}$, a complex number in which $b \neq 0$ could be a limit; however, in this text only real numbers will be considered; i.e., when the word "number" is used, it shall be presumed to be a real number.

THEOREM I. (Limit theorems for sequences) *Let k, L, and M be numbers and, if $\{a_i\}$ and $\{b_i\}$ both converge to L and M, respectively, then*

(i) $\{a_i\} = k, k, k, \ldots$ *and* $\lim\limits_{n \to \infty} a_n = k$.

(ii) $\lim\limits_{n \to \infty} (a_n \pm b_n) = L \pm M$

(iii) $\lim\limits_{n \to \infty} a_n b_n = LM$

(iv) $\lim\limits_{n \to \infty} k a_n = kL$

(v) $\lim\limits_{n \to \infty} \dfrac{a_n}{b_n} = \dfrac{L}{M}$, *provided* $M \neq 0$; *that is,* $b_n \neq 0$ *for sufficiently large* n, *i.e., for* $n \geqslant n_0$.

(vi) (Domination principle) *If* $\lim\limits_{n \to \infty} a_n = L$, *and* $\lim\limits_{n \to \infty} c_n = L$, *and there exists a sequence* $\{b_n\}$ *such that for some* $n \geqslant n_0$, $a_n < b_n < c_n$, *then* $\lim\limits_{n \to \infty} b_n = L$.

(vii) *If* $\lim\limits_{n \to \infty} a_n = L$, *then* $\lim\limits_{n \to \infty} |a_n| = |L|$.

THEOREM II. (Uniqueness of limits) *If* $\lim\limits_{n \to \infty} a_n = L_1$ *and* $\lim\limits_{n \to \infty} a_n = L_2$, *then* $L_1 = L_2$.

3.2 The Limit of a Function

If the values y of a function f exist arbitrarily close to a number L for corresponding values of x close to a number c, we say that *L is the limit of* $f(x) = y$ *as x approaches* c, even though $f(c)$ need not exist.

This statement may be described in terms of the rectangular coordinate system as follows (Fig. 3.1): Given a function $f(x) = y$, choose an arbitrarily small interval of length $2\epsilon, \epsilon > 0$, with the point $(0, L)$ at its center such that $L - \epsilon < f(x) < L + \epsilon$, i.e., $L - \epsilon < y < L + \epsilon$. Then there corresponds an interval on the X-axis which is the *inverse image of* the interval $(L - \epsilon, L + \epsilon)$ such that for $L - \epsilon, L$, and $L + \epsilon$ there correspond $c - \delta, c$, and $c + \delta'$, where $\delta > 0$ and $\delta' > 0$, even though $f(c)$ may not exist. Thus, c is contained in the interval $(c - \delta, c + \delta')$, which is the inverse image of $(L - \epsilon, L + \epsilon)$. Now, we want an interval along the X-axis with c at its *center*, and which is also contained in $(c - \delta, c + \delta')$. Notice in Fig. 3.1, that c is not the center point of the interval $(c - \delta, c + \delta')$. In this case, $\delta < \delta'$; hence, if we choose an interval about c of length $2\delta'$, it will contain points to the left of $c - \delta$ which are outside of our domain, i.e., the interval, $(c - \delta', c - \delta)$. However, the interval of length 2δ will satisfy these conditions. It is obvious that, in

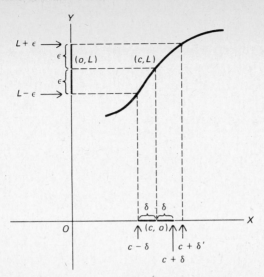

Fig. 3.1

general, the *smaller* of any two δ's will serve. If this construction exists, we say that L *is the limit of* $f(x)$. We also note that the expression $L - \epsilon < y < L + \epsilon$ means the same as $|f(x) - L| < \epsilon$; similarly, $c - \delta < x < c + \delta$ implies $|c - x| < \delta$. If $f(c)$ does not exist, we write $0 < |x - c| < \delta$, which means that $c - \delta < x < c$ or $c < x < c + \delta$, i.e., $x \neq c$. A formal statement of the foregoing conditions is given in Definition III, which follows.

DEFINITION III. *The value $f(x)$ of the function f has the constant L as a limit as x approaches c, if for every positive number ϵ there exists a positive number δ such that $|f(x) - L| < \epsilon$ for all x having the property that $0 < |x - c| < \delta$. When L exists, we write*

$$\lim_{x \to c} f(x) = L$$

If $x > c$, we write $x \longrightarrow c^+$ and say that L *is the right-hand limit*; on the other hand, if $x < c$, we write $x \longrightarrow c^-$, and say that L *is the left-hand limit*. In either case, L is called a *one-sided* limit. The statement that L is the limit of $f(x)$ as x approaches c implies that $\lim_{x \to c^+} f(x) = \lim_{x \to c^-} f(x)$. Alternatively, for the right-hand limit, if $\lim_{x \to c^+} f(x) = L$, then $|f(x) - L| < \epsilon$, whenever $c < x < c + \delta$; similarly, for the left-hand limit, $\lim_{x \to c^-} f(x) = L$ implies $|f(x) - L| < \epsilon$, whenever $c - \delta < x < c$.

A function f is said to be *continuous* on an interval $a \leqslant x \leqslant b$, if it is continuous at each point x_0 in this interval. If $f(x)$ is continuous at $x = c$, then the limit $L = f(c)$, i.e., $\lim_{x \to c} f(x) = f(c)$. A function is said to be *discontinuous* at a point, if it is not continuous at that point.

THEOREM III. (Limit theorems for functions)   *If* $\lim_{x \to c} f(x) = L$ *and* $\lim_{x \to c} g(x) = M$, *then*

(i) $\lim\limits_{x \to c} [f(x) \pm g(x)] = \lim\limits_{x \to c} f(x) \pm \lim\limits_{x \to c} g(x) = L \pm M$

(ii) $\lim\limits_{x \to c} f(x)\, g(x) = [\lim\limits_{x \to c} f(x)]\, [\lim\limits_{x \to c} g(x)] = LM$

(iii) $\lim\limits_{x \to c} f(x)/g(x) = \lim\limits_{x \to c} f(x)/\lim\limits_{x \to c} g(x) = L/M, M \neq 0$

(iv) $\lim\limits_{x \to c} \sqrt[n]{f(x)} = \sqrt[n]{\lim\limits_{x \to c} f(x)} = \sqrt[n]{L}$, *if* $L > 0$, *and* n *is a positive integer.*

(v) *For* $f(x) = k$, $\lim\limits_{x \to c} f(x) = \lim\limits_{x \to c} k = k$

(vi) $\lim\limits_{x \to c} [k \pm f(x)] = \lim\limits_{x \to c} k \pm \lim\limits_{x \to c} f(x) = k \pm L$

(vii) $\lim\limits_{x \to c} kf(x) = (\lim\limits_{x \to c} k)\, [\lim\limits_{x \to c} f(x)] = k \lim\limits_{x \to c} f(x) = kL$

(viii) $\lim\limits_{x \to c} k/f(x) = \lim\limits_{x \to c} k/\lim\limits_{x \to c} f(x) = k/\lim\limits_{x \to c} f(x) = k/L, L \neq 0$

3.3   Infinite and Nonexisting Limits

There are three possibilities involving the concept of infinity.

1. $\lim\limits_{x \to c} f(x) = \infty$, which means that as x approaches c, $f(x)$ grows large *without bound.* If $f(x)$ grows large without bound in the *negative* direction, we write, $f(x) \longrightarrow -\infty$; correspondingly, if $f(x)$ grows large without bound in the *positive* direction, $f(x) \longrightarrow +\infty$. An example of this case is

$$\lim_{x \to 0} \frac{1}{x^2} = \infty$$

2. $\lim\limits_{x \to \infty} f(x) = L$, which means that $f(x)$ has a *finite* limit as x becomes increasingly large. In this case, we have $|f(x) - L| < \epsilon$ whenever $|x| > N$, where $N \longrightarrow \infty$. As an example,

$$\lim_{x \to \infty} \frac{1}{x} = 0$$

3. $\lim\limits_{x \to c} f(x)$ does not exist. Such a case is $\lim\limits_{x \to 0} \log_{10} x$.

3.4 Limits of Composite Functions

If f and g are two functions such that $f \circ g = h$ and if $f[g(x)] = h(x)$, then $\lim\limits_{x \to c} f[g(x)] = f[\lim\limits_{x \to c} g(x)] = \lim\limits_{x \to c} h(x)$.

3.5 Limits of Special Functions

The following three limits are very important in the differential calculus.

$$\lim_{x \to 0} \frac{\sin x}{x} = 1 \qquad (x \text{ is measured in radians.}) \tag{1}$$

$$\lim_{x \to 0} \frac{\cos x - 1}{x} = 0 \qquad (x \text{ is measured in radians.}) \tag{2}$$

$$\lim_{x \to 0} (1 + x)^{1/x} = e, \text{ where } e = 2.71828\ldots \tag{3}$$

The number e is the base of the exponential function e^x and of the natural logarithm $\log_e x$. The exponential function is also written $\exp x$, and $\log_e x$ is most often written as just $\log x$ or $\ln x$. Limits (1), (2), and (3) above will be discussed in more detail in later chapters.

PROBLEMS WITH SOLUTIONS

1. Show that the sequence $1, \dfrac{1}{2}, \dfrac{1}{3}, \ldots, \dfrac{1}{n}, \ldots$ converges.

   Solution. Here $u_n = \dfrac{1}{n}$; we guess that the limit L is 0. For every positive ϵ, the condition $\left| \dfrac{1}{n} - 0 \right| < \epsilon$ implies that $\dfrac{1}{n} < \epsilon$, or $n > \dfrac{1}{\epsilon}$.

   Hence, a number $N = \dfrac{1}{\epsilon}$ exists such that $\left| \dfrac{1}{n} - 0 \right| < \epsilon$ for every integer $n \geqslant N$. The limit is therefore 0.

   From the theorems on limits of sequences, it follows that as $n \longrightarrow \infty$, $k\dfrac{1}{n} \longrightarrow 0$ (k constant), $\dfrac{1}{n^2} \longrightarrow 0$, $\dfrac{1}{n} + \dfrac{1}{n^2} \longrightarrow 0$, etc.

2. Show that the sequence for which $f(n) = \dfrac{1}{\sqrt[q]{n}} = \dfrac{1}{n^{1/q}}$ converges, where q is a positive integer.

   Solution. We guess that the limit L is 0. For every positive ϵ, $\left| \dfrac{1}{n^{1/q}} - 0 \right| < \epsilon$ implies that $\dfrac{1}{n^{1/q}} < \epsilon$, or $n^{1/q} > \dfrac{1}{\epsilon}$; thus, $n < 1/\epsilon^q$.

   Hence, a number $N = 1/\epsilon^q$ exists such that $\left| \dfrac{1}{n^{1/q}} - 0 \right| < \epsilon$ for every integer $n > N$. The limit is therefore 0.

3. Show that the sequence for which $f(n) = 1/n^{p/q}$ converges, where p and q are positive integers.

    Solution. This sequence converges to 0 by an argument similar to that used in Problem 2; hence, $N = 1/\epsilon^{q/p}$.

4. Show that the sequence $1, -1, 1, -1, \ldots, (-1)^n, \ldots$ ($n = 0, 1, 2, \ldots$) diverges.

    Solution. By Definition II, $|a_m - a_p|$ is always 0 or 2, depending on the parity of m and p, i.e., there exists no ϵ.

5. Show that the sequence $\{u_n\} = \left\{ \dfrac{n}{n+1} \right\}$ converges, i.e., the sequence,

    $$\frac{1}{2}, \frac{2}{3}, \frac{3}{4}, \ldots, \frac{n}{n+1}, \ldots$$

    Solution. We guess that this sequence converges to the limit 1. Note that $\dfrac{n}{n+1} < 1$ for every positive integer n; therefore, the condition $\left| \dfrac{n}{n+1} - 1 \right| < \epsilon$ implies that either $\dfrac{n}{n+1} - 1 < \epsilon$ or $-\dfrac{n}{n+1} + 1 < \epsilon$. Since the first equation provides a solution for $-n$ and the second equation provides a solution for $+n$, we choose the second equation, which yields $n > (1 - \epsilon)/\epsilon$. Hence, a number $N = (1 - \epsilon)/\epsilon$ exists such that $\left| \dfrac{n}{n+1} - 1 \right| < \epsilon$ for every integer $n \geqslant N$. The limit is therefore 1.

6. Show that the sequence $s_n = \dfrac{1}{n^2} (1 + 2 + \cdots + n)$ converges.

    Solution. To see this, it is first necessary to show that

    $$1 + 2 + \cdots + n = \frac{n(n+1)}{2}$$

    (This is usually done by *mathematical induction.*)

    Therefore, $s_n = \dfrac{n^2 + n}{2n^2} = \dfrac{1}{2} + \dfrac{1}{2n}$, and we see from Theorem I (i) and (iv) that $\dfrac{1}{2} \longrightarrow \dfrac{1}{2}$ and $\dfrac{1}{2n} \longrightarrow 0$ (see Problem 1.); thus, $\dfrac{1}{2} + \dfrac{1}{2n} \longrightarrow \dfrac{1}{2} + 0 = \dfrac{1}{2}$. Alternatively, we can write $|s_n - L| = \left| \left(\dfrac{1}{2} + \dfrac{1}{2n} \right) - \dfrac{1}{2} \right| < \epsilon$, which implies that $\dfrac{1}{2n} < \epsilon$, or $n > \dfrac{1}{2\epsilon}$.

Hence, a number $N = \dfrac{1}{2\epsilon}$ exists such that $\left| \left(\dfrac{1}{2} + \dfrac{1}{2n} \right) - \dfrac{1}{2} \right| < \epsilon$ for every integer $n \geqslant N$. The limit is, therefore, $\dfrac{1}{2}$.

7. Show that the sequence $s_n = \dfrac{1}{n^3}(1^2 + 2^2 + \cdots + n^2)$ converges.

   Solution. By mathematical induction, it can be shown that $1^2 + 2^2 + \cdots + n^2 = \dfrac{n(n+1)(2n+1)}{6}$; therefore, $s_n = \dfrac{2n^3 + 3n^2 + n}{6n^3} = \dfrac{1}{3} + \dfrac{1}{2n} + \dfrac{1}{6n^2}$. Since $\dfrac{1}{2n} \longrightarrow 0$ and $\dfrac{1}{6n^2} \longrightarrow 0$, $s_n \longrightarrow \dfrac{1}{3}$.

8. Show that the sequence $\dfrac{\cos 1}{1}, \dfrac{\cos 2}{2}, \ldots, \dfrac{\cos n}{n}, \ldots$ converges.

   Solution. This sequence converges to 0 by the domination principle (Theorem I (vi)), since $-\dfrac{1}{n} < \dfrac{\cos n}{n} < \dfrac{1}{n}$ for every positive integer n.

9. Show that the sequence for which $u_n = \dfrac{3n + 2}{n + 1}$ converges.

   Solution. First note that $\dfrac{3n + 2}{n + 1} = \dfrac{3 + \dfrac{2}{n}}{1 + \dfrac{1}{n}}$. Now, $3 + \dfrac{2}{n} \longrightarrow 3$ and $1 + \dfrac{1}{n} \longrightarrow 1$; thus, $\dfrac{3 + 2/n}{1 + 1/n} \longrightarrow 3$.

10. Show that the sequence for which $f(n) = \sqrt{n + 2} - \sqrt{n}$ converges.
    Solution. We write,

    $$f(n) = \left(\sqrt{n + 2} - \sqrt{n} \right) \dfrac{\sqrt{n + 2} + \sqrt{n}}{\sqrt{n + 2} + \sqrt{n}}$$

    $$= \dfrac{(n + 2) - n}{\sqrt{n + 2} + \sqrt{n}}$$

    $$= \dfrac{2}{\sqrt{n + 2} + \sqrt{n}}$$

    Now, $\dfrac{2}{\sqrt{n + 2} + \sqrt{n}} < \dfrac{2}{\sqrt{n} + \sqrt{n}} = \dfrac{1}{\sqrt{n}}$. Therefore, by the domination principle, $f(n) \longrightarrow 0$, since $0 < \dfrac{2}{\sqrt{n+2} + \sqrt{n}} < \dfrac{1}{\sqrt{n}}$, and $\lim\limits_{n \to \infty} 0 =$

$\lim\limits_{n \to \infty} \dfrac{1}{\sqrt{n}} = 0$.

11. Show that the sequence for which $u_n = n^{n-1}$ diverges.
Solution. Now, $n^{n-1} > n$ for $n > 2$. Also, the sequence for which $u_n = (n-1)^n$ diverges, since $(n-1)^n > n$ for $n > 2$.

12. Show that the sequence for which $u_n = \log_{10}(n+1)$ diverges.
Solution. Since, if $n = 10^k - 1$, k a positive integer, then $\log_{10}(n+1) = \log_{10} 10^k = k$ and $k \longrightarrow \infty$.

13. Given $f(x) = 2 + x$, find $\lim\limits_{x \to 1} f(x)$.

Solution. In the definition of the limiting value of f at 1, it is assumed that the limit L is known. We must, therefore, guess at the number L. Try $L = 3$ and let ϵ be a positive number. We must now show the existence of a δ such that $|(2 + x) - 3| < \epsilon$ for all $x \neq 1$ having the property $|x - 1| < \delta$. Now, $|(2 + x) - 3| = |x - 1|$; hence, $\delta = \epsilon$ will satisfy this requirement. Therefore, $\lim\limits_{x \to 1}(2 + x) = 3$.

14. Given $f(x) = a + bx$, find $\lim\limits_{x \to 1} f(x)$.
Solution. We guess that the limit $L = a + b$. For every positive ϵ, a δ must exist such that $|a + bx - (a + b)| < \epsilon$ for all $x \neq 1$ having the property $|x - 1| < \delta$. Now, $|a + bx - (a + b)| = |b| \, |x - 1| < \epsilon$, or $|x - 1| < \epsilon/|b|$; thus, take $\delta = \epsilon/|b|$.

15. Given $f(x) = \dfrac{1}{x}$, find $\lim\limits_{x \to 1/2} \dfrac{1}{x}$.

Solution. Since $f\left(\dfrac{1}{2}\right) = 2$, we may guess that the limit $L = 2$; from this, it follows that $f(x)$ is continuous at $x = \dfrac{1}{2}$. If this is true, then for every $\epsilon > 0$, there exists a $\delta > 0$ such that whenever $\left| \dfrac{1}{x} - 2 \right| < \epsilon$, $\left| x - \dfrac{1}{2} \right| < \delta$. Thus, we have $-\epsilon < \dfrac{1}{x} - 2 < \epsilon$, or $2 - \epsilon < \dfrac{1}{x} < 2 + \epsilon$, which yields $\dfrac{1}{2 + \epsilon} < x < \dfrac{1}{2 - \epsilon}$. Then, for $x - \dfrac{1}{2}$, we have $\dfrac{1}{2 + \epsilon} - \dfrac{1}{2} < x - \dfrac{1}{2} < \dfrac{1}{2 - \epsilon} - \dfrac{1}{2}$. Since x is contained in the open interval $\left(\dfrac{1}{2 + \epsilon}, \dfrac{1}{2 - \epsilon} \right)$, then, by continuity, $\dfrac{1}{2}$ must lie in this interval, and the length (absolute value) of δ must be equal to half of an interval with $\frac{1}{2}$ at its center, i.e., $\delta = \left| \dfrac{1}{2 - \epsilon} - \dfrac{1}{2} \right|$ or $\delta = \left| \dfrac{1}{2 + \epsilon} - \dfrac{1}{2} \right|$. Clearly, these two δ-values are different; therefore, we choose the *smaller* of them, which is $\delta = \left| \dfrac{1}{2 + \epsilon} - \dfrac{1}{2} \right|$. (Refer to Section 3.2.) Since $\delta > 0$, we must have $\delta = \dfrac{1}{2} - \dfrac{1}{2 + \epsilon} = \dfrac{\epsilon}{4 + 2\epsilon}$; thus, $\left| x - \dfrac{1}{2} \right| < \dfrac{\epsilon}{4 + 2\epsilon}$ and $\lim\limits_{x \to 1/2} \dfrac{1}{x} = 2$.

16. Given f such that $f(x) = \begin{cases} x^2, & x \neq 0 \\ 1, & x = 0 \end{cases}$. Find $\lim\limits_{x \to 0} f(x)$.

Solution. Evidently $\lim\limits_{x \to 0} x^2 = 0$, since for small values of x (x close to zero), x^2 is even smaller than x (for values of $x < 1$, $x^2 < x$). Thus, for every $\epsilon > 0$, a $\delta > 0$ must exist such that $|x^2 - 0| < \epsilon$ for all $x \neq 0$ having the property $|x - 0| < \delta$. We have $|x^2 - 0| = |x^2| = x^2 < \epsilon$; hence, the value $\sqrt{\epsilon}$ can be used for δ. But, since $f(0) = 1$, f is discontinuous at 0.

17. Find $\lim\limits_{x \to \infty} \dfrac{1}{x}$.

Solution. We guess that the limit $L = 0$. Then, for every $\epsilon > 0$, there must exist a number N such that $\left| \dfrac{1}{x} - 0 \right| < \epsilon$ for every $x > N$. We may take $\dfrac{1}{x} > 0$ so that $\left| \dfrac{1}{x} - 0 \right| = \left| \dfrac{1}{x} \right| = \dfrac{1}{x} < \epsilon$. Hence, $x > 1/\epsilon$ will be satisfied by every $x \geq N$, if we take $N = 1/\epsilon$.

PROBLEMS WITH ANSWERS

18. Given $f(x) = 4 - \dfrac{3}{x}$, find:

a. $\lim\limits_{x \to 2} 4 - \dfrac{3}{x}$ *Ans.*   a. $\dfrac{5}{2}$

b. $\lim\limits_{x \to 3} 4 - \dfrac{3}{x}$ b. 3

  c. $\lim\limits_{x\to\infty} 4 - \dfrac{3}{x}$

                                                                     c. 4

19. Let f be determined by $f(x) = \begin{cases} |x|, & x \neq 0 \\ 1, & x = 0 \end{cases}$ Find:

   a. $\lim\limits_{x\to 1} f(x)$ *Ans.*   a. 1

   b. $\lim\limits_{x\to 0} f(x)$ b. 0

20. Find $\lim\limits_{x\to 0^+} \dfrac{2}{1 - 3^{1/x}}.$ *Ans.*   0

21. Find $\lim\limits_{x\to 0^-} \dfrac{2}{1 - 3^{1/x}}.$ *Ans.* 2

22. Find $\lim\limits_{x\to 0} \dfrac{1}{2 + 3^{1/x}}.$ *Ans.*   Limit does not exist.

Answer Problems 25 through 29 "true" or "false."

23. $f(x) = \dfrac{1}{x}$ is discontinuous at $x = 0$.   *Ans.*   true

24. $f(x) = 2/(1 - 3^{1/x})$ is discontinuous at $x = 0$. (See Problems 20 and 21.)   *Ans.*   true

25. $f(x) = \begin{cases} x^2, & x \leqslant 0 \\ x, & x > 0 \end{cases}$, is continuous at $x = 0$.   *Ans.*   true

26. $f(x) = \begin{cases} \sqrt{1 - x^2}, 0 \leqslant x < \sqrt{2}/2 \\ x, x > \sqrt{2}/2, \end{cases}$ is continuous at $x = \sqrt{2}/2$.   *Ans.*   true

27. Test the following sequences for convergence and state the limit for those that converge.

   a. $a_n = \begin{cases} \dfrac{1}{2}n, & n \text{ even} \\[2mm] \dfrac{1}{3}n, & n \text{ odd} \end{cases}$ *Ans.*   Converges to 0.

b. $a_n = \dfrac{n^3 + 3m + 500}{n^2 + 1}$ *Ans.*   Diverges.

c. $a_n = \dfrac{3n^2 + 100}{6n^2 + 25}$ *Ans.*   Converges to $\dfrac{1}{2}$.

d. $a_n = \sqrt[n]{p}, p > 1$ and n is an integer greater than 1.
(Hint: Write $p^{1/n}$.)   *Ans.*   Converges to 1.

e. $a_n = \sqrt[n]{p}, p < 1$ *Ans.*   Converges to 1.

f. $a_n = p^n, p > 1$ *Ans.*   Diverges.

g. $a_n = p^n, p < 1$ *Ans.*   Converges to 0.

h. $a_n = \sqrt[n]{n}$ *Ans.*   Converges to 1.

i. $a_n = \sqrt{n + 1} - \sqrt{n}$ *Ans.*   Converges to 0.

j. $a_n = \dfrac{n}{\alpha_n}, \alpha > 1$ *Ans.*   Converges to 0.

k. $a_n = \dfrac{\alpha^n}{n!}, \alpha > 1$ *Ans.*   Converges to 0.

l. $a_n = \left(\sqrt{n + 1} - \sqrt{n}\right)\left(\sqrt{n + \dfrac{1}{2}}\right)$ *Ans.*   Converges to 0.

m. $a_n = \sqrt[3]{n + 1} - \sqrt[3]{n}$ *Ans.*   Converges to 0.

n. $a_n = n!$ *Ans.*   Diverges.

CHAPTER 4
SUMMATION AND INTEGRATION

4.1 Approximating Areas

Simple areas like those of rectangles, circles, and triangles are easy to compute, but there are many irregularly shaped regions whose areas cannot be computed so easily. A method for computing such a region is by decomposing it into smaller regions whose areas can be computed. For example, the area bounded by a continuous curve crossing the X-axis at two points can be approximated if this region is partitioned into rectangles (see Fig. 4.1). Notice that the smaller the rectangles are made, the more numerous they are, hence the better the approximation of the area. This method of approximating area is discussed below, but later the method, called **integration**, which provides an accurate way to measure areas, will be presented.

Let the function f be continuous and nonnegative on the *closed* interval $a \leqslant x \leqslant b$ (Fig. 4.2). Partition this interval into n open subintervals bounded by the points $a = x_0 < x_1 < x_2 < \cdots < x_k < \cdots <$

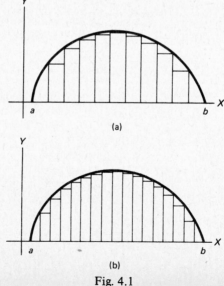

(a)

(b)

Fig. 4.1

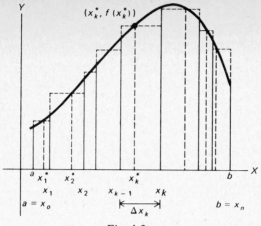

Fig. 4.2

$x_n = b$. Then choose a point x_k^* in the kth interval $[x_{k-1}, x_k]$, where $x_{k-1} < x_k^* < x_k$ and $k = 1, 2, \ldots, n$. Corresponding to x_k^*, we can find an ordinate y_k^* on the Y-axis such that the point (x_k^*, y_k^*) is on the curve $y = f(x)$ due to the continuity of $f(x)$. Hence, $y_k^* = f(x_k^*)$ is the *distance* from $f(x)$ to the X-axis at the point (x_k^*, y_k^*). In Fig. 4.2, we see that the X-axis has been partitioned as described above, but that each partitioning interval forms the base of a rectangle, and the height $f(x_k^*)$ of each rectangle is constructed from the point x_k^*, or $(x_k^*, 0)$, which is contained in each interval, and where k is zero or a positive integer. Therefore, each rectangle has area $f(_k^*)(x_k - x_{k-1})$, in particular the kth rectangle. The sum of all such rectangles is an *approximation* of the area A enclosed by the curve $y = f(x)$, the horizontal line $y = 0$ (the X-axis), and the vertical lines $x = a$ and $x = b$.

In order to simplify the formula to be written down, the symbol Σ (Greek sigma) is introduced. The following examples illustrate the use of this notation.

$$\sum_{k=1}^{3} \frac{1}{k} = \frac{1}{1} + \frac{1}{2} + \frac{1}{3}$$

This is read: *The sum of the reciprocals of the integers* 1, 2, *and* 3.

$$\sum_{k=1}^{n} k^2 = 1^2 + 2^2 + \cdots + n^2$$

(The sum of squares from 1 to n)

$$\sum_{k=1}^{n} f(k) = f(1) + f(2) + \cdots + f(n)$$

Thus, the sum,

$$A \cong \sum_{k=1}^{n} f(x_k^*)(x_k - x_{k-1})$$

represents an approximation of the **area under the curve.** Set $\Delta x_k = (x_k - x_{k-1})$ and let $\|\Delta x_k\|$, called the *norm* of Δx_k, be the largest of the numbers Δx_k, i.e., $\|\Delta x_k\| = \max \Delta x_k$. Then, the area A under the curve $y = f(x)$ from a to b is *exactly,*

$$A = \lim_{\substack{n \to \infty \\ \|\Delta x_k\| \to 0}} \sum_{k=1}^{n} f(x_k^*) \Delta x_k \tag{1}$$

Other notations for (1) are: A_a^b, read *the area from a to b*, and $\int_a^b f(x) \, dx$, read *the integral of f with respect to x from a to b*. Some authors refer to this as the definite integral of f from a to b; a and b are the lower and upper limits of integration respectively. The area A is

$$\int_a^b f(x) \, dx \tag{2}$$

Formula (2) is often used to represent concepts other than area.

The process of evaluating the limit by (1) is called *integration by the method of summation.* If $[a,b]$ is partitioned into subintervals of equal length Δx, then the right-hand member of (1) can be written in the form,

$$\lim_{n \to \infty} \sum_{k=1}^{n} f(x_k^*) \Delta x \tag{3}$$

We may choose $x_k^* = x_k$, but in any case, the limit is given by (2), and, since (2) is a limit formula, the following theorem results directly from Theorem III in Chapter 3.

THEOREM 1. *If f and g are continuous on a closed interval* $[a,b]$, *and k is a number, then*

(i) $\int_a^b kf(x)\,dx = k\int_a^b f(x)\,dx$ (k constant)

(ii) $\int_a^b [f(x) + g(x)]\,dx = \int_a^b f(x)\,dx + \int_a^b g(x)\,dx$

(iii) $\int_a^b f(x)\,dx = -\int_b^a f(x)\,dx$

(iv) $\int_a^c f(x)\,dx = \int_a^b f(x)\,dx + \int_b^c f(x)\,dx$, *where* $a < b < c$

(v) $\int_a^a f(x)\,dx = 0$

(vi) *For* $f(x) \leqslant 0, a \leqslant x \leqslant b, \int_a^b f(x)\,dx \leqslant 0$.

If $f(x)$ is sometimes negative and sometimes positive on $[a,b]$, then $\int_a^b f(x)\,dx$ adds numerically the areas below and above the X-axis. Also, because the definite integral is independent of the particular variable used, each of the variables x, t, u, etc. is called a *dummy variable of integration*, i.e., $\int_a^b f(x)\,dx = \int_a^b f(t)\,dt = \int_a^b f(u)\,du = \cdots$.

Example 1. Find the area bounded by $y = x^2$, $y = 0$, and $x = b$ (Fig. 4.3).

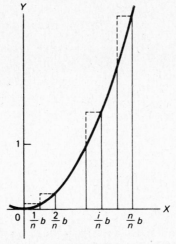

Fig. 4.3

Solution.

$$A = \lim_{n \to \infty} \sum_{i=1}^{n} \left(\frac{i}{n}b\right)^2 \left(\frac{1}{n}b\right)$$

$$= \lim_{n \to \infty} \frac{b^3}{n^3} \left(1^2 + 2^2 + \cdots + n^2\right)$$

$$= \lim_{n \to \infty} \frac{b^3}{n^3} \left[\frac{n(n+1)(2n+1)}{6}\right]^\dagger$$

$$= \lim_{n \to \infty} b^3 \left(\frac{2n^3 + 3n^2 + n}{6n^3}\right)$$

$$= \lim_{n \to \infty} b^3 \left(\frac{1}{3} + \frac{1}{2n} + \frac{1}{6n^2}\right)$$

$$= \frac{b^3}{3} \text{ sq. units}$$

Similarly, the area bounded by $y = x^2$, $y = 0$, and $x = a$ is $a^3/3$, and the area bounded by $y = x^2$, $y = 0$, $x = a$, and $x = b$, where $a < b$, is $\frac{1}{3}(b^3 - a^3)$. The principal difficulty here and in Example 1 arises in deriving the identity $1^2 + 2^2 + \cdots + n^2 = [n(n+1)(2n+1)]/6$.

Example 2. Find the area enclosed by $y = x^3 + 1$ and $y = x + 1$.
Solution. The required area A is the shaded area in Fig. 4.4; hence, (the area bounded by $y = x^3 + 1$, $y = 0$, and $x = 0$ *minus* the area bounded by $y = x + 1$, $y = 0$, and $x = 0$) *plus* (the area bounded by $y = x + 1$, $y = 0$, $x = 0$, and $x = 1$ *minus* the area bounded by $y = x^3 + 1$, $y = 0$, $x = 0$, and $x = 1$). Because of symmetry, both shaded areas are equal; thus, the area $A = 2$ (the area of triangle *ABC minus* the area bounded by $y = x^3 + 1$, $y = 1$, $x = 0$, and $x = 1$); thus,

$$A = 2 \left[\frac{1}{2} - \lim_{n \to \infty} \sum_{i=1}^{n} \left(\frac{k}{n}\right)^3 \frac{1}{n}\right]$$

$$= 1 - 2 \lim_{n \to \infty} \frac{n^2(n+1)^2}{4n^4}$$

$$= 1 - 2 \left(\frac{1}{4}\right)$$

$$= \frac{1}{2} \text{ sq. unit}$$

---

†For this derivation, see Richard Courant, *Differential and Integral Calculus* (2d ed.; New York: John Wiley & Sons, Inc., 1937), Vol. 1, p. 27.

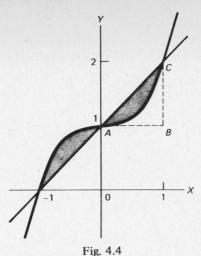

Fig. 4.4

To solve this problem we need to know that

$$1^3 + 2^3 + \cdots + n^3 = \frac{n^2(n+1)^2}{4}$$

It was recognized early that it was very difficult to evaluate directly $\lim_{n \to \infty} \sum_{k=1}^{n} f(x_k) \Delta x$ for even the simplest functions. As late as the mid-seventeenth century, mathematicians were still struggling with this problem for $f(x) = x^k$, k a positive integer. The heart of the difficulty is to find

$$\lim_{n \to \infty} \frac{x^{k+1}}{n^{k+1}} (1^k + 2^k + \cdots + n^k)$$

In 1657, Pascal solved the problem by showing that $1^k + 2^k + \cdots + n^k$ was a polynomial function of degree $k + 1$ in n. For $k = 1$ and 2, see Chapter 3, Problems 6 and 7. These and a few other cases are treated in Problems 1 through 5 below.

In general, only the simplest problems can be done by the method of summation. This method, however, is of extreme value in setting up a given problem. In Chapter 6, we shall see how $\int_a^b f(x) dx$ can be computed in another way.

PROBLEMS WITH SOLUTIONS

In Problems 1 through 12, find the indicated integrals by the method of summation.

1.
$$\int_0^x 1 \, du = \lim_{n \to \infty} \sum_{k=1}^n 1 \frac{x}{n}$$

$$= \lim_{n \to \infty} \left(\frac{x}{n} + \frac{x}{n} + \cdots + \frac{x}{n} \right)$$

$$= \lim_{n \to \infty} \frac{n}{n} x = x$$

2.
$$\int_0^x u \, du = \lim_{n \to \infty} \sum_{k=1}^n \left(\frac{k}{n} x \right) \times \frac{x}{n}$$

$$= \lim_{n \to \infty} \frac{x^2}{n^2} (1 + 2 + \cdots + n)$$

$$= \lim_{n \to \infty} x^2 \left(\frac{n^2 + n}{2n^2} \right) = \frac{x^2}{2}$$

3.
$$\int_0^x u^2 \, du = \lim_{n \to \infty} \sum_{k=1}^n \left(\frac{k}{n} x \right)^2 \times \frac{x}{n}$$

$$= \lim_{n \to \infty} \frac{x^3}{n^3} (1^2 + 2^2 + \cdots + n^2)$$

$$= \lim_{n \to \infty} \frac{x^3}{n^3} \times \frac{n(n+1)(2n+1)}{6} = \frac{x^3}{3}$$

4.
$$\int_0^x u^3 \, du = \lim_{n \to \infty} \sum_{k=1}^n \left(\frac{k}{n} x \right)^3 \times \frac{x}{n}$$

$$= \lim_{n \to \infty} \frac{x^4}{n^4} (1^3 + 2^3 + \cdots + n^3)$$

$$= \lim_{n \to \infty} \frac{x^4}{n^4} \times \frac{n^2 (n+1)^2}{4} = \frac{x^4}{4}$$

5.
$$\int_0^x u^4 \, du = \lim_{n \to \infty} \sum_{k=1}^n \left(\frac{k}{n} x \right)^4 \times \frac{x}{n}$$

$$= \lim_{n \to \infty} \frac{x^5}{n^5} (1^4 + 2^4 + \cdots + n^4)$$

$$= \lim_{n \to \infty} \frac{x^5}{n^5} \times \frac{6n^5 + 15n^4 + 10n^3 - n}{30} = \frac{x^5}{5}$$

6.
$$\int_1^2 x^2 \, dx = \frac{7}{3}$$
(by Problem 3)

7.
$$\int_{-1}^1 x^4 \, dx = \frac{2}{5}$$
(by Problem 5)

8.
$$\int_a^b 0 \, dx = 0. \quad \int_a^b dx = 0(b - a) = 0$$

9.
$$\int_{-1}^1 x \, dx = 0$$
(by Problem 2)

10.
$$\int_{-1}^1 x^3 \, dx = 0$$
(by Problem 4)

11.
$$\int_1^4 (2x + 3) \, dx = 2 \int_1^4 x \, dx + 3 \int_1^4 dx = 15 + 9 = 24$$

12.
$$\int_{-1}^3 (4 + 2x + x^2 - x^3) \, dx =$$

$$4 \int_{-1}^3 dx + 2 \int_{-1}^3 x \, dx + \int_{-1}^3 x^2 \, dx - \int_{-1}^3 x^3 \, dx = 16 + 8 + \frac{28}{3} - 20 = \frac{40}{3}$$

In Problems 13 through 20, set up the integral as the limit of a sum, but do not attempt to evaluate the limit.

13.
$$\int_0^3 \sqrt{x} \, dx = \lim_{n \to \infty} \sum_{k=1}^n \sqrt{k \frac{3}{n}} \times \frac{3}{n} = \lim_{n \to \infty} \frac{3^{3/2}}{n} \sum_{k=1}^n \sqrt{k}$$

14.
$$\int_1^3 \frac{x}{1 + x^2} \, dx = \lim_{n \to \infty} \sum_{k=1}^n \left[\frac{\left(1 + k \frac{2}{n}\right)}{1 + \left(1 + k \frac{2}{n}\right)^2} \right] \frac{2}{n}$$

15.
$$\int_0^2 3^x \, dx = \lim_{n \to \infty} \sum_{k=1}^n (3^{2 \, k/n}) \frac{2}{n}$$

16.
$$\int_0^{\pi/2} \sin x \, dx = \lim_{n\to\infty} \sum_{k=1}^{n} \left(\sin k \frac{\pi}{2n} \right) \frac{\pi}{2n}$$

17.
$$\int_2^{10} \log_{10} x \, dx = \lim_{n\to\infty} \sum_{k=1}^{n} \left[\log_{10} \left(2 + k \frac{8}{n} \right) \right] \frac{8}{n}$$

18.
$$\int_0^{2\pi} \cos x \, dx = \lim_{n\to\infty} \sum_{k=1}^{n} \cos \left(k \frac{2\pi}{n} \right) \frac{2\pi}{n}$$

19.
$$\int_0^1 \frac{1}{x} \, dx = \lim_{n\to\infty} \sum_{k=1}^{n} \left(\frac{1}{k\dfrac{1}{n}} \right) \frac{1}{n} = \lim_{n\to\infty} \sum_{k=1}^{n} \frac{1}{k}$$

20.
$$\int_a^b x^k \, dx = \lim_{n\to\infty} \sum_{k=1}^{n} \left[a + k \frac{(b-a)}{n} \right]^{*k} \left(\frac{b-a}{n} \right)$$

Express Problems 21 through 25 in the form of a definite integral. (The answer is not unique.)

21.
$$\lim_{n\to\infty} \sum_{k=1}^{n} \left(2 + k\frac{1}{n} \right)^{1/2} \times \frac{1}{n} = \int_0^1 \sqrt{2+x} \, dx = \int_2^3 \sqrt{x} \, dx$$

22.
$$\lim_{n\to\infty} \sum_{k=1}^{n} \left\{ \left(3 - k\frac{1}{n} \right)^{2/3} \right\} \frac{1}{n} = \int_0^1 (3-x)^{2/3} \, dx = \int_3^4 (x)^{2/3} \, dx$$

23.
$$\lim_{n\to\infty} \sum_{k=1}^{n} \frac{1}{n+k} = \lim_{n\to\infty} \sum_{k=1}^{n} \frac{1}{1 + k\dfrac{1}{n}} \times \frac{1}{n} = \int_0^1 \frac{1}{1+x} \, dx$$

24.
$$\lim_{n\to\infty} \sum_{k=1}^{n} \left(5 + k\frac{2}{n} \right)^2 \times \frac{2}{n} = \int_5^7 x^2 \, dx$$

25.
$$\lim_{n\to\infty} \sum_{k=1}^{n} \cos \frac{\pi}{2} \left(1 + k\frac{1}{n} \right) \times \frac{\pi}{4n} = \int_{\pi/4}^{\pi/2} \cos 2x \, dx$$

CHAPTER 5
DIFFERENTIATION

5.1 The Derivative

We have seen that the notion of integration arose in connection with the problem of finding the area enclosed by a curve. Differentiation, or the process of finding the derivative of a function, grew out of the geometric concept of the tangent to a curve and the physical concept of velocity.

Consider the graph of a continuous curve f, and let $P(x_0, f(x_0))$ and $Q(x_0 + h, f(x_0 + h))$ be two points on the graph of (Fig. 5.1). The slope

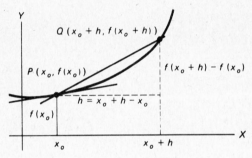

Fig. 5.1

m of the secant line PQ is $[f(x_0 + h) - f(x_0)]/h$. It may happen that as Q approaches P along the curve the secant line approaches some limiting position. Let us state this rather hazy geometric language in precise algebraic form: (1) since f is continuous, $f(x_0 + h) \longrightarrow f(x_0)$ as $h \longrightarrow 0$, and, therefore, $f(x_0 + h) - f(x_0) \longrightarrow 0$ as $h \longrightarrow 0$, and (2) it may happen nevertheless, that

$$m(x_0) = \lim_{h \to 0} \frac{f(x_0 + h) - f(x_0)}{h} \tag{1}$$

exists. If this limit exists, we call $m(x_0)$ *the slope of the line tangent to the curve at* P. The tangent line at P is defined as *the line through* P *with slope* $m(x_0)$. The equation of the tangent line is, therefore,

$$y - y_0 = m(x - x_0) \tag{2}$$

If limit (1) does not exist (is infinite), or if

$$\lim_{h \to 0} \frac{h}{f(x_\circ + h) - f(x_\circ)} = 0$$

and if $x_\circ$ belongs to the domain of f, the tangent line is vertical at $x = x_\circ$ (Fig. 5.2).

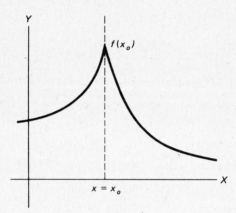

Fig. 5.2

Example 1. Find the equation of the line tangent to $y = x^2$ at $x = \frac{1}{2}$ (Fig. 5.3).

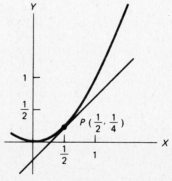

Fig. 5.3

Solution. We first find m, if it exists.

$$m = \lim_{h \to 0} \frac{f\left(\frac{1}{2} + h\right) - f\left(\frac{1}{2}\right)}{h}$$

$$= \lim_{h \to 0} \frac{\left(\frac{1}{2} + h\right)^2 - \left(\frac{1}{2}\right)^2}{h}$$

$$= \lim_{h \to 0} \frac{\frac{1}{4} + h + h^2 - \frac{1}{4}}{h}$$

$$= \lim_{h \to 0} (1 + h) = 1$$

The equation of the tangent line is $y - 1/4 = 1(x - 1/2)$, or $y = x - 1/4$.

Example 2. Find the slope of the line tangent to the curve $y = x^2$ at $P(x_0, x_0^2)$.
Solution.

$$m(x_0) = \lim_{h \to 0} \frac{f(x_0 + h) - f(x_0)}{h}$$

$$= \lim_{h \to 0} \frac{(x_0 + h)^2 - x_0^2}{h}$$

$$= \lim_{h \to 0} \frac{x_0^2 + 2x_0 h + h^2 - x_0^2}{h}$$

$$= \lim_{h \to 0} (2x_0 + h) = 2x_0$$

The slope of the tangent at P is said to be *the slope of the curve at P provided that there is only one tangent at P.* In general, if

$$\lim_{h \to 0} \frac{f(x + h) - f(x)}{h} \tag{3}$$

exists, it is called the *derivative of f evaluated at x.* The derivative of f is a function designated by f', and $f'(x)$ is the value of f' at x. The process of determining f' is called **differentiation**. There are many notations for the derivative of $f;$ some of these are $f', y', \dfrac{dy}{dx}, Df, D_x f, Dy,$ or $D_x y$.

And there are many notations for the *value* of the derivative of f and some of these are $f'(x), y'(x), Df(x), D_x f(x), Dy(x), D_x y(x)$, etc., as well as (3). The use of the subscript x in $D_x f$ and the dx in dy/dx, etc. is to indicate that differentiation has taken place with respect to the variable x in case confusion might arise from the presence of other variables.

It is also common practice to write Δx for h and Δy (or Δf) for $f(x + \Delta x) - f(x)$. Thus, the derivative of f evaluated at x is also written,

$$f'(x) = \lim_{\Delta x \to 0} \frac{f(x + \Delta x) - f(x)}{\Delta x} = \lim_{\Delta x \to 0} \frac{\Delta y}{\Delta x} = \frac{dy}{dx}$$

In discussing the slope of a plane curve, we usually use the same scale units on the two axes. Whatever the units of x and y are, $\dfrac{\Delta y}{\Delta x}$ is the *average rate of change* in y per unit change in x in the interval from x to $x + \Delta x$. Similarly, $\dfrac{dy}{dx}$ is the *instantaneous rate of change* in y per unit change in x at the point $(x, f(x))$. If y is distance and x is time, then $\dfrac{\Delta y}{\Delta x}$ and $\dfrac{dy}{dx}$ are to be interpreted as *average velocity* and *instantaneous velocity*, respectively.

Note that since $\Delta y \longrightarrow 0$ as $\Delta x \longrightarrow 0$, $\lim\limits_{\Delta x \to 0} \Delta y / \Delta x$ is initially of the form $0/0$. The principal difficulties in computing dy/dx directly from the limit lie in determining appropriate algebraic operations which reduce $\Delta y/\Delta x$ to a real-valued limit. The limits in Examples 1 and 2 above do not have the troublesome h, or Δx, in the denominator.

Example 3. Find $y'(x)$ where $y = \dfrac{1}{x}$.

Solution.

$$y'(x) = \lim_{\Delta x \to 0} \frac{\Delta y}{\Delta x} = \lim_{\Delta x \to 0} \frac{f(x + \Delta x) - f(x)}{\Delta x}$$

$$= \lim_{\Delta x \to 0} \frac{\dfrac{1}{x + \Delta x} - \dfrac{1}{x}}{\Delta x}$$

$$= \lim_{\Delta x \to 0} \frac{\dfrac{x - (x + \Delta x)}{x(x + \Delta x)}}{\Delta x}$$

$$= \lim_{\Delta x \to 0} \frac{-\Delta x}{x(x + \Delta x)\Delta x}$$

$$= \lim_{\Delta x \to 0} \frac{-1}{x(x + \Delta x)}$$

In the last expression, we can take the limit, but, in the earlier examples, we could not do this, because Δx was a factor of the denominator. The limit yields the answer

$$y'(x) = -\frac{1}{x^2}$$

Because of the usual difficulties encountered in computing $f'(x)$ directly from the definition, general rules of differentiation are given (Section 5.4) before special interpretations and problems are considered.

5.2 Higher Derivatives

If the limits involved exist, the higher derivatives f'', f''', ... are defined by

$$f''(x) = \frac{d}{dx}\left(\frac{dy}{dx}\right) = \frac{d^2y}{dx^2} = \lim_{\Delta x \to 0} \frac{f'(x + \Delta x) - f'(x)}{\Delta x}$$

$$f'''(x) = \frac{d}{dx}\left(\frac{d^2y}{dx^2}\right) = \frac{d^3y}{dx^3} = \lim_{\Delta x \to 0} \frac{f''(x + \Delta x) - f''(x)}{\Delta x}$$

$$\cdots$$

$$f^{(n)} = \frac{d^n y}{dx^n} = \lim_{\Delta x \to 0} \frac{f^{(n-1)}(x + \Delta x) - f^{(n-1)}(x)}{\Delta x}$$

Example. Given $f(x) = x^7 + 3e^{x^2} - \sin 2x$, find $f'(x)$ and $f''(x)$.

Solution. $f'(x) = 7x^6 + 6xe^{x^2} - 2\cos 2x$

$$f''(x) = 42x^5 + 12x^2 e^{x^2} + 6e^{x^2} + 4\sin 2x$$

5.3 Existence of Derivative

A *necessary* and *sufficient* condition for differentiability is that

(1) $\lim\limits_{\Delta x \to 0} \dfrac{\Delta y}{\Delta x}$ exist. This is more stringent than the corresponding necessary and sufficient condition for continuity that (2) $\lim\limits_{\Delta x \to 0} \Delta y = 0$

(i.e., $\lim\limits_{x \to a} f(x) = f(a)$), since conceivably (2) might hold, but (1) might fail. This is the case: *not all continuous functions have derivatives.*

PROBLEMS WITH SOLUTIONS

In Problems 1 through 6:

a. Find the slope of secant through $P(1, f(1))$, $Q(1 + \Delta x, f(1 + \Delta x))$.

b. Find the slope of tangent at P.

c. Find the equation of tangent at P.

d. Sketch $f(x)$ and the tangent to $f(x)$ at P.

1. $f(x) = x^2 + 2x + 1$. (Fig. 5.4)

2. $f(x) = \dfrac{1}{2} x^2$. (Fig. 5.5)

3. $f(x) = x - x^2$. (Fig. 5.6)

4. $f(x) = 2 - x^2$. (Fig. 5.7)

5. $f(x) = x^3$. (Fig. 5.8)

6. $f(x) = \dfrac{1}{x + 1}$. (Fig. 5.9)

Solutions.

1.

a. $\dfrac{(1 + \Delta x)^2 + 2(1 + \Delta x) + 1 - (1^2 + 2 + 1)}{\Delta x}$

b. $\displaystyle\lim_{\Delta x \to 0} \dfrac{1 + 2\Delta x + (\Delta x)^2 + 2 + 2\Delta x + 1 - 4}{\Delta x} = 4$

c. $y = 4x$

d.

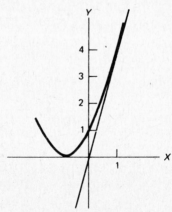

Fig. 5.4

2.

a. $\dfrac{\dfrac{1}{2}(1 + \Delta x)^2 - \dfrac{1}{2}}{\Delta x}$

b. $\lim\limits_{\triangle x \to 0} \dfrac{\triangle x + \frac{1}{2}(\triangle x)^2}{\triangle x} = 1$

c. $y = x - \dfrac{1}{2}$

d.

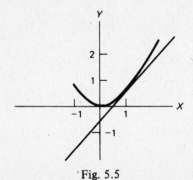

Fig. 5.5

3. a. $\dfrac{(1 + \triangle x) - (1 + \triangle x)^2}{\triangle x}$

b. $\lim\limits_{\triangle x \to 0} \dfrac{1 + \triangle x - 1 - 2\triangle x - (\triangle x)^2}{\triangle x} = -1$

c. $y = -1 (x - 1)$

d.

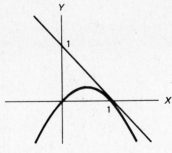

Fig. 5.6

4. a. $\dfrac{2 - (1 + \triangle x)^2 - (2 - 1)}{\triangle x}$

b. $\lim\limits_{\triangle x \to 0} \dfrac{-2\triangle x - (\triangle x)^2}{\triangle x} = -2$

c. $y - 1 = -2(x - 1)$

d.

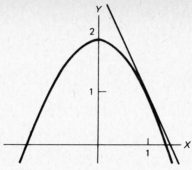

Fig. 5.7

5. a. $\dfrac{(1 + \triangle x)^3 - 1}{\triangle x}$

b. $\lim\limits_{\triangle x \to 0} \dfrac{3\triangle x + 3(\triangle x)^2 + (\triangle x)^3}{\triangle x} = 3$

c. $y - 1 = 3(x - 1)$

d.

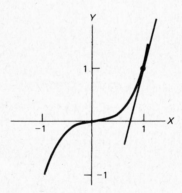

Fig. 5.8

6. a. $\dfrac{\dfrac{1}{1 + \triangle x + 1} - \dfrac{1}{2}}{\triangle x}$

b. $\lim\limits_{\triangle x \to 0} \dfrac{\dfrac{2 - (2 + \triangle x)}{2(2 + \triangle x)}}{\triangle x} =$

$$\lim_{\Delta x \to 0} \frac{-\Delta x}{2(2 + \Delta x)\Delta x} = -\frac{1}{4}$$

c. $y - \dfrac{1}{2} = -\dfrac{1}{4}(x - 1)$

d.

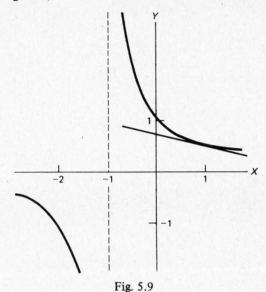

Fig. 5.9

PROBLEMS WITH ANSWERS

In Problems 7 through 12, find the slope of the tangent line.

7. $y = x^2 + x + 1$ at $(0,1)$.   *Ans.* 1

8. $y = 10x^2 - 1000x$ at $(1000; 9,000,000)$.   *Ans.* 19,000

9. $y = x^2$ at (x_0, x_0^2).   *Ans.* $2x_0$

10. $y = x^3$ at (x_0, x_0^3).   *Ans.* $3x_0^2$

11. $y = x^4$ at (x_0, x_0^4).   *Ans.* $4x_0^3$

12. $y = \dfrac{1}{x}$ at $\left(x_0, \dfrac{1}{x_0}\right)$.   *Ans.* $-\dfrac{1}{x_0^2}$

In Problems 13 through 20, given $f(x)$, find $f'(x)$ (k a constant).

13. $f(x) = kx^2$.   *Ans.* $f'(x) = 2kx$

14. $f(x) = kx^3$.   *Ans.* $f'(x) = 3kx^2$

15. $f(x) = kx$.   *Ans.* $f'(x) = k$

16. $f(x) = k$.   *Ans.* $f'(x) = 0$

17. $f(x) = 4x^2 - 7x + 3$.   *Ans.* $f'(x) = 8x - 7$

18. $f(x) = \dfrac{k}{x}$.   *Ans.* $f'(x) = -\dfrac{k}{x^2}$

19. $f(x) = \dfrac{k}{x^2}$.   *Ans.* $f'(x) = \dfrac{-2k}{x^3}$

20. $f(x) = (3 - 5x)^2$.   *Ans.* $f'(x) = 50x - 30$

5.4   Derivation of Rules of Differentiation

To establish rules by which functions can be differentiated at sight, the defining Δ-process is employed. This process, as it is applied again and again in the following cases, should be thoroughly mastered and the results should be memorized. The differentiation is with respect to x unless otherwise noted.

I. *The derivative of a constant is zero.*

$$y = c$$

$$y + \Delta y = c$$

$$\Delta y = 0$$

$$\frac{\Delta y}{\Delta x} = 0$$

$$\lim_{\Delta x \to 0} \frac{\Delta y}{\Delta x} = \frac{dy}{dx} = 0$$

II. *The derivative of the independent variable with respect to itself is unity.*

$$y = x$$

$$y + \Delta y = x + \Delta x$$

$$\Delta y = (x + \Delta x) - x = \Delta x$$

$$\frac{\Delta y}{\Delta x} = 1$$

$$\lim_{\Delta x \to 0} \frac{\Delta y}{\Delta x} = \frac{dy}{dx} = 1$$

III. *The derivative of ax*

$$y = ax$$

$$y + \Delta y = a(x + \Delta x)$$

$$\Delta y = ax + a\Delta x - ax = a\Delta x$$

$$\frac{\Delta y}{\Delta x} = a$$

$$\lim_{\Delta x \to 0} \frac{\Delta y}{\Delta x} = \frac{dy}{dx} = a$$

IV. *The derivative of x^n*

$$y = x^n$$

$$y + \Delta y = (x + \Delta x)^n$$

$$= x^n + nx^{n-1} \Delta x + \frac{n(n-1)}{2!} x^{n-2} (\Delta x)^2 + \cdots + (\Delta x)^n$$

(See the binomial theorem, Section 1.2.)

$$\Delta y = nx^{n-1} \Delta x + \frac{n(n-1)}{2!} x^{n-2} (\Delta x)^2 + \cdots + (\Delta x)^n$$

$$\frac{\Delta y}{\Delta x} = nx^{n-1} + \frac{n(n-1)}{2!} x^{n-2} \Delta x + \cdots + (\Delta x)^{n-1}$$

$$\lim_{\Delta x \to 0} \frac{\Delta y}{\Delta x} = \frac{dy}{dx} = nx^{n-1}$$

Even though this rule has been developed by the *binomial theorem* for a positive whole number n, the result itself holds for any constant value of n as can be shown by other methods. The student should apply the rule for any number n.

V. *The derivative of a (finite) sum is the sum of the derivatives*

Let u and v be functions of x, and consider

$$y = u + v$$

$$y + \Delta y = u + \Delta u + v + \Delta v$$

$$\Delta y = \Delta u + \Delta v$$

$$\frac{\Delta y}{\Delta x} = \frac{\Delta u}{\Delta x} + \frac{\Delta v}{\Delta x}$$

$$\lim_{\Delta x \to 0} \frac{\Delta y}{\Delta x} = \lim_{\Delta x \to 0} \left(\frac{\Delta u}{\Delta x} + \frac{\Delta v}{\Delta x} \right) = \frac{dy}{dx} = \frac{du}{dx} + \frac{dv}{dx}$$

VI. *The product rule for derivatives*

Let u and v be functions of x, then

$$y = uv$$

$$y + \Delta y = (u + \Delta u)(v + \Delta v) = uv + u\Delta v + v\Delta u + \Delta u \Delta v$$

$$\Delta y = u\Delta v + v\Delta u + \Delta u \Delta v$$

$$\frac{\Delta y}{\Delta x} = u\frac{\Delta v}{\Delta x} + v\frac{\Delta u}{\Delta x} + \frac{\Delta u \Delta v}{\Delta x}$$

$$\lim_{\Delta x \to 0} \frac{\Delta y}{\Delta x} = \frac{dy}{dx} = u\frac{dv}{dx} + v\frac{du}{dx}$$

Since this rule involves the derivatives of u and v separately, it cannot be applied in the event u and/or v does not have a derivative.

VII. *The quotient rule for derivatives*

$$y = \frac{u}{v}$$

$$y + \Delta y = \frac{u + \Delta u}{v + \Delta v}$$

$$\Delta y = \frac{u + \Delta u}{v + \Delta v} - \frac{u}{v} = \frac{v\Delta u - u\Delta v}{v(v + \Delta v)}$$

$$\frac{\Delta y}{\Delta x} = \frac{v\dfrac{\Delta u}{\Delta x} - u\dfrac{\Delta v}{\Delta x}}{v(v + \Delta v)}$$

$$\lim_{\Delta x \to 0} \frac{\Delta y}{\Delta x} = \frac{dy}{dx} = \frac{v\dfrac{du}{dx} - u\dfrac{dv}{dx}}{v^2}$$

It follows from this that $\dfrac{d}{dx}\left(\dfrac{1}{u}\right) = -\dfrac{1}{u^2}\dfrac{du}{dx}$; hence, $\dfrac{d}{dx}\left(\dfrac{1}{x}\right) = -\dfrac{1}{x^2}$.

VIII. *The derivative rule for inverse functions*

Since $\dfrac{\Delta y}{\Delta x} = \dfrac{1}{\dfrac{\Delta x}{\Delta y}}$

therefore, $\dfrac{dy}{dx} = \dfrac{1}{\dfrac{dx}{dy}}$

where dx/dy is the derivative of the inverse of $y = f(x)$.

IX. *The chain rule for derivatives*

Suppose $y = f(u)$, where $u = g(x)$. Now,

$$\frac{\Delta y}{\Delta x} = \frac{\Delta y}{\Delta u}\frac{\Delta u}{\Delta x}, \Delta u \neq 0$$

Therefore, taking limits, we have the very important relation,

$$\frac{dy}{dx} = \frac{dy}{du}\frac{du}{dx}$$

X. *The differentiation of parametric equations*

If $x = f(t)$ and $y = g(t)$, then dx/dt and dy/dt can be determined using rule IX; thus,

$$\frac{dy}{dx} = \frac{dy}{dt}\frac{dt}{dx} = \frac{\dfrac{dy}{dt}}{\dfrac{dx}{dt}}$$

XI. *The derivative of the nth power of a function of x*

Using rules IV and IX,

$$y = u^n$$

$$\frac{dy}{dx} = nu^{n-1}\frac{du}{dx}$$

XII. *The derivative of logarithms to bases a and e*

$$y = \log_a u$$

$$y + \Delta y = \log_a (u + \Delta u)$$

$$\Delta y = \log_a (u + \Delta u) - \log_a u = \log_a \left(\frac{u + \Delta u}{u} \right)$$

$$\frac{\Delta y}{\Delta u} = \frac{1}{\Delta u} \log_a \left(1 + \frac{\Delta u}{u} \right) = \frac{1}{u} \log_a \left(1 + \frac{\Delta u}{u} \right)^{u/\Delta u}$$

Now, the $\lim_{k \to 0} (1 + k)^{1/k} = e \cong 2.718$.

Hence,

$$\frac{dy}{du} = \frac{1}{u} \lim_{\Delta u \to 0} \log_a \left(1 + \frac{\Delta u}{u} \right)^{u/\Delta u}$$

$$= \frac{1}{u} \log_a e$$

$$\frac{dy}{dx} = \frac{1}{u} \log_a e \frac{du}{dx}$$

If base e logarithms are used, $\log_e e = 1$, and this reduces to

$$\frac{d(\log_e u)}{dx} = \frac{1}{u} \frac{du}{dx}$$

One reason for using base e logarithms (natural) is that they simplify the differentiation of the logarithmic function. In the future, where no base is mentioned, base e will always be understood.

XIII. *The derivative of a^u*

$$y = a^u$$

$$\log y = u \log a \qquad\qquad (1)$$

$$\frac{1}{y} \frac{dy}{dx} = \log a \frac{du}{dx}$$

By differentiating both sides of (1),

$$\frac{dy}{dx} = a^u \log a \frac{du}{dx}$$

If $a = e$, we have, as a very important special case,

$$\frac{d(e^u)}{dx} = e^u \frac{du}{dx}$$

XIV. *The derivative of u^v*

$$y = u^v$$

$$\log y = v \log u$$

$$\frac{1}{y}\frac{dy}{dx} = v \times \frac{1}{u}\frac{du}{dx} + \log u \frac{dv}{dx}$$

$$\frac{dy}{dx} = vu^{v-1}\frac{du}{dx} + u^v \log u \frac{dv}{dx}$$

It is worth noting that the two parts of this formula are the expressions for the derivative of u^v, v being thought of as a constant, and for the derivative of u^v, u being thought of as a constant, respectively (see XI and XIII).

An alternative method is to write

$$y = u^v = \exp(\log u^v) = \exp(v \log u)$$

Differentiating the expression on the right,

$$\frac{dy}{dx} = \exp(v \log u)\, D_x (v \log u)$$

$$= u^v \left(v \times \frac{1}{u} \times \frac{du}{dx} + \frac{dv}{dx} \times \log u \right)$$

XV. *The derivative of sin u*

$$y = \sin u$$

$$y + \Delta y = \sin(u + \Delta u) = \sin u \cos \Delta u + \cos u \sin \Delta u$$

$$\Delta y = \sin u (\cos \Delta u - 1) + \cos u \sin \Delta u$$

$$\frac{\Delta y}{\Delta u} = \sin u \left(\frac{\cos \Delta u - 1}{\Delta u} \right) + \cos u \frac{\sin \Delta u}{\Delta u}$$

Now, $\lim\limits_{\Delta u \to 0} \dfrac{\cos \Delta u - 1}{\Delta u} \doteq 0$, and $\lim\limits_{\Delta u \to 0} \dfrac{\sin \Delta u}{\Delta u} = 1$, provided radian measure is used; hence,

$$\frac{dy}{du} = \cos u$$

$$\frac{dy}{dx} = \cos u \frac{du}{dx}$$

One reason why radian measure is used is that it simplifies the differentiation of the trigonometric functions. Where trigonometric functions are involved, radian measure will be used unless otherwise noted.

XVI. *The derivative of cos u*

To differentiate cosine, we note that

$$y = \cos u = \sin\left(\frac{\pi}{2} - u\right)$$

Hence, by XV,

$$\frac{dy}{dx} = \cos\left(\frac{\pi}{2} - u\right)\frac{d}{dx}\left(\frac{\pi}{2} - u\right) = -\cos\left(\frac{\pi}{2} - u\right)\frac{du}{dx}$$

Thus, $\dfrac{dy}{dx} = -\sin u\,\dfrac{du}{dx}$

XVII. *The derivative of tan u*

$$y = \tan u = \frac{\sin u}{\cos u}$$

$$\frac{dy}{du} = \frac{\cos^2 u + \sin^2 u}{\cos^2 u} = \sec^2 u$$

$$\frac{dy}{dx} = \sec^2 u\,\frac{du}{dx}$$

XVIII. *The derivative of cot u*

$$y = \cot u = \frac{1}{\tan u}$$

Differentiating this as a quotient, we get

$$\frac{dy}{du} = \frac{0 - \sec^2 u}{\tan^2 u} = -\csc^2 u$$

$$\frac{dy}{dx} = -\csc^2 u\,\frac{du}{dx}$$

XIX. *The derivative of sec u*

$$y = \sec u = \frac{1}{\cos u}$$

$$\frac{dy}{du} = \frac{0 + \sin u}{\cos^2 u}$$

$$= \sec u \tan u$$

$$\frac{dy}{dx} = \sec u \tan u \frac{du}{dx}$$

XX. The derivative of csc u

$$y = \csc u = \frac{1}{\sin u}$$

$$\frac{dy}{du} = -\frac{\cos u}{\sin^2 u}$$

$$= -\csc u \cot u$$

$$\frac{dy}{dx} = -\csc u \cot u \frac{du}{dx}$$

XXI. The derivative of sin⁻¹ u

$$y = \sin^{-1} u$$

$$u = \sin y$$

$$\frac{du}{dy} = \cos y$$

Now, $\cos y$ will be positive if y is a first or fourth quadrantal angle and negative if the terminal side of the angle y lies in the second or third quadrant; hence,

$$\frac{dy}{du} = \frac{1}{\cos y} = \pm \frac{1}{\sqrt{1 - u^2}}$$

$$\frac{dy}{dx} = \pm \frac{1}{\sqrt{1 - u^2}} \frac{du}{dx}$$

$$\frac{dy}{dx} = \frac{1}{\sqrt{1 - u^2}} \frac{du}{dx}, \quad -\frac{\pi}{2} \leqslant \mathrm{Sin}^{-1} u \leqslant \frac{\pi}{2}$$

The notation $\mathrm{Sin}^{-1} u$ is used to indicate the *principal values* of $\sin^{-1} u$; those authors who use $\arcsin u$ as the notation for the inverse $\sin u$ generally adopt $\mathrm{Arcsin}\, u$ to indicate principal values.

XXII. *The derivative of* $\cos^{-1} u$

$$y = \cos^{-1} u$$

$$u = \cos y$$

$$\frac{du}{dy} = -\sin y$$

$$= \mp \sqrt{1 - u^2}$$

The minus sign is to be chosen if y is in the first or second quadrant, and the positive sign chosen if y is in the third or fourth quadrant.

$$\frac{dy}{du} = \frac{\mp 1}{\sqrt{1 - u^2}}$$

$$\frac{dy}{dx} = \frac{\mp 1}{\sqrt{1 - u^2}} \frac{du}{dx}$$

$$\frac{dy}{dx} = \frac{-1}{\sqrt{1 - u^2}} \frac{du}{dx}, \quad 0 \leqslant \cos^{-1} u \leqslant \pi$$

XXIII. *The derivative of* $\tan^{-1} u$

$$y = \tan^{-1} u$$

$$u = \tan y$$

$$\frac{du}{dy} = \sec^2 y = 1 + u^2$$

$$\frac{dy}{du} = \frac{1}{1 + u^2}$$

$$\frac{dy}{dx} = \frac{1}{1 + u^2} \frac{du}{dx}$$

XXIV. *The derivative of* $\cot^{-1} u$

$$y = \cot^{-1} u$$

$$u = \cot y$$

$$\frac{du}{dy} = -\csc^2 y = -(1 + u^2)$$

$$\frac{dy}{du} = \frac{-1}{1 + u^2}$$

$$\frac{dy}{dx} = \frac{-1}{1 + u^2} \frac{du}{dx}$$

XXV. *The derivative of* $sec^{-1} u$

$$y = \sec^{-1} u$$

$$u = \sec y$$

$$\frac{du}{dy} = \sec y \tan y = \pm u\sqrt{u^2 - 1}$$

$$\frac{dy}{du} = \frac{\pm 1}{u\sqrt{u^2 - 1}}$$

$$\frac{dy}{dx} = \frac{\pm 1}{u\sqrt{u^2 - 1}} \frac{du}{dx}$$

Here the plus sign is to be used if the angle y is in the first or third quadrant, and the minus sign used if the angle is in the second or fourth quadrant.

$$\frac{dy}{dx} = \frac{1}{u\sqrt{u^2 - 1}} \frac{du}{dx}, \ -\pi \leqslant \mathrm{Sec}^{-1} u \leqslant -\frac{\pi}{2} \text{ and } 0 \leqslant \mathrm{Sec}^{-1} u \leqslant \frac{\pi}{2}$$

XXVI. *The derivative of* $csc^{-1} u$

$$y = \csc^{-1} u$$

$$u = \csc y$$

$$\frac{du}{dy} = -\csc y \cot y = \mp u\sqrt{u^2 - 1}$$

$$\frac{dy}{du} = \frac{\mp 1}{u\sqrt{u^2 - 1}}$$

$$\frac{dy}{dx} = \frac{\mp 1}{u\sqrt{u^2 - 1}} \frac{du}{dx}$$

Here the minus sign holds when y is in the first or third quadrant, and the plus sign is used when y is in the second or fourth quadrant.

$$\frac{dy}{dx} = \frac{-1}{u\sqrt{u^2 - 1}} \frac{du}{dx}, \ -\pi \leqslant \mathrm{Csc}^{-1} u \leqslant -\frac{\pi}{2} \text{ and } 0 \leqslant \mathrm{Csc}^{-1} u \leqslant \frac{\pi}{2}$$

This completes the list of the fundamental elementary functions; their derivatives should be memorized. The rules and the results will be needed for further work in differentiation.

5.5 Table of Rules of Differentiation

We summarize the rules of differentiation in the following condensed table of derivatives.

THE FUNCTION

THE DERIVATIVE

1. $y = x^n$

$$\frac{dy}{dx} = nx^{n-1}$$

2. $y = u + v$

$$\frac{dy}{dx} = \frac{du}{dx} + \frac{dv}{dx}$$

3. $y = uv$

$$\frac{dy}{dx} = u\frac{dv}{dx} + v\frac{du}{dx}$$

4. $y = \dfrac{u}{v}$

$$\frac{dy}{dx} = \frac{v\dfrac{du}{dx} - u\dfrac{dv}{dx}}{v^2}$$

5. $y = f(u), u = g(x)$

$$\frac{dy}{dx} = \frac{dy}{du}\frac{du}{dx}$$

6. $y = u^n$

$$\frac{dy}{dx} = nu^{n-1}\frac{du}{dx}$$

7. $y = \log_a u$

$$\frac{dy}{dx} = \frac{1}{u}\log_a e\frac{du}{dx}$$

8. $y = \log u$

$$\frac{dy}{dx} = \frac{1}{u}\frac{du}{dx}$$

9. $y = a^u$

$$\frac{dy}{dx} = a^u \log a\frac{du}{dx}$$

10. $y = e^u$

$$\frac{dy}{dx} = e^u\frac{du}{dx}$$

11. $y = u^v$

$$\frac{dy}{dx} = vu^{v-1}\frac{du}{dx} + u^v \log u\frac{dv}{dx}$$

12. $y = \sin u$

$$\frac{dy}{dx} = \cos u\frac{du}{dx}$$

13. $y = \cos u$

$$\frac{dy}{dx} = -\sin u\frac{du}{dx}$$

14. $y = \tan u$

$$\frac{dy}{dx} = \sec^2 u\frac{du}{dx}$$

15. $y = \cot u$

$$\frac{dy}{dx} = -\csc^2 u\frac{du}{dx}$$

16. $y = \sec u$

$$\frac{dy}{dx} = \sec u \tan u\frac{du}{dx}$$

17. $y = \csc u$

$$\frac{dy}{dx} = -\csc u \cot u\frac{du}{dx}$$

18. $y = \operatorname{Sin}^{-1} u$

$$\frac{dy}{dx} = \frac{1}{\sqrt{1-u^2}}\frac{du}{dx}$$

19. $y = \operatorname{Cos}^{-1} u$

$$\frac{dy}{dx} = \frac{-1}{\sqrt{1-u^2}}\frac{du}{dx}$$

20. $y = \operatorname{Tan}^{-1} u$

$$\frac{dy}{dx} = \frac{1}{1+u^2}\frac{du}{dx}$$

21. $y = \text{Cot}^{-1} u$ $\qquad\qquad \dfrac{dy}{dx} = \dfrac{-1}{1+u^2} \dfrac{du}{dx}$

22. $y = \text{Sec}^{-1} u$ $\qquad\qquad \dfrac{dy}{dx} = \dfrac{1}{u\sqrt{u^2-1}} \dfrac{du}{dx}$

23. $y = \text{Csc}^{-1} u$ $\qquad\qquad \dfrac{dy}{dx} = \dfrac{-1}{u\sqrt{u^2-1}} \dfrac{du}{dx}$

5.6 Implicit Differentiation

Where y is given explicitly as a function of x, there is no difficulty in obtaining $\dfrac{dy}{dx}$, since if $y = f(x)$, then $\dfrac{dy}{dx} = f'(x)$. But, in the event y is given implicitly as a function of x, i.e., $F(x,y) = 0$, then, instead of first trying to solve for y, it is generally preferable to differentiate immediately as the equation stands, and then later to solve for $\dfrac{dy}{dx}$ (in terms of x and y). Such differentiation is termed **implicit differentiation**.

Example 1. Given $x^5 + x^2 y^3 - y^6 + 7 = 0$, find $\dfrac{dy}{dx}$.

Solution. Differentiating implicitly, we get

$$5x^4 + 2xy^3 + 3x^2 y^2 \frac{dy}{dx} - 6y^5 \frac{dy}{dx} = 0$$

Solving this for $\dfrac{dy}{dx}$ gives the answer

$$\frac{dy}{dx} = \frac{5x^4 + 2xy^3}{6y^5 - 3x^2 y^2}$$

Example 2. Given $y = e^{xy} + \sin x$, find $\dfrac{dy}{dx}$.

Solution. Implicit differentiation yields

$$\frac{dy}{dx} = ye^{xy} + xy'e^{xy} + \cos x$$

Hence,

$$\frac{dy}{dx} = \frac{ye^{xy} + \cos x}{1 - xe^{xy}}$$

PROBLEMS WITH SOLUTIONS

Solutions

1. $y = 7 + \sqrt{2} - 3x^0 + 5c.$ $\qquad$ $\dfrac{dy}{dx} = 0$

2. $y = 6x - 4a.$ $\qquad$ $y'(x) = 6$

3. $y = \sqrt{5}\ x^{3/2} + x^{1/2} - kx^{-2}.$ $\qquad$ $D_x y(x) = \dfrac{3\sqrt{5}}{2}\ x^{1/2} + \dfrac{1}{2\sqrt{x}}$
$$+ 2kx^{-3}$$

4. $s = \sqrt{3t} + \dfrac{9}{t^3}.$ $\qquad$ $\dfrac{ds}{dt} = \dfrac{\sqrt{3}}{2\sqrt{t}} - \dfrac{27}{t^4}$

5. $r = 1 + \dfrac{1}{\theta} + \dfrac{1}{\theta^2}.$ $\qquad$ $\dfrac{dr}{d\theta} = -\dfrac{1}{\theta^2} - \dfrac{2}{\theta^3}$

6. $w = (4z + 3)(z^2 - 7).$ $\qquad$ $\dfrac{dw}{dz} = (4z + 3)(2z) + 4(z^2 - 7)$

7. $y = (1 - x)(1 + x^2)(2 - x^3).$ $\qquad$ $y'(x) = (1 - x)(1 + x^2)(-3x^2)$
$$+ (1 - x)(2 - x^3)(2x)$$
$$- (1 + x^2)(2 - x^3)$$

8. $y = \dfrac{4x - 1}{x^2 - x^5}.$ $\qquad$ $y'(x) =$
$$\dfrac{(x^2 - x^5)4 - (4x - 1)(2x - 5x^4)}{(x^2 - x^5)^2}$$

9. $y = (5 - 3x^2)^9.$ $\qquad$ $y'(x) = 9(5 - 3x^2)^8(-6x)$

10. $y = (1 - 2x)(x^2 - 2)^3.$ $\qquad$ $y'(x) = (1 - 2x)6x(x^2 - 2)^2$
$$- 2(x^2 - 2)^3$$

11. $y = 2^{(x+1)}.$ $\qquad$ $y'(x) = 2^{(x+1)} \log 2$

12. $y = e^{2-x} - \log \sin x.$ $\qquad$ $y'(x) = -e^{2-x} - \dfrac{\cos x}{\sin x}$

13. $y = \tan \dfrac{x^2}{2} - \sec^2 \left(\dfrac{\pi}{3} - x \right)$ $y'(x) = x \sec^2 \dfrac{x^2}{2}$

$$+ 2 \sec^2 \left(\dfrac{\pi}{3} - x \right) \tan \left(\dfrac{\pi}{3} - x \right)$$

14. $y = 3 \cos^{-1} 2x.$ $y'(x) = \dfrac{-6}{\sqrt{1 - 4x^2}}$

15. $y = \tan^{-1} \dfrac{x}{a}.$ $y'(x) = \dfrac{a}{a^2 + x^2}$

In Problems 16 through 20, find the first and second derivatives.

Solutions

16. $y = x^3 - \cos 3x - \log x^5.$ $y' = 3x^2 + 3 \sin 3x - \dfrac{5}{x}$

$$y'' = 6x + 9 \cos 3x + \dfrac{5}{x^2}$$

17. $y = \dfrac{e^x + e^{-x}}{e^x - e^{-x}}.$ $y' = \dfrac{-4}{(e^x - e^{-x})^2}$

$$y'' = \dfrac{8(e^x + e^{-x})}{(e^x - e^{-x})^3}$$

18. $y = \dfrac{x}{2} \sqrt{1 - x^2} + \dfrac{1}{2} \sin^{-1} x.$ $y' = \sqrt{1 - x^2}$

$$y'' = \dfrac{-x}{\sqrt{1 - x^2}}$$

19. $y = \log \tan \left(\dfrac{\pi}{4} + \dfrac{x}{2} \right).$ $y' = \dfrac{1}{2 \cos \left(\dfrac{\pi}{4} + \dfrac{x}{2} \right) \sin \left(\dfrac{\pi}{4} + \dfrac{x}{2} \right)}$

$$y'' = \dfrac{\cos 2 \left(\dfrac{\pi}{4} + \dfrac{x}{2} \right)}{4 \left[\cos \left(\dfrac{\pi}{4} + \dfrac{x}{2} \right) \sin \left(\dfrac{\pi}{4} + \dfrac{x}{2} \right) \right]^2}$$

20. $y = \log \dfrac{x - a}{x + a}.$ $y' = \dfrac{2a}{x^2 - a^2}$

$$y'' = \dfrac{-4a}{(x^2 - a^2)^3}$$

CHAPTER 6

THE RELATIONSHIP BETWEEN INTEGRATION AND DIFFERENTIATION

6.1 Antidifferentiation

We can begin with a function f and compute its derivative f' defined by

$$f'(x) = \lim_{\Delta x \to 0} \frac{f(x + \Delta x) - f(x)}{\Delta x}$$

For example, if $f(x) = x^2$, then $f'(x) = 2x$, as we have seen in Chapter 5. The process seems to be reversible: given $f'(x)$, we might be able to find $f(x)$. If $f'(x) = 2x$ is given, then from our previous knowledge of differentiation, we come up with the suggestion that $f(x) = x^2$. But we should note that $f(x) = x^2 + 17$ would also differentiate into $f'(x) = 2x$, since the derivative of the constant 17 (and indeed of any constant) is zero. Therefore, if $f'(x) = 2x$, then $f(x) = x^2 + C$, where C is any real number. Because of the additive constant C, this reverse operation, called **antidifferentiation**, is not quite the inverse of differentiation. It is a bit more general. The **antiderivative** is also called the **indefinite integral**.

Example. Given $f'(x) = -4x + 3$, find $f(x)$.
Solution. For the $-4x$ term, we should start with an x^2, but differentiating x^2 would give us $2x$, and we have $-4x$. Therefore, we should start with $-2x^2$ to get $-4x$ by differentiating. For the 3 term, we get $3x$; therefore, $f(x) = -2x^2 + 3x + C$.

6.2 Integration

In Chapter 4, the definite integral was treated by the summation method; thus,

$$\int_a^b f(x)\,dx = \lim_{\substack{n \to \infty \\ |\Delta x_k| \to 0}} \sum_{k=1}^n f(x^*)\,\Delta x_k$$

The evaluation of this limit is very difficult and, from this point of view, the process of finding a definite integral of any but the very simplest of functions becomes an onerous task.

6.3 Differentiation

In Chapter 5, we dealt with the derivative,

$$f'(x) = \lim_{\Delta x \to 0} \frac{f(x + \Delta x) - f(x)}{\Delta x}$$

The problem of finding an antiderivative has also been considered.

6.4 Integration and Differentiation Related

There is no immediate and obvious connection between the integral and the derivative, but consider the following special cases.

$$f(u) = \int_0^x u^2 \, du = \frac{x^3}{3} \quad \text{(the definite integral by summation)}$$

$$f(x) = \frac{x^3}{3}, \text{ where } f'(x) = x^2 \quad \text{(the derivative)}$$

$$f'(x) = x^2, \text{ where } f(x) = \frac{x^3}{3} + C \quad \text{(the antiderivative, or indefinite integral)}$$

Clearly, there is a connection in this case and, indeed, there is a relationship between integration and differentiation for any continuous and differentiable function. Theorems I through VII below lead to Theorems VIII and IX which explain this relationship, and, by means of which, an integral, definite or indefinite, may be calculated by antidifferentiation instead of by summation.

THEOREM I. (Intermediate value theorem) *If f is continuous on the closed interval* $[a, b]$, *then f assumes an absolute minimum m and an absolute maximum M somewhere on* $[a,b]$. *Further, if c is a number such that* $m < c < M$, *then* x_0 *exists such that* $f(x_0) = c$; *that is, f assumes each value in between its minimum and its maximum at least once on* $[a,b]$ (Fig. 6.1).

THEOREM II. *Let f be continuous on the closed interval* $[a,b]$ *and let f be a maximum (or minimum) at an interior point* x_0, *i.e., where* $a < x_0 < b$. *If* $f'(x_0)$ *exists, then* $f'(x_0) = 0$ (Fig. 6.2).

THEOREM III. (Rolle's theorem) *Let f be continuous on the closed interval* $[a,b]$ *and let f be differentiable in the open interval* (a,b).

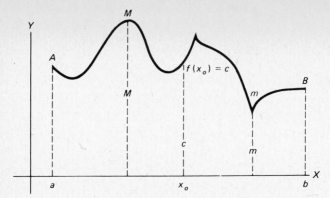

Fig. 6.1

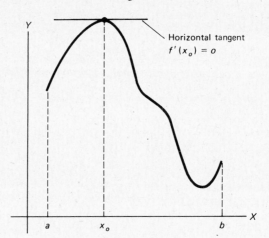

Fig. 6.2

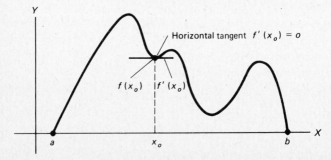

Fig. 6.3

If $f(a) = f(b) = 0$, *then there is at least one point x_0 in (a,b) such that* $f'(x_0) = 0$ (Fig. 6.3).

THEOREM IV. (Mean value theorem for derivatives) *Let f be continuous on the closed interval $[a,b]$, and let f be differentiable on the open interval (a,b). Then there exists at least one point x_0 in (a,b) such that* (Fig. 6.4)

$$f'(x_0) = \frac{f(b) - f(a)}{b - a}$$

Note that Rolle's theorem is a special case of Theorem IV.

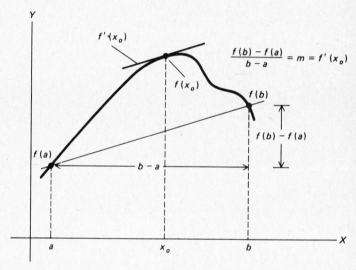

Fig. 6.4

THEOREM V. *If $f(x)$ has its derivative equal to zero at every point on an interval I, then $f(x) = C$ (C = constant) on I.*

THEOREM VI. *If f and g are two functions whose derivatives are each equal to $\phi(x)$ on I, then they differ at most by an additive constant; that is, $f(x) = g(x) + C$ on I.*

THEOREM VII. (Mean value theorem for integrals) *If f is continuous on the closed interval $[a,b]$, then there exists a point x_0 in (a,b) such that* (Fig. 6.5)

$$\int_a^b f(x)\,dx = f(x_0)(b - a)$$

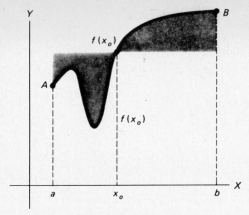

Fig. 6.5

THEOREM VIII. (Fundamental theorem of calculus, first form) *If f is continuous on a closed interval* [a,b] *and if f′ is integrable there, then*

$$\int_a^b f'(x)\,dx = f(b) - f(a)$$

THEOREM IX. (Fundamental theorem of calculus, second form) *If f is continuous on a closed interval* [a,b] *and if*

$$F(x) = \int_a^x f(t)\,dt, x \in [a,b]$$

then F′(x) = f(x).

Theorems VI and VIII tell us how to find an integral without using the summation method. In particular, to evelute $\int_a^b f(x)\,dx$, first, find any function G such that $G'(x) = f(x)$, where G is an *antiderivative* of f. Then,

$$\int_a^b f(x)\,dx = G(b) - G(a)$$

Similarly, to solve $\int_a^x f(t)\,dt$, we find an *antiderivative* G of f such that $G'(t) = f(t)$; then

$$\int_a^x f(t)\,dt = G(x) - G(a)$$

If only an antiderivative is to be found, we write

$$\int^x f(t)\,dt = G(x)$$

$$\int f(x)\,dx = G(x)$$

$$\int f(x)\,dx = G(x) + C$$

where C is the constant of integration.

Example 1. Find $\int_a^b x^2\,dx$.

Solution. We recall from our knowledge of differentiation that $D_x x^3 = 3x^2$ and an antiderivative of x^2 is $x^3/3$; hence,

$$\int_a^b x^2\,dx = \frac{x^3}{3}\bigg|_a^b = \frac{b^3}{3} - \frac{a^3}{3}$$

where $\dfrac{x^3}{3}\bigg|_a^b$ is read: x^3 *over three evaluated at* b, *the upper limit of integration, minus* x^3 *over three evaluated at* a, *the lower limit of integration.*

We may use any other antiderivative of x^2, such as $\dfrac{x^3}{3} + C$; thus,

$$\int_a^b x^2\,dx = \frac{x^3}{3} + C\bigg|_a^b = \left(\frac{b^3}{3} + C\right) - \left(\frac{a^3}{3} + C\right)$$

$$= \frac{b^3}{3} - \frac{a^3}{3}\ \text{(as before)}$$

Example 2. Evaluate $\int_a^x t^2\,dt$.

Solution.

$$\int_a^x t^2\,dt = \frac{t^3}{3}\bigg|_a^x = \frac{x^3}{3} - \frac{a^3}{3}$$

Example 3. Evaluate $\int 4x^5 \, dx$.

Solution.

$$\int 4x^5 \, dx = \frac{4}{6} x^6 + C = \frac{2}{3} x^6 + C$$

because $D_x \left(\frac{2}{3} x^6 + C \right) = 4x^5$.

The basic problem of integral calculus is to evaluate $\int f(x) \, dx$ for a given function f by *antidifferentiation*; hence, we must be thoroughly familiar with the rules of differentiation (Chapter 5).

PROBLEMS WITH ANSWERS

In Problems 1 through 8, given $f'(x)$, find the antiderivative (indefinite integral).

1. $f'(x) = 0$.   *Ans.* $f(x) = C$

2. $f'(x) = k$.   *Ans.* $f(x) = kx + C$

3. $f'(x) = x$.   *Ans.* $f(x) = \dfrac{x^2}{2} + C$

4. $f'(x) = x^2$.   *Ans.* $f(x) = \dfrac{x^3}{3} + C$

5. $f'(x) = x^3$.   *Ans.* $f(x) = \dfrac{x^4}{4} + C$

6. $f'(x) = \dfrac{1}{x^2}$.   *Ans.* $f(x) = -\dfrac{1}{x} + C$

7. $f'(x) = 3x^2 + 5x - 9$.   *Ans.* $f(x) = x^3 + \dfrac{5}{2}x^2 - 9x + C$

8. $f'(x) = 5x^3 - \dfrac{7}{x^2}$.   *Ans.* $f(x) = \dfrac{5}{4}x^4 + \dfrac{7}{x} + C$

In Problems 9 and 10 find the integral (by summation) of $f(u)$ from 0 to x. (See Chapter 4.)

9. $f(u) = 0$.   *Ans.* $\displaystyle\int_0^x 0 \, du = \lim_{n \to \infty} (0 + 0 + \cdots + 0) \frac{1}{n} x = 0$

10. $f(u) = k$ (k = const).   *Ans.* $\int_0^x k \, du = \lim\limits_{n \to \infty} (k + k + \cdots + k) \dfrac{1}{n} x$

$$= \lim\limits_{n \to \infty} nk \dfrac{1}{n} x = kx$$

In Problems 11 through 16 find the integral of $f(u)$ from a to x. (See Chapter 4.)

11. $f(u) = 0$.   *Ans.* $\int_a^x 0 \, du = 0$

12. $f(u) = k$.   *Ans.* $\int_a^x k \, du = kx - ka$

13. $f(u) = ku$ (k const).   *Ans.* $\int_a^x ku \, du = k\left(\dfrac{x^2}{2} - \dfrac{a^2}{2}\right)$

14. $f(u) = ku^2$.   *Ans.* $\int_a^x ku^2 \, du = k\left(\dfrac{x^3}{3} - \dfrac{a^3}{3}\right)$

15. $f(u) = ku^3$.   *Ans.* $\int_a^x ku^3 \, du = k\left(\dfrac{x^3}{4} - \dfrac{a^3}{4}\right)$

16. $f(u) = 3x^2 + 5x - 9$.   *Ans.* $\int_a^x (3x^2 + 5x - 9) \, dx$

$$= \left(x^3 + \dfrac{5}{2}x^2 - 9x\right) - \left(a^3 + \dfrac{5}{2}a^2 - 9a\right)$$

CHAPTER 7
RULES OF INTEGRATION

7.1 Indefinite Integrals

Now that we have seen the rules of differentiation, we can proceed to the study of integration by the method of finding antiderivatives. The symbol,

$$\int f(x)dx \tag{1}$$

for the integral comes about in the following way. When a function, say f, is to be differentiated, we must specify the variable with respect to which the differentiation is to take place. The derivative of $f(x)$ with respect to x is $\dfrac{df}{dx}$; the derivative with respect to t is $\dfrac{df}{dt}$. This could be written $\dfrac{df}{dx} \times \dfrac{dx}{dt}$. So, in the reverse process of integration, we must integrate with respect to a certain variable. We write

$$\frac{dy}{dx} = f(x) \tag{2}$$

$$dy = f(x)\,dx \tag{3}$$

Integrating both sides of (3), we get

$$y = \int dy = \int f(x)dx \tag{4}$$

where dx in (4) tells us that the integration of $f(x)$ is to take place with respect to x. The function f in (4) is called the **integrand**. Note that $\int x^3\,dx = \frac{1}{4}x^4 + C$ and that $\int t^3\,dt = \frac{1}{4}t^4 + C$, but that $\int x^3\,dt$ cannot be performed until we know what function x is of t. This corresponds directly with the parallel situation in differentiation where $\dfrac{d(x^3)}{dx} = 3x^2$,

92

but $\dfrac{d(x^3)}{dt} = 3x^2 \dfrac{dx}{dt}$, and this cannot be completed until we know the

relation between x and t. For example, let $x = e^{2t}$, then $\dfrac{d(x^3)}{dt} =$

$\dfrac{d(e^{6t})}{dt} = 6e^{6t}$. Similarly, $\int x^3 \, dt$ would become $\int e^{6t} \, dt = \frac{1}{6} e^{6t} + C$, as

the student can quickly check by differentiation.

In the development of the formulas for differentiation, we have generally worked with the function u. We shall do likewise in the formulas for integration. The integrals I through XVII are the direct consequence of our knowledge of the derivatives developed in Chapter 5. They should be memorized.

I. $\displaystyle\int 0 \, dx = C$

II. $\displaystyle\int k \, dx = k \int dx = kx + C$

III. $\displaystyle\int x^n \, dx = \frac{x^{n+1}}{n+1} + C, n \neq -1$

IV. $\displaystyle\int u^n \, du = \frac{u^{n+1}}{n+1} + C, n \neq -1$

V. $\displaystyle\int \frac{1}{u} \, du = \log u + C$

VI. $\displaystyle\int [u(x) \pm v(x)] \, dx = \int u(x) \, dx \pm \int v(x) \, dx$

(The integral of a (finite) sum is the sum of the integrals.)

VII. $\displaystyle\int e^u \, du = e^u + C$

VIII. $\displaystyle\int a^u \, du = \frac{a^u}{\log a} + C$

IX. $\displaystyle\int \sin u \, du = -\cos u + C$

X. $\displaystyle\int \cos u \, du = \sin u + C$

XI. $\displaystyle\int \sec^2 u \, du = \tan u + C$

XII. $\displaystyle\int \csc^2 u \, du = -\cot u + C$

XIII. $\displaystyle\int \sec u \tan u \, du = \sec u + C$

XIV. $\int \csc u \cot u \, du = - \csc u + C$

XV. $\int \dfrac{du}{\sqrt{a^2 - u^2}} = \sin^{-1} \dfrac{u}{a} + C$

XVI. $\int \dfrac{du}{a^2 + u^2} = \dfrac{1}{a} \tan^{-1} \dfrac{u}{a} + C$

XVII. $\int \dfrac{du}{u\sqrt{u^2 - a^2}} = \dfrac{1}{a} \sec^{-1} \dfrac{u}{a} + C$

The following integrals are met so frequently that they too should be memorized. It will be noted that they do not come from differentiation formulas which we have considered as the standard elementary ones to be committed to memory, but they are, nevertheless, very important and a knowledge of them will be of tremendous aid in the integration of more complicated functions.

XVIII. $\int \tan u \, du = \log \sec u + C$

XIX. $\int \cot u \, du = \log \sin u + C$

XX. $\int \sec u \, du = \log (\sec u + \tan u) + C$

XXI. $\int \csc u \, du = \log (\csc u - \cot u) + C$

XXII. $\int \dfrac{du}{u^2 - a^2} = \dfrac{1}{2a} \log \dfrac{u - a}{u + a} + C$

XXIII. $\int \dfrac{du}{a^2 - u^2} = \dfrac{1}{2a} \log \dfrac{a + u}{a - u} + C$

XXIV. $\int \dfrac{du}{\sqrt{u^2 \pm a^2}} = \log (u + \sqrt{u^2 \pm a^2}) + C$

XXV. $\int \sqrt{a^2 - u^2} \, du = \dfrac{u}{2} \sqrt{a^2 - u^2} + \dfrac{a^2}{2} \sin^{-1} \dfrac{u}{a} + C$

XXVI. $\int \sqrt{u^2 \pm a^2} \, du = \dfrac{u}{2} \sqrt{u^2 \pm a^2} \pm \dfrac{a^2}{2} \log (u + \sqrt{u^2 \pm a^2}) + C$

It would be a simple matter to verify these integrals by differentiating the answers, thus obtaining the integrands. But we should derive these answers by way of beginning a discussion of the general methods of integration. To do this, we begin with XVIII, offhand, we know of no function that will differentiate into tan u; but write

$$\int \tan u \, du = \int \frac{\sin u}{\cos u} \, du \qquad (5)$$

and then examine V, which says that the integral of a fraction whose numerator is the differential du of the denominator u is $\log u$. The denominator of (5) is $\cos u$, whose differential is $-\sin u\,du$. Hence, we could write

$$\int \tan u\, du = -\int \frac{-\sin u\, du}{\cos u} = -\log \cos u + C \tag{6}$$

Since $-\log \cos u = \log (\cos u)^{-1} = \log \frac{1}{\cos u} = \log \sec u$, the form of the answer given in XVIII follows.

Similarly for XIX, we have

$$\int \cot u\, du = \int \frac{\cos u\, du}{\sin u} = \log \sin u + C \quad \text{(by V)} \tag{7}$$

To derive XX, we write

$$\int \sec u\, du = \int \frac{du}{\cos u} = \int \frac{\cos u\, du}{\cos^2 u} = \int \frac{d(\sin u)}{1 - \sin^2 u} \tag{8}$$

which is of the form $\int \frac{dv}{1 - v^2}$ (if we set $v = \sin u$), and this, in turn, is of type XXIII; therefore,

$$\int \sec u\, du = \frac{1}{2} \log \frac{1 + \sin u}{1 - \sin u} + C \tag{9}$$

This is a perfectly good standard form for $\int \sec u\, du$. By noting that $\frac{1 + \sin u}{1 - \sin u} = (\sec u + \tan u)^2$, a trigonometric identity, the final form XX is obtained.

There is still another useful form for this integral:

$$\int \sec u\, du = \log \tan \left(\frac{\pi}{4} + \frac{u}{2} \right) + C$$

This may be derived from (9) by the change in variable $u = v - \frac{\pi}{2}$ (in the answer).

The integral in XXI, $\int \csc u \, du$, is derived in a manner exactly analogous to that used for $\int \sec u \, du$ and is left as an exercise for the student.

For formula XXII, we make use of *partial fractions* which might be called a process inverse to that of reducing a fraction to a common denominator. This method is explained in some detail in Section 7.4; at the present, it will be sufficient to note that

$$\frac{1}{u^2 - a^2} = \frac{\dfrac{1}{2a}}{u - a} - \frac{\dfrac{1}{2a}}{u + a}$$

By reducing the right-hand members to a common denominator, we obtain the left-hand member of the equation; hence,

$$\int \frac{du}{u^2 - a^2} = \frac{1}{2a} \int \frac{du}{u - a} - \frac{1}{2a} \int \frac{du}{u + a} \tag{10}$$

But, each of these last two integrals is of form V; therefore,

$$\int \frac{du}{u^2 - a^2} = \frac{1}{2a} \log (u - a) - \frac{1}{2a} \log (u + a) + C$$

$$= \frac{1}{2a} \log \frac{u - a}{u + a} + C \tag{11}$$

Since XXIII is handled in identically the same way, we leave it to the student as an exercise.

To obtain one of the integrals in XXIV, $\int \dfrac{du}{\sqrt{u^2 + a^2}}$, we make use of the device known as the method of substitution (or transformation). (See Section 7.3 for a fuller discussion.) Let $u = a \tan v$; then $du = a \sec^2 v \, dv$, and our integral becomes

$$\int \frac{du}{\sqrt{u^2 + a^2}} = \int \frac{a \sec^2 v \, dv}{\sqrt{a^2 \tan^2 v + a^2}}$$

$$= \int \frac{a \sec^2 v \, dv}{a \sec v} = \int \sec v \, dv$$

$$= \log (\sec v + \tan v) + C \text{ (by XX.)} \tag{12}$$

But, if $u = a \tan v$, $\sec v = \dfrac{\sqrt{u^2 + a^2}}{a}$, and (12) becomes

$$\int \frac{du}{\sqrt{u^2 + a^2}} = \log \left(\frac{u}{a} + \frac{\sqrt{u^2 + a^2}}{a} \right) + C$$

$$= \log \left(\frac{u + \sqrt{u^2 + a^2}}{a} \right) + C$$

$$= \log (u + \sqrt{u^2 + a^2}) - \log a + C \qquad (13)$$

But a is a constant; hence, we might combine $- \log a + c$, calling it a new constant k. This completes the derivation of the one integral in XXIV and the other, using the minus sign, follows the same pattern.

To perform the integration in XXV, another substitution is made setting $u = a \sin v$ and $du = a \cos v \, dv$; the integral becomes

$$\int \sqrt{a^2 - u^2} \, du = \int a^2 \cos^2 v \, dv$$

$$= a^2 \int \left(\frac{1 + \cos 2v}{2} \right) dv$$

$$= \frac{a^2}{2} \int (1 + \cos 2v) \, dv$$

$$= \frac{a^2}{2} \left(v + \frac{1}{2} \sin 2v \right) + C$$

$$= \frac{a^2}{2} (v + \sin v \cos v) + C$$

$$= \frac{a^2}{2} \left(\sin^{-1} \frac{u}{a} + \frac{u}{a} \frac{\sqrt{a^2 - u^2}}{a} \right) + C \qquad (14)$$

This is formula XXV.

Finally, integrals XXVI are best treated by a process called *integration by parts* which is explained in Section 7.2.

7.2 Integration by Parts

This method makes use of the simple form of the differential of a product:

$$d(uv) = u \, dv + v \, du \qquad (1)$$

Transposing, we have

$$u \, dv = d(uv) - v \, du \qquad (2)$$

Upon integrating (2), we obtain

$$\int u \, dv = uv - \int v \, du \qquad (3)$$

Hence, the integral $\int u \, dv$ is made to depend upon the integral $\int v \, du$, which may be more readily handled or may be recognizable as one of the standard forms. We illustrate the method first with a simple example.

Example 1. Find $\int xe^x \, dx$.

Solution. This does not immediately come under any of the standard cases.

Let us choose $u = x$ and $dv = e^x \, dx$. Then $du = dx$ and $v = \int e^x \, dx = e^x$. (We shall add our constant of integration later.) Then our integral becomes, integrating by parts

$$\int u \, dv = uv - \int v \, du$$

$$\int xe^x \, dx = xe^x - \int e^x \, dx$$

This last integral is one of the standard forms; hence,

$$\int xe^x \, dx = xe^x - e^x + C$$

We shall now derive formula XXVI using the plus sign. First we transform the integral by setting $u = a \tan z$ and du becomes $du = a \sec^2 z \, dz$. The integral, therefore, becomes

$$\int \sqrt{u^2 + a^2} \, du = \int a \sec z \, a \sec^2 z \, dz \qquad (4)$$

$$= a^2 \int \sec^3 z \, dz \qquad (5)$$

This last integral, we integrate by parts, setting $u = \sec z$ and $dv = \sec^2 z \, dz$; hence, $du = \sec z \tan z \, dz$ and $v = \int \sec^2 z \, dz = \tan z$. Thus, (5) becomes, upon applying the integration by parts formula (3)

$$a^2 \int \sec^3 z \, dz = a^2 \sec z \tan z - a^2 \int \sec z \tan^2 z \, dz \tag{6}$$

$$= a^2 \sec z \tan z - a^2 \int \sec z \, (\sec^2 z - 1) \, dz$$

$$= a^2 \sec z \tan z + a^2 \int \sec z \, dz - a^2 \int \sec^3 z \, dz$$

$$= a^2 \sec z \tan z + a^2 \log (\sec z + \tan z) - a^2 \int \sec^3 z \, dz$$

$$\text{(by XX)} \tag{7}$$

The last member in (7) is the very thing we seek, namely, $\int \sec^3 z \, dz$; but it has the factor $-a^2$. Hence, if we bring it over to the left-hand side of the equation, we get

$$2a^2 \int \sec^3 z \, dz = a^2 \sec z \tan z + a^2 \log (\sec z + \tan z) \tag{8}$$

Dividing by 2,

$$a^2 \int \sec^3 z \, dz = \frac{a^2}{2} \sec z \tan z + \frac{a^2}{2} \log (\sec z + \tan z) \tag{9}$$

Finally, transforming this back into the variable u, we have XXVI;

$$\int \sqrt{u^2 + a^2} \, du = \frac{u}{2} \sqrt{u^2 + a^2} + \frac{a^2}{2} \log (u + \sqrt{u^2 + a^2}) + C$$

The student should carry through XXVI when the minus sign is used.

This completes the derivation of the standard formulas of integration I through XXVI, and we now further illustrate the principle of integration by parts.

Example 2. Find $\int x^2 \sin x \, dx$.

Solution. Set $u = x^2$ and $dv = \sin x \, dx$.
The object is so to choose u and dv as *first* to make dv integrable and
second to make $\int v \, du$ simpler than the original $\int u \, dv$. Here, $du = 2x \, dx$
and $v = -\cos x$; then

$$\int x^2 \sin x \, dx = -x^2 \cos x + \int 2x \cos x \, dx$$

Note that this last integral is of the same type, except that we have x instead of x^2 as a multiplier of the trigonometric part. We therefore apply
the rule once more to $\int x \cos x \, dx$, setting $u = x$ and $dv = \cos x \, dx$,
which gives $du = dx$ and $v = \sin x$. Our integral becomes

$$\int x^2 \sin x \, dx = -x^2 \cos x + 2 \left(x \sin x - \int \sin x \, dx \right)$$

$$= -x^2 \cos x + 2x \sin x + 2 \cos x + C$$

Example 3. Find $\int e^x \sin x \, dx$.

Solution. Set $u = e^x$ and $dv = \sin x \, dx$; then $du = e^x \, dx$ and $v = -\cos x$.

$$\int e^x \sin x \, dx = -e^x \cos x + \int e^x \cos x \, dx$$

For the last integral, set $u = e^x$ and $dv = \cos x \, dx$; then $du = e^x \, dx$ and
$v = \sin x$; hence,

$$\int e^x \sin x \, dx = -e^x \cos x + \left(e^x \sin x - \int e^x \sin x \, dx \right)$$

Transposing the last integral

$$2 \int e^x \sin x \, dx = -e^x \cos x + e^x \sin x$$

Finally,

$$\int e^x \sin x \, dx = \frac{e^x}{2} (\sin x - \cos x) + C$$

Example 4. Find $\int x^3 \sqrt{x^2 - 2} \, dx$.

Solution. Set $u = x^2$ and $dv = x\sqrt{x^2 - 2} \, dx$. Then $du = 2x \, dx$ and $v = \frac{1}{3}(x^2 - 2)^{3/2}$, by IV; thus,

$$\int x^3 \sqrt{x^2 - 2} \, dx = \frac{x^2}{3}(x^2 - 2)^{3/2} - \frac{2}{3}\int x(x^2 - 2)^{3/2} \, dx$$

This last integral is again one of type IV; hence,

$$\int x^3 \sqrt{x^2 - 2} \, dx = \frac{x^2}{3}(x^2 - 2)^{3/2} - \frac{2}{15}(x^2 - 2)^{5/2} + C$$

Only by solving many problems will the student gain proficiency in this method. It is not an easy matter to choose the parts u and dv without wide experience. It should be said that a particular problem may not yield readily to this method; in this case, other methods should be tried. Among these is the one of *substitutions* already used in deriving several of the standard formulas.

7.3 Integration by Substitution

No general rule can be laid down as to which substitution (transformation) will reduce a given integral to one of recognizable form. But, in certain cases, particular substitutions automatically suggest themselves. Since the Pythagorean theorem for a right triangle states that the square on the hypotenuse equals the sums of squares of the two sides, an appropriate trigonometric substitution might simplify a given integral involving such quantities as $\sqrt{u^2 - a^2}$, $\sqrt{u^2 + a^2}$, and $\sqrt{a^2 - u^2}$ (Fig. 7.1); for example,

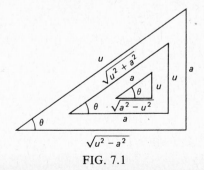

FIG. 7.1

If $\sqrt{u^2 - a^2}$ is present, try $u = a \csc \theta$.
If $\sqrt{u^2 + a^2}$ is present, try $u = a \tan \theta$.
If $\sqrt{a^2 - u^2}$ is present, try $u = a \sin \theta$.

But success is not assured even when such a substitution is used, and it may be necessary to apply other methods.

Example 1. Find $\int \dfrac{x^2 \, dx}{\sqrt{4 - x^2}}$.

Solution. Set $x = 2 \sin \theta$, and $dx = 2 \cos \theta \, d\theta$.

(The student should draw a figure and note that $\dfrac{x}{\sqrt{4 - x^2}} = \tan \theta$.)

$$\int \frac{x^2 \, dx}{\sqrt{4 - x^2}} = 4 \int \sin \theta \, \tan \theta \, \cos \theta \, d\theta$$

$$= 4 \int \sin^2 \theta \, d\theta$$

This last integral is readily handled by writing

$$\sin^2 \theta = \frac{1 - \cos 2\theta}{2}$$

Therefore,

$$\int \frac{x^2 \, dx}{\sqrt{4 - x^2}} = 2 \int (1 - \cos 2\theta) \, d\theta$$

$$= 2\theta - \sin 2\theta + C$$
$$= 2\theta - 2 \sin \theta \cos \theta + C$$
$$= 2 \sin^{-1} \frac{x}{2} - \frac{x}{2} \sqrt{4 - x^2} + C$$

Example 2. Find $\int \dfrac{x \, dx}{(x^2 + 9)^{5/2}}$.

Solution. This requires no substitution, since it is already in standard form IV; hence,

$$\int \frac{x \, dx}{(x^2 + 9)^{5/2}} = \frac{1}{2} \int (x^2 + 9)^{-5/2} \, (2x \, dx)$$

$$= \frac{1}{2} \frac{(x^2 + 9)^{-3/2}}{-\frac{3}{2}} + C$$

$$= -\frac{1}{3}(x^2 + 9)^{-3/2} + C$$

Example 3. Find $\int \sin^3 x \, dx$.

Solution. Write this as

$$\int (1 - \cos^2 x) \sin x \, dx = \int (\sin x - \cos^2 x \sin x) \, dx$$

$$= -\cos x + \frac{1}{3} \cos^3 x + C$$

Example 4. Find $\int \dfrac{dx}{(5 + x^2)^{3/2}}$.

Solution. Set $x = \sqrt{5} \tan \theta$ and $dx = \sqrt{5} \sec^2 \theta \, d\theta$.

$$\int \frac{dx}{(5 + x^2)^{3/2}} = \int \left(\frac{1}{5^{3/2}} \cos^3 \theta \right) (5^{1/2} \sec^2 \theta \, d\theta)$$

$$= \frac{1}{5} \int \cos \theta \, d\theta$$

$$= \frac{1}{5} \sin \theta + C$$

$$= \frac{1}{5} \frac{x}{\sqrt{5 + x^2}} + C$$

Again, some other substitution may be suggested by the particular form of the function to be integrated.

Example 5. Find $\int \dfrac{dx}{x + 5 - \sqrt{x + 5}}$.

Solution. Here it seems reasonable to think of $(x + 5)$ as a new variable. Because of the square root term we set $\sqrt{x + 5} = z$; then $x + 5 = z^2$ and $dx = 2z \, dz$. Our integral becomes

$$\int \frac{dx}{x + 5 - \sqrt{x + 5}} = \int \frac{2z \, dz}{z^2 - z}$$

$$= 2 \int \frac{dz}{z - 1} = 2 \log (z - 1) + C$$

$$= 2 \log (\sqrt{x + 5} - 1) + C$$

Example 6. Find $\int \dfrac{dx}{\sqrt{x^2 - 3x + 2}}$.

Solution. If this integral could be put into the form $\int \dfrac{du}{\sqrt{u^2 \pm a^2}}$, it could be integrated by XXIV. That this is possible will be seen after a little reflection upon the process of completing the square. We write

$$\int \frac{dx}{\sqrt{x^2 - 3x + 2}} = \int \frac{dx}{\sqrt{x^2 - 3x + \frac{9}{4} + 2 - \frac{9}{4}}}$$

$$= \int \frac{dx}{\sqrt{\left(x - \frac{3}{2}\right)^2 - \frac{1}{4}}}$$

Now set $x - \frac{3}{2} = z$, and we get

$$= \int \frac{dz}{\sqrt{z^2 - \frac{1}{4}}} = \log \left(z + \sqrt{z^2 - \frac{1}{4}}\right) + C$$

$$= \log (x - \frac{3}{2} + \sqrt{x^2 - 3x + 2}) + C$$

The principle of completing the square should be applied to integrals of the form XV, XVI, and XXII–XXVI, which contain a linear (first-degree) term u, in addition to the quadratic term u^2 and the constant a^2. We illustrate with another example.

Example 7. Find $\int \dfrac{(3x - 5)\,dx}{x^2 + 2x + 9}$.

Solution. We first write this in the form

$$\frac{3}{2} \int \frac{(2x + 2)\,dx}{x^2 + 2x + 9} - 8 \int \frac{dx}{x^2 + 2x + 9} \tag{a}$$

The factor in the denominators is the same as in the original integral. The numerator of the first integral in (a) is made (chosen) to be the derivative of this denominator. This is done to put that integral into standard form $V, \int \dfrac{du}{u}$. The multiplicative factor $3/2$ will result in giving us the $3x$ term that is needed. Finally, $-8 \int \dfrac{dx}{x^2 + 2x + 9}$ is added so that we get our -5 term in the original integral. A partial answer is, therefore,

$$\frac{3}{2} \log (x^2 + 2x + 9) - 8 \int \frac{dx}{x^2 + 2x + 9} \tag{b}$$

We now complete the square in the integral remaining in (b), and this integral becomes

$$\int \frac{dx}{x^2 + 2x + 9} = \int \frac{dx}{(x + 1)^2 + 8} \tag{c}$$

Next let $x + 1 = z$ and $dx = dz$, then (c) becomes

$$\int \frac{dz}{z^2 + 8}$$

which is of form XVI and equals $\dfrac{1}{2\sqrt{2}} \tan^{-1} \dfrac{z}{2\sqrt{2}}$.

The final answer then is

$$\int \frac{3x - 5}{x^2 + 2x + 9} = \frac{3}{2} \log (x^2 + 2x + 9) - \frac{1}{2\sqrt{2}} \tan^{-1} \frac{x + 1}{2\sqrt{2}} + C$$

Example 8. Find $\int \dfrac{(x + 1)\, dx}{2x^2 - 4x - 5}$.

Solution. We treat this in the same way we treated the problem in Example 7.

$$\int \frac{(x + 1)\, dx}{2x^2 - 4x - 5} = \frac{1}{4} \int \frac{(4x - 4)\, dx}{2x^2 - 4x - 5} + 2 \int \frac{dx}{2x^2 - 4x - 5}$$

$$= \frac{1}{4} \log (2x^2 - 4x - 5) + 2 \int \frac{dx}{2(x^2 - 2x - \frac{5}{2})}$$

This last integral equals

$$\int \frac{dx}{(x-1)^2 - \frac{7}{2}} \tag{a}$$

Now set $x - 1 = z$ and $dx = dz$, then (a) becomes

$$\int \frac{dz}{z^2 - \frac{7}{2}}$$

This is of type XXII and equals

$$\frac{1}{2\sqrt{\frac{7}{2}}} \log \frac{z - \sqrt{\frac{7}{2}}}{z + \sqrt{\frac{7}{2}}} = \frac{\sqrt{14}}{14} \log \frac{\sqrt{2}(x-1) - \sqrt{7}}{\sqrt{2}(x-1) + \sqrt{7}}$$

Hence, finally,

$$\int \frac{(x+1)\,dx}{2x^2 - 4x - 5} = \frac{1}{4} \log (2x^2 - 4x - 5) + \frac{\sqrt{14}}{14} \log \frac{\sqrt{2}(x-1) - \sqrt{7}}{\sqrt{2}(x-1) + \sqrt{7}} + C$$

Because of the complicated factors involved there seems to be no point in combining these terms.

The important fact to be gained from the last two examples is that integrals of the type $\int \frac{(ax+b)\,dx}{px^2 + qx + r}$ always reduce by the methods used in these examples to integrals of types V, XVI, and/or XXII, or XXIII (logarithms and arc tangents).

Example 9. $\int x(7 + 3x)^{1/3}\,dx$.

Solution. The substitution $7 + 3x = z^3$ seems a reasonable one, since this will rationalize the integrand; $dx = z^2\,dz$.

$$\int x(7 + 3x)^{1/3}\,dx = \int \frac{z^3 - 7}{3} \cdot z \cdot z^2\,dz$$

$$= \int \left(\frac{z^6}{3} - \frac{7}{3}z^3 \right)\,dz$$

$$= \frac{z^7}{21} - \frac{7}{12}z^4 + C$$

$$= \frac{(7 + 3x)^{7/3}}{21} - \frac{7}{12}(7 + 3x)^{4/3} + C$$

Again, we repeat that the student will gain knowledge of integration by means of substitutions only after solving many problems and making many false starts; what seems to be the "obvious" substitution to make often proves to be of no aid at all in a particular problem.

There is a third standard method in use in integration beyond the method of integration by parts and the method of substitutions. This method is discussed in the next section.

7.4 Integration by Partial Fractions

This method is particularly useful in the integration of rational fractions (the quotient of two polynomials) where the denominator has real factors. Suppose the integrand is of the form $\dfrac{ax + b}{x^2 + px + q} \equiv \dfrac{ax + b}{(x - \alpha)(x - \beta)}$, where the factors of $x^2 + px + q$ are $(x - \alpha)$ and $(x - \beta)$. We suppose that $ax + b$ is not one of these factors. Algebraically, it is evident that constants A and B exist such that

$$\frac{ax + b}{(x - \alpha)(x - \beta)} \equiv \frac{A}{x - \alpha} + \frac{B}{x - \beta} \tag{1}$$

If the right-hand members are reduced to a common denominator, (1) becomes

$$\frac{ax + b}{(x - \alpha)(x - \beta)} \equiv \frac{A}{x - \alpha} + \frac{B}{x - \beta} \equiv \frac{A(x - \beta) + B(x - \alpha)}{(x - \alpha)(x - \beta)} \tag{2}$$

$$\equiv \frac{(A + B)x + (-A\beta - B\alpha)}{(x - \alpha)(x - \beta)}$$

Now, A and B can be determined so that $A + B = a$ and $A\beta + B\alpha = -b$. In solving these simultaneous equations by determinants, we find

$$A = \frac{\begin{vmatrix} a & 1 \\ -b & \alpha \end{vmatrix}}{\begin{vmatrix} 1 & 1 \\ \beta & \alpha \end{vmatrix}} \qquad B = \frac{\begin{vmatrix} 1 & a \\ \beta & -b \end{vmatrix}}{\begin{vmatrix} 1 & 1 \\ \beta & \alpha \end{vmatrix}} \qquad \begin{vmatrix} 1 & 1 \\ \beta & \alpha \end{vmatrix} = \alpha - \beta \neq 0$$

We suppose that α is distinct from β; hence,

$$\int \frac{(ax + b)\,dx}{(x - \alpha)(x - \beta)} = \int \frac{A\,dx}{x - \alpha} + \int \frac{B\,dx}{x - \beta}$$

$$= A \log (x - \alpha) + B \log (x - \beta) + C$$

Example 1. $\int \dfrac{(2x-3)\,dx}{x^2-x-42}$

Solution. $\dfrac{2x-3}{x^2-x-42} = \dfrac{A}{x-7} + \dfrac{B}{x+6}$

$$2x-3 = A(x+6) + B(x-7)$$

Hence, $A = \frac{11}{13}$, $B = \frac{15}{13}$, and

$$\int \frac{(2x-3)\,dx}{x^2-x-42} = \frac{11}{13}\int \frac{dx}{x-7} + \frac{15}{13}\int \frac{dx}{x+6}$$

$$= \frac{11}{13}\log(x-7) + \frac{15}{13}\log(x+6) + C$$

Sometimes it is convenient to write the arbitrary constant of integration as a logarithm, i.e., write $C = \log k$. Our answer then becomes

$$\log(x-7)^{11/13} + \log(x+6)^{15/13} + \log k = \log k(x-7)^{11/13}(x+6)^{15/13}$$

Example 2. $\int \dfrac{6x^3 + 17x^2 + 13x - 6}{3x^3 + 2x^2 - x}\,dx$

Solution. In case the numerator is of the same degree as the denominator (or of higher degree), the first thing to do is to divide out until the numerator is of lower degree. Upon dividing, we find

$$\frac{6x^3 + 17x^2 + 13x - 6}{3x^3 + 2x^2 - x} = 2 + \frac{13x^2 + 15x - 6}{3x^3 + 2x^2 - x}$$

We now factor the denominator, and write

$$\frac{13x^2 + 15x - 6}{x(3x-1)(x+1)} = \frac{A}{x} + \frac{B}{3x-1} + \frac{C}{x+1}$$

$$13x^2 + 15x - 6 = A(3x-1)(x+1) + Bx(x+1) + Cx(3x-1)$$

Setting, $x = 0$, $x = \frac{1}{3}$ and $x = -1$, respectively, we compute $A = 6$, $B = 1$, and $C = -2$; hence,

$$\int \frac{6x^3 + 17x^2 + 13x - 6}{3x^3 + 2x^2 - x}\,dx = \int \left(2 + \frac{6}{x} + \frac{1}{3x-1} - \frac{2}{x+1}\right)dx$$

$$= 2x + 6\log x + \frac{1}{3}\log(3x-1)$$

$$- 2\log(x+1) + \log c$$

$$= 2x + \log \frac{cx^6 (3x - 1)^{1/3}}{(x + 1)^2}$$

In the case of a repeated root, there is a slight modification of the process. It is evident that $\dfrac{ax + b}{(x - \alpha)(x - \alpha)(x - \alpha)}$ could not be written as the sum of three fractions each with constant numerator and linear denominator; that is,

$$\frac{ax + b}{(x - \alpha)(x - \alpha)(x - \alpha)} \neq \frac{A}{x - \alpha} + \frac{B}{x - \alpha} + \frac{C}{x - \alpha}$$

Since each of the right-hand members is of the same type, we write

$$\frac{ax + b}{(x - \alpha)(x - \alpha)(x - \alpha)} = \frac{A}{x - \alpha} + \frac{B}{(x - \alpha)^2} + \frac{C}{(x - \alpha)^3}$$

from which $ax + b \equiv A(x - \alpha)^2 + B(x - \alpha) + C$.

The numbers, A, B, and C can now be determined in the usual way, and the integration reduces to the types $\displaystyle\int \frac{du}{u}$ and $\displaystyle\int u^n \, du$.

Example 3. Find $\displaystyle\int \frac{(x^2 - 5)\,dx}{x(x - 1)^2}$

Solution. Write

$$\frac{x^2 - 5}{x(x - 1)^2} = \frac{A}{x} + \frac{B}{x - 1} + \frac{C}{(x - 1)^2}$$

$$x^2 - 5 = A(x - 1)^2 + Bx(x - 1) + Cx$$

$$A = -5, B = 6, C = -4$$

Hence, $\displaystyle\int \frac{(x^2 - 5)\,dx}{x(x - 1)^2} = -5 \int \frac{dx}{x} + 6 \int \frac{dx}{x - 1} - 4 \int \frac{dx}{(x - 1)^2}$

$$= -5 \log x + 6 \log (x - 1) + \frac{4}{x - 1} + \log c$$

$$= \log \frac{c(x - 1)^6}{x^5} + \frac{4}{x - 1}$$

In case the denominator cannot be broken up wholly into real linear factors, then another algebraic modification is necessary. For each nonrepeated, irreducible, quadratic factor appearing in the denominator, we must have a partial fraction of the form $\dfrac{Ax + B}{x^2 + px + q}$; where the numerator is a linear function. An example will make it clear why this is so.

Example 4. Find $\displaystyle\int \dfrac{(3x^2 - x - 3)dx}{(x + 1)(x^2 + x + 1)}$.

Solution. $\dfrac{3x^2 - x - 3}{(x + 1)(x^2 + x + 1)} = \dfrac{A}{x + 1} + \dfrac{Bx + C}{x^2 + x + 1}$

$$3x^2 - x - 3 = A(x^2 + x + 1) + (Bx + C)(x + 1) \qquad\qquad (a)$$

Note that if the second partial fraction did not have a linear numerator $Bx + C$, but, instead, had just a constant D, then we should in general be unable to determine the two constants A and D satisfying the three conditions implied in (a). First let us try using just A and D.

$3x^2 - x - 3 = A(x^2 + x + 1) + D(x + 1)$, or (1) $3 = A$, (2) $-1 = A + D$, and (3) $-3 = A + D$; it is obvious that conditions (2) and (3) are inconsistent. Returning to (a), we compute $A = 1, B = 2$, and $C = -4$, and our integral becomes

$$\int \frac{(3x^2 - x - 3)dx}{(x + 1)(x^2 + x + 1)} = \int \frac{dx}{x + 1} + \int \frac{(2x - 4)dx}{x^2 + x + 1}$$

$$= \log(x + 1) + \int \frac{(2x + 1)dx}{x^2 + x + 1} - 5\int \frac{dx}{x^2 + x + 1}$$

$$= \log(x + 1) + \log(x^2 + x + 1) -$$

$$\frac{10}{\sqrt{3}} \tan^{-1} \frac{2x + 1}{\sqrt{3}} + C \quad (b)$$

The last integral in (b) is treated in the manner of Example 7, Section 7.3.

If repeated quadratic factors are present, the situation is still more complex arithmetically, but the problem presents no theoretical difficulties.

Example 5. Find $\displaystyle\int \dfrac{(x^3 + x - 3)dx}{(x^2 + 2x + 2)^2}$.

Solution. We write

$$\frac{x^2 + x - 3}{(x^2 + 2x + 2)^2} = \frac{Ax + B}{(x^2 + 2x + 2)} + \frac{Cx + D}{(x^2 + 2x + 2)^2}$$

Here, $A = 1, B = -2, C = 3, D = 1$. Hence,

$$\int \frac{(x^3 + x - 3)\,dx}{(x^2 + 2x + 2)^2} = \int \frac{(x - 2)\,dx}{x^2 + 2x + 2} + \int \frac{(3x + 1)\,dx}{(x^2 + 2x + 2)^2}$$

The first integral on the right is solved by the method shown in Example 7 in Section 7.3; thus,

$$\int \frac{(x - 2)\,dx}{x^2 + 2x + 2} = \frac{1}{2} \log (x^2 + 2x + 2) - 3 \tan^{-1} (x + 1)$$

The second integral must be broken up into two parts, as follows:

$$\int \frac{(3x + 1)\,dx}{(x^2 + 2x + 2)^2} = \frac{3}{2} \int \frac{(2x + 2)\,dx}{(x^2 + 2x + 2)^2} - 2 \int \frac{dx}{(x^2 + 2x + 2)^2}$$

$$= -\frac{3}{2} (x^2 + 2x + 2)^{-1} - 2 \int \frac{dx}{(x^2 + 2x + 2)^2}$$

This last integral is best handled by Formula 33 in the table of integrals (Appendix B). Using this with $a = 1, n = 2$ we get

$$-2 \int \frac{dx}{(x^2 + 2x + 2)^2} = -2 \int \frac{dx}{[(x + 1)^2 + 1]^2}$$

$$= -2 \left[\frac{1}{2} \frac{x + 1}{(x + 1)^2 + 1} + \frac{1}{2} \int \frac{dx}{(x + 1)^2 + 1} \right]$$

$$= -\frac{x + 1}{(x + 1)^2 + 1} - \tan^{-1} (x + 1)$$

Combining these partial results, we get as the solution to our problem:

$$\int \frac{(x^3 + x - 3)\,dx}{(x^2 + 2x + 2)^2}$$
$$= \frac{1}{2} \log(x^2 + 2x + 2) - \frac{x + \frac{5}{2}}{x^2 + 2x + 2} - 4 \tan^{-1} (x + 1) + C$$

We summarize the general principles of integration by partial fractions. First, divide out until the numerator is of lower degree than the denominator. There are then four cases.

Case I.   Nonrepeated linear factors

For each such factor, write a term $\dfrac{A}{x - \alpha}$; each such partial fraction integrates into a logarithm.

Case II.   Repeated linear factors

For each r-fold linear factor, write the sum

$$\frac{A_1}{x - \alpha} + \frac{A_2}{(x - \alpha)^2} + \cdots + \frac{A_r}{(x - \alpha)^r}$$

The first of these will integrate into a logarithm; all of the others are of type $\int u^n \, du$.

Case III.   Nonrepeated quadratic factors

For each such factor, write $\dfrac{Ax + B}{x^2 + px + q}$. The integral of each such fraction is accomplished after the methods of Examples 7 and 8 in Section 7.3 (by formulas V, XVI, XXII, and XXIII).

Case IV.   Repeated quadratic factors

For each r-fold quadratic factor write the sum

$$\frac{A_1 x + B_1}{x^2 + px + q} + \frac{A_2 x + B_2}{(x^2 + px + q)^2} + \cdots + \frac{A_r x + B_r}{(x^2 + px + q)^r}$$

The first of these is integrated as in Case III; each of the others is integrated as follows: Write

$$\int \frac{(A_k x + B_k)\,dx}{(x^2 + px + q)^k} = \frac{A_k}{2} \int \frac{(2x + p)\,dx}{(x^2 + px + q)^k,}$$

$$+ \left(B_k - \frac{pA_k}{2} \right) \int \frac{dx}{(x^2 + px + q)^k}$$

The first integral on the right is of the type $\int u^n \, du$. The second integral yields (after completing the square) to repeated applications of Formula 33 in the table of integrals (Appendix B).

Since every polynomial is factorable into linear and quadratic factors, the method of partial fractions will integrate any rational function (quotient of two polynomials).

7.5 Tables of Integrals

Many integrals have been computed and catalogued in so-called tables of integrals. When a given integral does not readily yield to any of the three standard methods:

> Integration by parts
> Integration by substitutions
> Integration by partial fractions

then it may be possible, by the use of these methods, to put the given integral into a form that can be found in a table of integrals. Therefore, we list a fourth method of integration:

> Integration by use of tables

A short table of integrals is provided in Appendix B.

PROBLEMS WITH ANSWERS

1. $\displaystyle\int \sin^3 2x \cos 2x\, dx.$ *Ans.* $\frac{1}{8} \sin^4 2x + C$

2. $\displaystyle\int \frac{e^x - e^{-x}}{e^x + e^{-x}}\, dx.$ *Ans.* $\log(e^x + e^{-x}) + C$

3. $\displaystyle\int 4xe^{2x^2}\, dx.$ *Ans.* $e^{2x^2} + C$

4. $\displaystyle\int \sec^2 (3x - 1)\, dx.$ *Ans.* $\frac{1}{3} \tan(3x - 1) + C$

5. $\displaystyle\int \csc 2x \cot 2x\, dx.$ *Ans.* $-\frac{1}{2} \csc 2x + C$

6. $\displaystyle\int \frac{dx}{x^2 + 4x + 9}.$ *Ans.* $\dfrac{\sqrt{5}}{5} \tan^{-1} \dfrac{x + 2}{\sqrt{5}} + C$

7. $\displaystyle\int \frac{dx}{\sqrt{1 - x - x^2}}.$ *Ans.* $\sin^{-1} \dfrac{2x + 1}{\sqrt{5}} + C$

8. $\int 3x^2 e^{2x}\, dx.$ *Ans.* $\frac{3}{2} e^{2x} \left(x^2 - x + \frac{1}{2} \right) + C$

9. $\int \dfrac{dx}{x + \sqrt{x}}\,.$ *Ans.* $2 \log (1 + \sqrt{x}\,) + C$

10. $\int \dfrac{dx}{1 + e^x}$ *Ans.* $x - \log (1 + e^x) + C$

11. $\int \tan^{-1} x\, dx.$ *Ans.* $x \tan^{-1} x - \frac{1}{2} \log (1 + x^2) + C$

12. $\int \dfrac{(x - 2)\, dx}{x^2 + 4x - 5}\,.$ *Ans.* $\frac{1}{2} \log x^2 + 4x + C$

CHAPTER 8

APPLICATIONS OF DIFFERENTIATION

8.1 Slopes

As we have already seen, $y' = \dfrac{dy}{dx} = m = \tan\theta =$ slope of the tangent to the curve $y = f(x)$. This was defined to be the slope of the curve itself in Section 5.1.

Example 1. Find the slope of the curve $y = 4x^3 - 3x + 1$ at the point for which $x = -1$.

Solution.
$$y' = 12x^2 - 3$$
$$y'(-1) = 9$$

Example 2. Find the points at which the tangent to the curve $x^2 + 3y^2 = 1$ makes $60°$ with the X-axis. (See Section 5.6, implicit differentiation.)

Solution.
$$2x + 6yy' = 0$$
$$y' = -\frac{x}{3y} = \pm\frac{x}{\sqrt{3}\sqrt{1 - x^2}}$$

Setting this equal to $\tan 60° = \sqrt{3}$, we have

$$\pm\frac{x}{\sqrt{3}\sqrt{1 - x^2}} = \sqrt{3}$$
$$\pm x = 3\sqrt{1 - x^2}$$
$$x^2 = 9(1 - x^2)$$
$$x = \pm\frac{3}{10}\sqrt{10}$$
$$y = \pm\frac{\sqrt{30}}{30}$$

Upon proper pairing of signs, we find that the points are

$$\left(\frac{3}{10}\sqrt{10}, \frac{\sqrt{30}}{30}\right) \text{ and } \left(-\frac{3}{10}\sqrt{10}, \frac{\sqrt{30}}{30}\right)$$

8.2 Tangents and Normals

To find the equation of the tangent to a curve, the point-slope form of the equation of a straight line is used, namely,

$$y - y_1 = m(x - x_1) \text{ (equation of tangent)} \tag{1}$$

where $m = y'$ evaluated at the point (x_1, y_1). Since a line perpendicular to this would have the slope $-\dfrac{1}{m}$, the equation of the normal becomes

$$y - y_1 = -\frac{1}{m}(x - x_1) \quad \text{(equation of normal)} \tag{2}$$

Example 1. Find (a) the equation of the tangent and (b) the equation of the normal to the curve $y^2 = 5x - 1$ at the point $(1, -2)$.

Solution. (a) The equation of the tangent will be of the form $y + 2 = m(x - 1)$, where $m = y'$ is evaluated at $(1, -2)$. Now $2yy' = 5$, or $y' = \dfrac{5}{2y}$; hence the equation becomes

$$y + 2 = -\tfrac{5}{4}(x - 1)$$

(b) The equation of the normal is

$$y + 2 = \tfrac{4}{5}(x - 1)$$

Refer to Fig. 8.1 for the following definitions:

Length of subtangent = ST

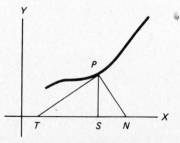

FIG. 8.1

Length of subnormal = *SN*
Length of tangent   = *PT*
Length of normal   = *PN*

If in the equation (1) we set $y = 0$ and solve for x, we get the co-ordinates of the tangent T, $T\left(x_1 - \dfrac{y_1}{m}, 0\right)$; similarly for the normal N, $N(x_1 + my_1, 0)$. With this information the lengths ST, SN, PT, and PN can be readily determined.

$$ST = \left| x_1 - \left(x_1 - \frac{y_1}{m}\right)\right| = \left| \frac{y_1}{m}\right|$$

$$SN = |(x_1 + my_1) - x_1| = |my_1|$$

$$PT = \left| \frac{y_1 \sqrt{1 + m^2}}{m}\right|$$

$$PN = |y_1 \sqrt{1 + m^2}|$$

Example 2. Find the lengths of the tangent and subnormal to the curve $y = x^5 - 2x + 3$ at $(1, 2)$.

Solution. $y' = 5x^4 - 2$; hence, $y'(1) = 3$.

Thus, $PT = \dfrac{2\sqrt{1 + 9}}{3} = \dfrac{2}{3}\sqrt{10}$ and $SN = 3 \times 2 = 6$

8.3 Angle between Two Curves

The angle between two curves will be the angle between the tangents. This angle is best given by the following formula:

$$\tan \theta_{12} = \frac{m_2 - m_1}{1 + m_1 m_2}$$

Angle θ_{12} is measured counterclockwise from the tangent to curve 1 *to* the tangent to curve 2.

Example 1. Find the angle at which the curves intersect in the first quadrant.

$$x^2 + y^2 = 9 \tag{1}$$

$$y^2 = 8x \tag{2}$$

Solution. The point of intersection is obtained by solving the two equations simultaneously. This yields $x^2 + 8x = 9$, or $x = 1, -9$. The point of intersection in the first quadrant, therefore, has coordinates $(1, 2\sqrt{2})$. For (1), $y' = -\dfrac{x}{y}$; for (2), $y' = \dfrac{4}{y}$. Hence, at $(1, 2\sqrt{2})$

$m_1 = -\frac{1}{4}\sqrt{2}$ and $m_2 = \sqrt{2}$; therefore,

$$\tan\theta_{12} = \frac{\sqrt{2} + \dfrac{1}{4}\sqrt{2}}{1 - \dfrac{1}{2}} = \frac{5}{2}\sqrt{2}$$

The angle of intersection is given by $\theta_{12} = \tan^{-1}\frac{5}{2}\sqrt{2}$.

Example 2. Show that (1) $x^2 - xy + y^2 - 3 = 0$ and (2) $x + y = 0$ intersect at right angles.

Solution. The points of intersection are readily found to be $(1, -1)$ and $(-1, 1)$. For (1), $y' = \dfrac{2x - y}{x - 2y}$ and at either point $m_1 = 1$. For (2), $y' = -1$ and $m_2 = -1$; hence,

$$\tan\theta_{12} = \frac{-1 - 1}{1 - 1} = \frac{-2}{0}$$

Therefore, the angle of intersection is $90°$. Or again, since $m_1 = -\dfrac{1}{m_2}$, the lines are perpendicular.

PROBLEMS WITH SOLUTIONS

1. a. Find the slope of the curve $y = x^2$ at $x = 1$.
    b. Find the slope of the curve $y = x^3 + x^2 + x + 1$ at $x = 1$.
    Solution.
    a. The slope is $y' = 2x$; thus, $y'(1) = 2$.
    b. Similarly, $y' = 3x^2 + 2x + 1$ and $y'(1) = 6$.

2. a. Find the points at which the tangent to the curve $y = x^2$ makes an angle of $45°$ with the X-axis.
    b. Find the points at which the tangent to the circle $x^2 + y^2 = 1$ makes an angle of $45°$ with the X-axis.
    Solution.
    a. Now, $y' = 2x$ and the angles $45°$ and $135°$ must both be used. For $45°$, $2x = 1$, and for $135°$, $2x = -1$. Therefore, the points are $\left(\frac{1}{2}, \frac{1}{4}\right)$ and $\left(-\frac{1}{2}, \frac{1}{4}\right)$.
    b. Here, $y' = \dfrac{-x}{y} = \dfrac{-x}{\sqrt{1 - x^2}} = 1$ or -1; thus, $x = \pm\dfrac{1}{\sqrt{2}} = \pm\dfrac{\sqrt{2}}{2}$. There-

fore, the points are $\left(\dfrac{\sqrt{2}}{2}, \dfrac{\sqrt{2}}{2}\right), \left(-\dfrac{\sqrt{2}}{2}, \dfrac{\sqrt{2}}{2}\right), \left(\dfrac{\sqrt{2}}{2}, -\dfrac{\sqrt{2}}{2}\right)$, and $\left(-\dfrac{\sqrt{2}}{2}, -\dfrac{\sqrt{2}}{2}\right)$.

3. a. Find the equations of the tangent and normal to the curve $y = x^2$ at the point $(2, 4)$.

   b. Find the equations of the tangent and normal to the curve $y = x^3$ at $(2, 8)$.

   Solution.

   a. Since $y' = 2x$, $y'(2) = 4$, the tangent is $y - 4 = 4(x - 2)$, or $4x - y = 4$. The normal is $y - 4 = -\frac{1}{4}(x - 2)$, or $x + 4y = 18$.

   b. Here, $y' = 3x^2$ and $y'(2) = 12$; therefore, the tangent is $y - 8 = 12(x - 2)$, or $12x - y = 16$. The normal is $y - 8 = -\frac{1}{12}(x - 2)$, or $x + 12y = 98$.

4. a. Find the lengths of the tangent, normal, subtangent, and subnormal to the curve $y = x^2$ at $(2, 4)$.

   b. Find the lengths of the tangent, normal, subtangent, and subnormal to the curve $y = x^3$ at $(-2, -8)$.

   Solution.

   a. $y' = 2x$ and $y'(2) = m = 4$. The required lengths are: tangent = $\dfrac{4\sqrt{17}}{4} = \sqrt{17}$, normal = $4\sqrt{17}$, subtangent = $\frac{4}{4} = 1$, and subnormal = $4 \times 4 = 16$.

   b. $y' = 3x^2$ and $y'(-2) = m = 12$. The required lengths are: tangent = $\left|\dfrac{-8\sqrt{145}}{12}\right| = \frac{2}{3}\sqrt{145}$, normal = $|-8\sqrt{145}| = 8\sqrt{145}$, subtangent = $\left|-\frac{8}{12}\right| = \frac{2}{3}$, and subnormal = $|-8 \times 12| = 96$.

5. Find the angle between the two lines $y = \dfrac{1}{\sqrt{3}}x$ and $y = \sqrt{3}x$.

   Solution. The slopes are $m_1 = \dfrac{1}{\sqrt{3}}$ and $m_2 = \sqrt{3}$. Since the angle has not been specified, we simply write $\tan \theta = \dfrac{\sqrt{3} - \dfrac{1}{\sqrt{3}}}{1 + 1} = \dfrac{1}{\sqrt{3}}$. Therefore, $\theta = 30°$ (and, of course, $150°$).

6. Find the angles at which the line $y = x$ intersects the curve $y = x^2$.
   Solution. The points of intersection are $(0, 0)$ and $(1, 1)$, and $y' = 2x$. The slope of the curve is therefore 0 at $(0, 0)$ and 2 at $(1, 1)$.

At $(0, 0)$, $m_1 = 0$, $m_2 = 1$, and $\tan \theta = \dfrac{1 - 0}{1 + 0} = 1$, hence $\theta = 45°$

(and $135°$). At $(1, 1)$, $m_1 = 1$, $m_2 = 2$, and $\tan \theta = \dfrac{2 - 1}{1 + 2} = \dfrac{1}{3}$, hence

$\theta = \text{Tan}^{-1} \dfrac{1}{3}$ (and its supplement).

7. Find angle ABC of a triangle with vertices $A(2, 1)$, $B(-4, 3)$, and $C(1, -5)$.

   Solution. The slopes are $m_{AB}, = -\dfrac{1}{3}$, $m_{BC} = -\dfrac{8}{5}$, and $m_{CA} = 6$. Angle $ABC = \theta_{12}$ is measured from $\overrightarrow{BC}$ to $\overrightarrow{BA}$; therefore,

$$\tan ABC = \frac{-\dfrac{1}{3} - \left(-\dfrac{8}{5}\right)}{1 + \left(-\dfrac{8}{5}\right)\left(-\dfrac{1}{3}\right)} = \frac{19}{23}$$

   Hence, angle $ABC = \text{Tan}^{-1} \dfrac{19}{23}$.

8. Find the angle of intersection of $y = \sin x$ and $y = \cos x$.

   Solution. The first quadrantal point of intersection is $(\pi/4, \sqrt{2}/2)$. For $y = \cos x$, $y' = -\sin x$ and $m_1 = y'(\pi/4) = -\sqrt{2}/2$; similarly, $m_2 = \sqrt{2}/2$; therefore,

$$\tan \theta = \frac{\dfrac{\sqrt{2}}{2} + \dfrac{\sqrt{2}}{2}}{1 - \left(\dfrac{\sqrt{2}}{2}\right)\left(\dfrac{\sqrt{2}}{2}\right)} = 2\sqrt{2}$$

   The acute angle is given by $\text{Tan}^{-1} 2\sqrt{2}$.

PROBLEMS WITH ANSWERS

9. For each curve, find the slope at the indicated point.

   a. $y = x^5 + x^3 + x + 1$, $(1, 4)$

   Ans. 9

   b. $y = x \log(x + 2)$, $(-1, 0)$

   Ans. -1

   c. $y = e^{x^2}$, $(0, 1)$

   Ans. 0

   d. $y = e^x \sin x$, $(0, 0)$

   Ans. 1

   e. $y = xe^{-x} \cos x$, $\left(\dfrac{\pi}{2}, 0\right)$

   Ans. $-\dfrac{\pi}{2} e^{-\pi/2}$

10. Find the points on the curve $y = x^3$ where the tangent line is (a) parallel and (b) perpendicular to the line $3x - 4y + 1 = 0$.

    Ans. (a) $\left(\dfrac{1}{2}, \dfrac{1}{8}\right)$, $\left(-\dfrac{1}{2}, -\dfrac{1}{8}\right)$

    (b) no such point

11. Find the equations of the tangent and normal to each of the following curves at the indicated point.

   a. $y = x + \dfrac{1}{x}$, $(1, 2)$

   Ans.   tangent: $y = 2$, normal: $x = 1$

   b. $y = \dfrac{x - 1}{x + 1}$, $(1, 0)$

   Ans.   tangent: $x - 2y - 1 = 0$, normal: $2x + y - 2 = 0$

   c. $y = \sin x + \cos x$, $(\pi, -1)$

   Ans.   tangent: $x + y + 1 - \pi = 0$, normal: $x - y - 1 - \pi = 0$

   d. $y = e^x \sin x$, $(0, 0)$

   Ans.   tangent: $x - y = 0$, normal: $x + y = 0$

   e. $y = x \log x$, (e, e)

   Ans.   tangent: $2x - y - e = 0$, normal: $x + 2y - 3e = 0$

12. Find the lengths of the tangent (T), normal (N), subtangent (ST), and subnormal (SN) to each of the following curves at the indicated point.

   a. $y = 2x^4 - x + 1$, $(1, 2)$

   Ans. $T = \dfrac{10}{7}\sqrt{2}$, $N = 10\sqrt{2}$, $ST = \dfrac{2}{7}$, $SN = 14$

   b. $y = \dfrac{x + 2}{x(x - 1)}$, $(2, 2)$

   Ans. $T = \dfrac{2}{5}\sqrt{29}$, $N = \sqrt{29}$, $ST = \dfrac{4}{5}$, $SN = 5$

   c. $y = \sin 2x$, $\left(\dfrac{\pi}{2}, 0\right)$

   Ans. $T = 0$, $N = 0$, $ST = 0$, $SN = 0$

   d. $y = e^{-x}$, $(0, 1)$

   Ans. $T = \sqrt{2}$, $N = \sqrt{2}$, $ST = 1$, $SN = 1$

   e. $y = \mathrm{Tan}^{-1} x$, $\left(1, \dfrac{\pi}{4}\right)$

   Ans. $T = \dfrac{\pi}{4}\sqrt{5}$, $N = \dfrac{\pi}{8}\sqrt{5}$, $ST = \dfrac{\pi}{2}$, $SN = \dfrac{\pi}{8}$

13. Find the equations of the tangent, the normal, the length ST of the subtangent, and the length SN of the subnormal to each curve for the indicated value of the parameter.

   a. $x = t^2 + 1$, $y = t + 3$, $t = 2$

   Ans. $x - 4y + 15 = 0$, $4x + y - 25 = 0$, $ST = 20$, $SN = \dfrac{5}{4}$

   b. $x = \dfrac{1}{t} + 1$, $y = t^2 + 1$, $t = 1$

   Ans. $2x + y - 6 = 0$, $x - 2y + 2 = 0$, $ST = 1$, $SN = 4$

c. $x = \sec\theta$, $y = \tan\theta$, $\theta = \dfrac{\pi}{4}$

   Ans. $\sqrt{2}\,x - y - 1 = 0$, $x + \sqrt{2}\,y - 2\sqrt{2} = 0$, $ST = \frac{1}{2}\sqrt{2}$, $SN = \sqrt{2}$

d. $x = \dfrac{3t}{1 + t^3}$, $y = \dfrac{3t^2}{1 + t^3}$, $t = 2$

   Ans. $4x - 5y + 4 = 0$, $15x + 12y - 26 = 0$, $ST = \frac{5}{3}$, $SN = \frac{16}{15}$

14. Find θ, the angle ABC, in each triangle.
    a. $A(-1, 1)\ B(1, -2)$, $C(3, 2)$

       Ans. $\theta = \text{Tan}^{-1}\,\frac{7}{4}$

    b. $A(0, 4)$, $B(-3, -1)$, $C(2, -4)$

       Ans. $\theta = 90°$

    c. $A(-1, 0)$, $B(0, 1)$, $C(5, 5)$

       Ans. $\theta = \text{Tan}^{-1}\left(-\frac{1}{9}\right)$

    d. $A(-3, 1)$, $B(4, 2)$, $C(3, 10)$

       Ans. $\theta = \text{Tan}^{-1}\,57$

15. Find θ, the smaller angle of intersection of the curves.
    a. $x = \sqrt{y}$, $x^2 + y^2 = \frac{3}{4}$

       Ans. $\theta = \text{Tan}^{-1}\,2\sqrt{2}$

    b. $\dfrac{x^2}{4} + y^2 = 1$, $x^2 - \dfrac{y^2}{2} = 1$

       Ans. $\theta = 90°$

    c. $y^2 = 2x^3$, $2x^2 + 3y^2 = 56$

       Ans. $\theta = 90°$

    d. $y = 2\,\text{Tan}^{-1}\,x$, $y = 3e^{-x} - 3$

       Ans. $\theta = 45°$

    e. $y = e^x$, $y = 2e^{-x}$

       Ans. $\theta = \text{Tan}^{-1}\,2\sqrt{2}$

8.4 Maxima and Minima

Consider a function f continuous on the closed interval $[a, b]$. Now examine Fig. 8.2, and observe that $f(a + h) < f(a)$ for sufficiently small positive h; hence, $f(a)$ is an endpoint *relative* **maximum**. We also see that $f(b - h) > f(b)$; thus, $f(b)$ is an endpoint *relative* **minimum**. We assume horizontal tangents at $x_1, x_3, x_5, x_7, x_9, x_{11}$ and vertical tangents at x_4 and x_6. Since $f(x_1) < f(x_1 - h)$ and $f(x_1) < f(x_1 + h)$ for sufficiently small positive h, $f(x_1)$ is a relative minimum. Relative minima also occur at x_4 and x_9. Since $f(x_3) > f(x_3 - h)$ and $f(x_3) > f(x_3 + h)$ for sufficiently small positive h, $f(x_3)$ is a relative maximum. Relative maxima also occur at x_5 and x_{11}.

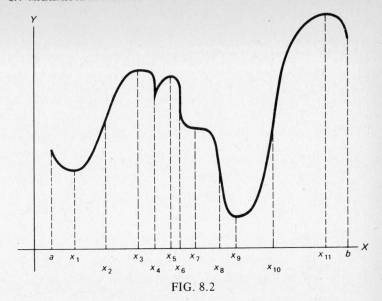

FIG. 8.2

Relative **extrema** (maximum or minimum) occur either at endpoints or at points $(x_1, x_3, x_5, x_9,$ and $x_{11})$ with horizontal tangents. There is a horizontal tangent at x_7, but f does not have an extremum there. Relative extrema may occur at points with vertical tangents, as at x_4. There is a vertical tangent at x_6, but there is no extremum there. Henceforth, we shall drop the word "relative."

A point with a horizontal tangent is called a **stationary point**. At such points $\dfrac{df}{dx} = 0$, if f is a differentiable function. At a point with a vertical tangent, $\dfrac{df}{dx}$ does not exist, hence we sometimes write $\dfrac{df}{dx} = \infty$, or $\dfrac{1}{\dfrac{df}{dx}} = 0$.

If we stay strictly in the closed interval $[a, b]$, then f does not have a derivative at a or b. Sometimes we say that f has a *right-hand* derivative at a, since

$$\lim_{h \to 0^+} \frac{f(a + h) - f(a)}{h}, h > 0$$

Similarly, f has a *left-hand* derivative at b.

The maximum value of f in $[a, b]$ is called the *absolute maximum*. This occurs at x_{11}. Similarly the *absolute minimum* occurs at x_9. A point where $f' = 0$ or $f' = \infty$ is called a **critical point**.

A curve is *concave downward* in an interval if it lies below its tangent lines as in the intervals $(x_2, x_4), (x_4, x_6), (x_7, x_8)$, and (x_{10}, b). A curve is *concave upward* in an interval if it lies above its tangents, as in the intervals $(a, x_2), (x_6, x_7)$, and (x_8, x_{10}). A point at which the sense of concavity changes is called an **inflection point**, or **point of inflection**, as at x_2, x_6, x_7, x_8, and x_{10}. If f and f' are both differentiable, then an inflection point of f is a stationary point of f'.

8.5 Tests for Extrema

First find the critical points: each must be tested. Let x_0 be the abscissa of a critical point and let h be a positive number such that no other critical point is in the interval $(x_0 - h, x_0 + h)$.

Test I. Using the first derivative
1. *If* $f'(x_0) = 0$:
    (i) $f(x_0)$ *is a maximum if* $f'(x_0 - h) > 0$ *and* $f'(x_0 + h) < 0$.
    (ii) $f(x_0)$ *is a minimum if* $f'(x_0 - h) < 0$ *and* $f'(x_0 + h) > 0$.
    (iii) $f(x_0)$ *is not an extremum if* $f'(x_0 - h)$ *and* $f'(x_0 + h)$ *have like signs.*
2. *If* $f'(x_0) = \infty$: same as 1.

Test II. Using the second derivative
If $f'(x_0) = 0$ *and* $f''(x)$ *exists*:
    (i) $f(x_0)$ *is a maximum if* $f''(x_0) < 0$.
    (ii) $f(x_0)$ *is a minimum if* $f''(x_0) > 0$.
    (iii) *Test II is not applicable if* $f''(x_0) = 0$ *or if* $f''(x_0)$ *does not exist.*

Test III. Using higher derivatives
If $f'(x_0) = f''(x_0) = \ldots = f^{(n-1)}(x_0) = 0$ *and if* $f^n(x_0) \neq 0$, *then*:
    (i) *For n even* $\begin{cases} f(x_0) \text{ is a maximum if } f^{(n)}(x_0) < 0. \\ f(x_0) \text{ is a minimum if } f^{(n)}(x_0) > 0. \end{cases}$
    (ii) *For n odd*, $f(x_0)$ *is not an extremum.*

Example 1. Examine $y = 2x^3 + 3x^2 - 12x - 15$ for extreme values (Fig. 8.3).

Solution.
$$y' = 6x^2 + 6x - 12$$
$$= 6(x + 2)(x - 1) = 0$$
$$x = -2, 1, \text{ critical values}$$

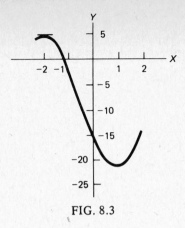

FIG. 8.3

The first derivative test for $x = -2$ yields

$$\left.\begin{array}{l} y' > 0 \text{ before} \\ y' < 0 \text{ after} \end{array}\right\} \therefore y(-2) = 5 \text{ (max.)}$$

For the point $x = 1$ we get

$$\left.\begin{array}{l} y' < 0 \text{ before} \\ y' > 0 \text{ after} \end{array}\right\} \therefore y(1) = -22 \text{ (min.)}$$

Example 2. Examine $y = 3x^4 - x^3 + 2$ for extreme values (Fig. 8.4).

Solution.
$$y' = 12x^3 - 3x^2$$
$$= 3x^2 (4x - 1) = 0$$
$$x = 0, \tfrac{1}{4}, \text{ critical values}$$

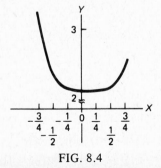

FIG. 8.4

The first derivative test for $x = 0$ yields

$$\left.\begin{array}{l} y' < 0 \text{ before} \\ y' < 0 \text{ after} \end{array}\right\} \therefore y(0) = 2 \text{ is neither a } maximum \text{ nor a } minimum$$

For the point $x = \frac{1}{4}$, we get

$$\left.\begin{array}{l} y' < 0 \text{ before} \\ y' > 0 \text{ after} \end{array}\right\} \therefore y\left(\frac{1}{4}\right) = \frac{511}{256} \text{ (min.)}$$

Using the tests involving higher derivatives, we find

$$y'(0) = y''(0) = 0, \text{ since } y'' = 36x^2 - 6x$$

$$y''' = 72x - 6, y'''(0) = -6$$

Since the first nonvanishing derivative at the point $x = 0$ is of odd order ($n = 3$), $y(0)$ is neither a *maximum* nor a *minimum*.

$$y''\left(\frac{1}{4}\right) = \frac{3}{4}, \therefore y\left(\frac{1}{4}\right) = \frac{511}{256} \text{ (min.)}$$

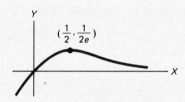

FIG. 8.5

Example 3. Examine $y = xe^{-2x}$ for extreme values (Fig. 8.5).

Solution.
$$y' = e^{-2x}(1 - 2x) = 0$$

$$x = \frac{1}{2}, \text{ critical value}$$

$$y'' = 4e^{-2x}(x - 1)$$

$$y''\left(\frac{1}{2}\right) = -2e^{-1} < 0$$

Therefore,

$$y\left(\frac{1}{2}\right) = \frac{1}{2e} \text{ (max.)}$$

Example 4. If a tin can is to be made from a given amount of metal so that it will have maximum volume, what must its relative dimensions be (Fig. 8.6)?

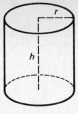

FIG. 8.6

Solution. The "given amount of metal" means that the total surface area is specified. Call it S and let r = radius and h = height of the can, then

$$S = 2\pi rh + 2\pi r^2 \tag{1}$$

Now the quantity to be maximized is the volume

$$V = \pi r^2 h \tag{2}$$

$$= \frac{Sr}{2} - \pi r^3 \text{ (because of (1))}$$

$$\frac{dV}{dr} = \frac{S}{2} - 3\pi r^2 = 0 \tag{3}$$

$$r = \pm \sqrt{\frac{S}{6\pi}}, \text{ the critical values (the minus sign has no meaning here.)}$$

$$\frac{d^2 V}{dr^2} = -6\pi r \text{ is negative for all (positive) values of } r. \tag{4}$$

$$r = \sqrt{\frac{S}{6\pi}} \text{ corresponds to a } \textit{maximum.}$$

$$h = \frac{S - 2\pi r^2}{2\pi r} = S \frac{\frac{2}{3}}{2\pi \sqrt{\frac{S}{6\pi}}}$$

$$= 2 \sqrt{\frac{S}{6\pi}} = 2r$$

Hence, the relative dimensions are $h = 2r$.

Example 5. Two corridors, each of width a, meet at right angles. Find the length of the longest pipe that can be passed horizontally around the corner.

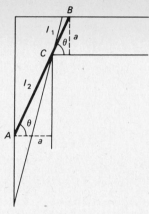

FIG. 8.7

Solution. In Fig. 8.7, we see that many lines (pipes) can be drawn connecting such points as A and B and touching corner C. The length of the *longest pipe* that will go around the corner is the length of the *shortest line ACB*. Call the length l and write $l = l_1 + l_2$; thus,

$$l = a \csc \theta + a \sec \theta \tag{1}$$

$$l' = \frac{dl}{d\theta} = a(-\csc \theta \cot \theta + \sec \theta \tan \theta) = 0 \tag{2}$$

Hence,

$\tan^3 \theta = 1, \theta = 45°$, and $225°$ are the critical values, but the one of
interest is $\theta = 45°$.

Testing at, say, $30°$ (before) and $60°$ (after), we have

$$l'(30°) = a\left(-2\sqrt{3} + \tfrac{2}{3}\right) < 0,$$

$$l'(60°) = a\left(2\sqrt{3} - \tfrac{2}{3}\right) > 0$$

Hence,

$l(45°) = 2a\sqrt{2}$, is the minimum length of line and the maximum
length of pipe that will go around the corner.

Example 6. Find the point where the slope of the curve whose parametric equations are $x = 2t^2 - 1, y = 3t^3 + t$ is a *minimum*.

Solution. The slope m will be given by $\dfrac{dy}{dx} = y'$.

$$\frac{dx}{dy} = 4t \tag{1}$$

$$\frac{dy}{dt} = 9t^2 + 1 \tag{2}$$

$$m = \frac{dy}{dx} = \frac{\dfrac{dy}{dt}}{\dfrac{dx}{dt}} = \frac{9t^2 + 1}{4t} = \frac{9}{4}t + \frac{1}{4t} \tag{3}$$

It is the slope that is to be minimized; therefore, we must find $\dfrac{dy'}{dx} = y''$

and set this equal to zero in order to obtain the critical points for slope.

$$y'' = \frac{d^2y}{dx^2} = \frac{d}{dx}\left(\frac{dy}{dx}\right) = \frac{d}{dt}\left(\frac{dy}{dx}\right) \cdot \frac{dy}{dx} \tag{4}$$

$$= \left(\frac{9}{4} - \frac{1}{4t^2}\right) \cdot \frac{1}{4t} = \frac{1}{16t^3}(9t^2 - 1) = 0$$

$$\therefore t = \pm\frac{1}{3}$$

There are two critical points, namely, for $t = \frac{1}{3}$, $P\left(-\frac{7}{9}, \frac{4}{9}\right)$ and, for $t = -\frac{1}{3}$, $Q\left(-\frac{7}{9}, -\frac{4}{9}\right)$. In testing, we vary x slightly and keep in mind that "before" and "after" refer to values of x, not t.

Testing at P, $\left.\begin{array}{l} y'' < 0 \text{ before} \\ y'' > 0 \text{ after} \end{array}\right\}$ $\therefore$ Slope is *minimum*

Testing at Q, $\left.\begin{array}{l} y'' > 0 \text{ before} \\ y'' < 0 \text{ after} \end{array}\right\}$ $\therefore$ Slope is *maximum*

The test used, though involving y'', is the *first derivative test; y''* is the *first derivative of y'* whose extreme values are being investigated.

PROBLEMS WITH SOLUTIONS

1. Examine $f(x) = x^4 - x^2$ for extreme values.
   Solution. Here $f'(x) = 4x^3 - 2x$, and the critical numbers are those numbers for which $4x^3 - 2x = 0$; namely, $-\sqrt{2}/2$, 0, and $\sqrt{2}/2$. The critical points are $(0, 0)$, $(\sqrt{2}/2, -1/4)$, and $(-\sqrt{2}/2, -1/4)$. Since $f''(x) = 12x^2 - 2$ yields $f''(-\sqrt{2}/2) > 0$, $f''(0) < 0$, and $f''(\sqrt{2}/2) > 0$, the function f has a minimum of $f(-\sqrt{2}/2) = -1/4$, a maximum of $f(0) = 0$, and a minimum of $f(\sqrt{2}/2) = -1/4$.

2. Show that $f(x) = x^5 + x^4$ has a maximum at the critical point $(-4/5, 256/3125)$ and a minimum at $(0, 0)$.
   Solution. $f'(x) = 5x^4 + 4x^3$; thus, $5x^4 + 4x^3 = x^3(5x + 4) = 0$ if $x = 0$ and $x = -4/5$. Thus, $f(-4/5) = 256/3125$ and $f(0) = 0$; therefore, the critical points are $(-4/5, 256/3125)$ and $(0, 0)$.

   We test these points for maximum and minimum as follows: $f''(x) = 20x^3 + 12x^2$; hence, $f''(-4/5) = -64 < 0$. Therefore, $(-4/5, 256/3125)$ is a *maximum*. Since $f''(0) = 0$, this test fails, and we must use the higher derivative test (Test III); hence, $f'''(0) = 0$, but $f^{(4)}(x) = 120x^2 + 24$, so that $f^{(4)}(0) = 24 > 0$; therefore, $(0, 0)$ is a minimum.

3. Examine the following for extrema:
   a. $x^{2/3} + y^{2/3} = 1$, with $y \geqslant 0$
   b. $y = x^{2/3}(2x - 5)$
   c. $y = x - \sin x$
   Solution.
   a. Differentiating implicitly with respect to x,

   $$\frac{2}{3}x^{-1/3} + \frac{2}{3}y^{-1/3}\,y' = 0, \text{ or } y' = -\frac{y^{1/3}}{x^{1/3}} = -\frac{\sqrt{1 - x^{2/3}}}{x^{1/3}}$$

   If $y' = 0$, then $\sqrt{1 - x^{2/3}} = 0$, or $x = \pm 1$; actually, $(1, 0)$ and $(-1, 0)$ are endpoints at which minima occur; hence, $(0, 1)$ corresponds to a maximum.
   b. $y' = \frac{10}{3}\frac{x - 1}{x^{1/3}}$; critical points: $(1, -3)$ and $(0, 0)$. Since $y'\left(\frac{1}{2}\right) < 0$ and $y'\left(\frac{3}{2}\right) > 0$, the function has a minimum of -3 at $x = 1$. At $x = 0$, y' does not exist, but, since $y \leqslant 0$ in the interval $-\frac{1}{2} \leqslant x \leqslant \frac{1}{2}$, the function has a maximum value of 0 at $x = 0$.
   c. $y' = 1 - \cos x$, $y'' = \sin x$, and $y''' = \cos x$. Accordingly, $y'(2k\pi) = 0$, $y''(2k\pi) = 0$, and $y'''(2k\pi) = 1$; therefore, f has no extrema.

4. An open cylindrical cup with a given volume is to be made out of the least possible amount of tin. Find the relative dimensions.
   Solution. The surface area $A_s = 2\pi rh + \pi r^2$ is to be minimized subject to the condition $V = \pi r^2 h$ is a constant. $A_s = 2\pi r(V/\pi r^2) + \pi r^2 = 2V/r + \pi r^2$. $A_s' = -2V/r^2 + 2\pi r$, $A_s'' = 4V/r^3 + 2\pi$. $A_s' = 0$ when $r = (V/\pi)^{1/3}$, $A_s''(V/\pi)^{1/3} > 0$, and A_s is minimum at $r = (V/\pi)^{1/3}$. Further,

   $$h = \frac{V}{\pi r^2} = \frac{V\pi^{2/3}}{\pi V^{2/3}} = \left(\frac{V}{\pi}\right)^{1/3} = r$$

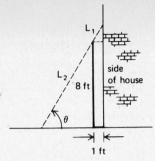

FIG. 8.8

5. A wall 8 ft. high runs parallel to the side of a house and 1 ft. from it. What is the length L of the shortest ladder that will reach the house from the ground outside the wall (Fig. 8.8)?

Solution.

$L = L_1 + L_2 = \sec \theta + 8\csc \theta$

$L' = \sec \theta \tan \theta - 8 \csc \theta \cot \theta = 0$

$\tan^3 \theta = 8, \tan \theta = 2, \theta = 63° 26'$

$L'(45°) = \sqrt{2} - 8\sqrt{2} < 0,$

$L'(70°) = (2.9238)(2.7475) - 8(1.0642)(0.36397) > 0$; therefore, L is least when $\tan \theta = 2$ and L (min.) $= 2.2359 + 8(1.1180) = 11.1799$ ft.

6. Find the dimensions of the right circular cone of maximum volume that can be inscribed in a sphere of radius r (Fig. 8.9).

Solution. Let V = volume, R = radius of the sphere, and r = radius of the base of the cone, then

$$V = \frac{1}{3}\pi r^2 (R + h)$$

$$= \frac{1}{3}\pi (R^2 - h^2)(R + h)$$

$$= \frac{1}{3}\pi (R^3 + R^2 h - Rh^2 - h^3)$$

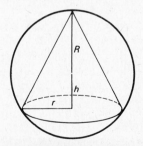

FIG. 8.9

$$V' = \tfrac{1}{3}\pi(R^2 - 2Rh - 3h^2)$$

$$= \frac{\pi}{3}(R - 3h)(R + h) = 0$$

Then, $h = \tfrac{1}{3}R, -R$, which is of no physical interest, but

$$V'' = \tfrac{1}{3}\pi(-2R - 6h), \text{ and}$$

$$V''\left(\tfrac{1}{3}R\right) = \tfrac{1}{3}\pi(-2R - 2R) < 0$$

Therefore, the dimensions sought are $h = \tfrac{1}{3}R$ and $r = \tfrac{2}{3}\sqrt{2}R$.

7. A rectangular piece of tin is to have a square cut out from each corner and the four projecting flaps turned up to form an open box. What is the length of a side of the square cut out if the box is to have a maximum volume (Fig. 8.10)?

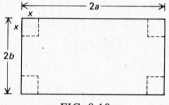

FIG. 8.10

Solution. Let V = volume, then

$$V = x(2a - 2x)(2b - 2x)$$

$$= 4x(a - x)(b - x)$$

$$= 4[abx - (a + b)x^2 + x^3]$$

$$V' = 4[ab - 2(a + b)x + 3x^2] = 0$$

$$x = \frac{1}{3}\left(a + b \pm \sqrt{a^2 - ab + b^2}\right)$$

We have no interest in the value of x computed by using the plus sign before the radical, but, using the minus sign and the second-derivative test, we have

$$V'' = 4[-2(a + b) + 6x], \text{ and at the critical point,}$$

$$V'' = 4(-2a - 2b + 2a + 2b - 2\sqrt{a - ab + b^2})$$

$$= -8\sqrt{a^2 - ab + b^2}$$

Therefore, the box has maximum volume.

8. A man is in a boat at P, 1 mi. from the nearest point A on shore. He wishes to go to B which is 1 mi. farther down the shore. If he can row 4 mi./hr. and walk 5 mi./hr., toward what point C should he row in order to reach B in least time?

Solution. Let $x = AC$ and let t be the total time from P to B, then

$$t = \frac{1}{4}(1 + x^2)^{1/2} + \frac{1}{5}(1 - x)$$

$$t' = \frac{x}{4\sqrt{1 + x^2}} - \frac{1}{5} = 0$$

$$x = \pm\frac{4}{3}, \text{ but choose } x = \frac{4}{3}.$$

Since $t'' = \frac{1}{4}\frac{1}{(1 + x^2)^{3/2}}$ and $t''\left(\frac{4}{3}\right) > 0$, $x = \frac{4}{3}$ yields a minimum t, but this is absurd, since if $x = \frac{4}{3}$, C is beyond B. Therefore, the critical value which is of interest is $x = 1$ (an endpoint) and the man should row directly to B for the minimum time.

9. We are given a straight fence 100 ft. long, and wish, by adding 200 ft. more, to form a rectangular enclosure whose boundary contains the original fence. How can this be done so as to enclose the greatest possible area?

Solution. Let x denote the length of the new fence to be aligned with the original 100 ft., then the area is

$$A = (100 + x)(50 - x)$$
$$= 5000 - 50x - x^2$$
$$A' = -50 - 2x = 0 \text{ and}$$
$$x = -25 \text{ ft.}$$

Since $A'' = -2 < 0$, $x = -25$ yields a minimum. This result is absurd, since 25 ft. of the original fence would then have to be removed, contrary to our hypothesis. The correct solution is $x = 0$, which yields an endpoint maximum.

10. Find the relative and absolute extrema of $f(x) = x^2 (9x^2 - 28x + 24)$ for $0 \leqslant x \leqslant 14/9$.

Solution. From $f'(x) = 12x(x - 1)(3x - 4)$, we obtain the critical values 0, 1, 4/3, and the endpoint 14/9. Routine testing will show that $f(0)$ is an absolute minimum, $f(4/3)$ is a relative minimum, $f(1)$ is a relative maximum, and $f(14/9)$ is an absolute (endpoint) maximum.

PROBLEMS WITH ANSWERS

11. Examine the following for extrema.

    a. $y = x^2 - 4x + 3$

    Ans. min. $(2, -1)$

    b. $y = x^3 - 2x^2$

    Ans. max. $(0, 0)$, min. $\left(\frac{4}{3}, -\frac{32}{27}\right)$

    c. $y = \sin^2 x$

    Ans. max. $\left(\frac{2n+1}{2}\pi, 1\right)$, min. $(n\pi, 0)$

    d. $y = \tan x - x$

    Ans. none

12. Examine the following for extrema.

    a. $y = x^4 + x^3 + 1$

    Ans. inflection point $(0, 1)$, min. $\left(-\frac{3}{4}, \frac{229}{256}\right)$

    b. $y = xe^{-x}$

    Ans. max. $(1, e^{-1})$

    c. $y = \dfrac{x}{1 + x^2}$

    Ans. min. $\left(-1, -\frac{1}{2}\right)$, max. $\left(1, \frac{1}{2}\right)$

    d. $y = \dfrac{1}{1 - x} - x$

    Ans. min. $(0, 1)$, max. $(2, -3)$

    e. $y = \dfrac{x}{\log x}$

    Ans. min. (e, e)

    f. $y = x^{1/x}$

    Ans. max. $(e, e^{1/e})$

13. Find all extrema and points of inflection.

    a. $y = 2x^2 - 2x^3$

    Ans. min. $(0, 0)$, max. $\left(\frac{2}{3}, \frac{8}{27}\right)$, inflection point $\left(\frac{1}{3}, \frac{4}{27}\right)$

    b. $y = 1 - \dfrac{x^2}{2} + \dfrac{x^4}{24}$

    Ans. max. $(0, 1)$, min. $\left(\pm\sqrt{6}, -\frac{1}{2}\right)$, inflection point $\left(\pm\sqrt{2}, \frac{1}{6}\right)$

    c. $y = (8 - x)^{2/3}$

    Ans. absolute min. (cusp) $(8, 0)$

    d. $y = \dfrac{3x^2}{x + 1}$

    Ans. min. $(0, 0)$, max. $(-2, -12)$

14. The perimeter of a sector of a circle is to be 16 ft. in length. What is the radius of the circle if the sector has maximum area?

$Ans.$ $r = 4$ ft.

15. The sum of the perimeters of a circle and a square is a constant. Find the relative dimensions of the circle and square if the total area is a minimum.

$Ans.$ The diameter of the circle equals the side of the square.

16. An irrigation ditch is to have a trapezoidal cross section of area A sq. units with (equal) sides at $45°$ from the vertical. What should the depth be for the minimum retaining surface?

$Ans.$ $\sqrt{\dfrac{A}{2\sqrt{2} - 2}}$ units

17. A picture 8 ft. high hangs on a wall with the bottom of the picture 7 ft. from the floor. How far back from the wall should a person, whose eyes are 5 ft. from the floor, stand so as to make the picture subtend the largest visual angle?

$Ans.$ $2\sqrt{5}$ ft.

18. What is the volume of the right circular cylinder of greatest volume that can be inscribed in a right circular cone of radius r and height h?

$Ans.$ $\dfrac{4}{27}\pi r^2 h$

19. In strip mining, the cost increases directly as the depth. Suppose that the value of a product increases as the square root of the depth and that, at a depth of 25 ft., the cost of mining is \$1.25 per cubic unit, and the value is \$5.00 per cubic unit. At what depth will the greatest profit be made, and what is that profit?

$Ans.$ depth 100 ft., profit \$5.00 per cu. unit

20. A letter Y 16 in. high and 12 in. wide at the top is to be made. Find the length of its stem if the total length of the stem and the two equal branches is to be a minimum.

$Ans.$ $16 - 2\sqrt{3}$ in.

21. The dimensions of a room are a, b, and c (height). A fly crawls by way of two walls from one corner of the floor to the corner diagonally opposite at the ceiling. If the fly follows the shortest path, how far from the floor will it be when it crosses from one wall to the other?

$Ans.$ $\dfrac{ac}{a + b}$

8.6 Straight-Line Motion

Various problems in kinematics can be solved by the use of the derivative. If a particle moves along a straight line so that its distance s from some fixed point is a function of time, then $s = f(t)$, the velocity $v = \dfrac{ds}{dt}$, and the acceleration $a = \dfrac{dv}{dt} = \dfrac{d^2 s}{dt^2}$.

Example 1. The height s feet after t seconds of a certain body thrown vertically upward is given by $s = 96t - 16t^2$. Find (a) the velocity v and acceleration a at any time t, (b) the initial velocity, (c) the maximum height reached by the body, (d) the velocity at the end of 1 sec., and (e) the time when the body returns to the ground and its velocity then.

Solution.

(a) $v = 96 - 32t$ ft./sec.; $a = -32$ ft./sec^2.

(b)   At $t = 0, v = 96$ ft./sec.

(c) The maximum height is attained when $\dfrac{ds}{dt} = v = 0$, or $96 - 32t = 0$,

   $t = 3$ sec.; maximum $s = 96 \times 3 - 16 \times 9 = 144$ ft.

(d) $v(1) = 64$ ft./sec.

(e) $s = 0 = 96t - 16t^2$, $t = 0$ (body started up)

               $t = 6$ (body returned to the ground)

   $v(6) = -96$ ft./sec.

Example 2. A particle moves in a straight line so that $s = A \cos(kt + \theta)$. Find the velocity v at any time t and show that the acceleration a is proportional to s.

Solution. $v = \dfrac{ds}{dt} = -kA \sin(kt + \theta)$

$$a = \dfrac{dv}{dt} = -k^2 A \cos(kt + \theta) = -k^2 s$$

Such a motion, with the acceleration proportional to displacement, is *simple harmonic motion*. The *amplitude* of the motion is A, the *period* is $\dfrac{1}{\text{frequency}} = \dfrac{2\pi}{k}$, and the *phase constant* is θ.

8.7 Curvilinear Motion

When a particle moves along a curve, the expressions for velocity (Fig. 8.11) and acceleration (Fig. 8.12) are a little more complicated.

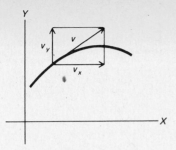

FIG. 8.11

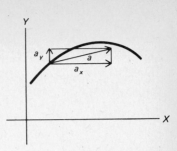

FIG. 8.12

Let the equation of the curve be given in *parametric* form $x = f(t)$, $y = g(t)$, where the parameter t represents time. The velocity v is a vector quantity and has components $v_x = \dfrac{dx}{dt}$ and $v_y = \dfrac{dy}{dt}$. (See Chapter 12.) The *magnitude* of v (speed) is

$$|v| = \sqrt{v_x^2 + v_y^2}$$

The *direction* of v is given by

$$\tan \theta = \frac{v_y}{v_x} = \frac{dy}{dx}$$

The velocity vector is tangent to the curve of motion.

Similarly, for the x- and y-components of acceleration, we have

$$a_x = \frac{dv_x}{dt} = \frac{d^2 x}{dt^2}, \text{ and } a_y = \frac{dv_y}{dt} = \frac{d^2 y}{dt^2}$$

The magnitude of a is

$$|a| = \sqrt{a_x^2 + a_y^2}$$

The direction of a is given by

$$\tan \phi = \frac{a_y}{a_x}$$

The acceleration vector is not, in general, tangent to the curve of motion.

Sometimes it is important to resolve the acceleration vector into components tangent and normal, respectively, to the curve. The tangential component a_T and the normal component a_N are given by

$$a_T = \frac{v_x a_x + v_y a_y}{|v|}$$

$$a_N = \frac{v_x a_y - v_y a_x}{|v|}$$

Another important concept is that of angular velocity. When a particle moves along the circumference of a circle, the central angle θ, measured from some fixed direction, is a function of time t. We define angular velocity ω as the rate of change of θ with respect to time t and write

$$\omega = \frac{d\theta}{dt}$$

Likewise, angular acceleration α is denoted by

$$\alpha = \frac{d\omega}{dt} = \frac{d^2\theta}{dt^2}$$

Or, again, we speak of the angular velocity and angular acceleration of a *vector* $\overrightarrow{OP}$ drawn from the origin O to a point P as P moves along a curve.

Example 1. A particle moves along the parabola $y = x^2$ with a constant speed of 10 ft./sec. Find v_y when v_x is 2 ft./sec. Also find the corresponding point on the curve and the x- and y-components of acceleration there.

Solution. $v^2 = v_x^2 + v_y^2$ (1)

$100 = 4 + v_y^2$, or $v_y = \pm 4\sqrt{6}$ ft./sec.

Differentiating the equation of the curve of motion with respect to time t, we get

$$\frac{dy}{dt} = 2x \frac{dx}{dt}$$

and

$$v_y = 2xv_x \pm 4\sqrt{6} = 4x   (2)$$

hence,

$$x = \pm\sqrt{6}, y = 6$$

Differentiating (1) and (2) with respect to t, we arrive at two equations in the unknowns a_x and a_y; namely,

$$0 = v_x a_x + v_y a_y \text{ and } a_y = 2x a_x + 2v_x^2$$

Solving these simultaneously, we get

$$a_x = \mp \frac{16}{25}\sqrt{6} \text{ ft./sec}^2. \text{ and } a_y = \frac{8}{25} \text{ ft./sec}^2$$

Example 2. A particle moves so that its x- and y-coordinates are given by $x = a \cos 2t$, $y = b \sin 2t$. Find (a) its velocity and acceleration x- and y-components and (b) the *magnitude* and *direction* of the velocity and acceleration vectors. (c) Show that the path is an ellipse. (d) Find t when a_y is a maximum. (Let t be both time in seconds and radian measure; x and y are to be measured in feet.)

Solution.

(a)
$$v_x = -2a \sin 2t \text{ ft./sec.}$$
$$v_y = 2b \cos 2t \text{ ft./sec.}$$
$$a_x = -4a \cos 2t \text{ ft./sec}^2.$$
$$a_y = -4b \sin 2t \text{ ft./sec}^2.$$

(b)
$$|v| = \sqrt{v_x^2 + v_y^2} = 2\sqrt{a^2 \sin^2 2t + b^2 \cos^2 2t}$$
$$|a| = \sqrt{a_x^2 + a_y^2} = 4\sqrt{a^2 \cos^2 2t + b^2 \sin^2 2t}$$

$$\tan \theta = \frac{v_y}{v_x} = -\frac{b}{a} \cot 2t$$

$$\tan \phi = \frac{a_y}{a_x} = \frac{b}{a} \tan 2t$$

(c)
$$\frac{x}{a} = \cos 2t \text{ and } \frac{y}{b} = \sin 2t$$

Hence,
$$\frac{x^2}{a^2} + \frac{y^2}{b^2} = 1 \text{ (the equation of an ellipse)}$$

(d)
$$\frac{da_y}{dt} = -8b \cos 2t = 0, t = \frac{\pi}{4}, \text{ and } \frac{3\pi}{4}$$

Using the second derivative test, we have

$$\frac{d^2 a_y}{dt^2} = 16b \sin 2t$$

At $t = \frac{\pi}{4}$, this is plus, therefore, a_y is a *minimum*; at $t = \frac{3\pi}{4}$, this is minus, therefore, a_y is a *maximum*. The particle is moving around the curve counterclockwise and has maximum a_y at the lower end of the minor axis. Note that here $a_x = 0$.

Example 3. A particle P moves around the circumference of a circle with constant angular velocity. Find $v_x, v_y, a_x, a_y, a_T,$ and a_N.

Solution. Let the equation of the circle be given in parametric form

$$x = r \cos \theta \text{ and } y = r \sin \theta$$

Then $v_x = -r \sin \theta \dfrac{d\theta}{dt}$ and $v_y = r \cos \theta \dfrac{d\theta}{dt}$

$$= -r\omega \sin \theta = r\omega \cos \theta$$

$$a_x = -r\omega^2 \cos \theta \text{ and } a_y = -r\omega^2 \sin \theta$$

$$a_T = \frac{v_x a_x + v_y a_y}{|v|}$$

$$= \frac{(-r\omega \sin \theta)(-r\omega^2 \cos \theta) + (r\omega \cos \theta)(-r\omega^2 \sin \theta)}{|v|} = 0$$

$$a_N = \frac{v_x a_y - v_y a_x}{|v|}$$

$$= \frac{(-r\omega \sin \theta)(-r\omega^2 \sin \theta) - (r\omega \cos \theta)(-r\omega^2 \cos \theta)}{\sqrt{r^2 \omega^2 \sin^2 \theta + r^2 \omega^2 \cos^2 \theta}} = r\omega^2$$

Thus all of the acceleration is directed toward the center. The projection of P upon a diameter executes simple harmonic motion since $a_x = \omega^2 x$; that is, the acceleration in the x direction is proportional to the displacement in that direction. (See Example 2, Section 8.6.)

Example 4. (a) Find the angular velocity and angular acceleration of a particle which moves along the curve $y^2 = x^3$ in such a way that v_x is constant and equal to 2 ft./sec., and (b) find ω and α at the point $(1, 1)$.
Solution.

(a) Now

$$\theta = \tan^{-1} \frac{y}{x} = \tan^{-1} x^{1/2}$$

$$\omega = \frac{d\theta}{dt} = \frac{1}{1 + x} \left(\frac{1}{2} x^{-1/2} \frac{dx}{dt} \right)$$

$$= \frac{1}{\sqrt{x}(1 + x)}$$

$$\alpha = \frac{d^2 \theta}{dt^2} = \frac{-\left(\frac{1}{2} x^{-1/2} + \frac{3}{2} x^{1/2} \right)}{x(1 + x)} \frac{dx}{dt}$$

$$= -\frac{1 + 3x}{x^{3/2}(1 + x)^2}$$

(b)

$$\omega|_{(1,1)} = \tfrac{1}{2} \text{ rad./sec.}$$

$$\alpha|_{(1,1)} = -1 \text{ rad./sec.}^2$$

PROBLEMS WITH SOLUTIONS

1. For a body moving in a straight line a distance s, where $s = t^3 - 3t^2 - 3t - 3$, find the time t at which the velocity v and acceleration a are both positive.

   Solution. $\qquad\qquad v(t) = 3t^2 - 6t - 3$
   $$= 3(t^2 - 2t - 1)$$

   If $v(t) = 0$, $t = 1 \pm \sqrt{2}$. Therefore, $v(t) > 0$ if $t < 1 - \sqrt{2}$ or if $t > 1 + \sqrt{2}$. Since $a(t) = 6t - 6 > 0$, if $t > 1$, the answer to the problem is $t > 1 + \sqrt{2}$.

2. For an object moving in a straight line on an inclined plane, the distance s(ft.) from an origin $(0, 0)$ at time t (sec.) is given by $s = 5t - 6t^2$, $0 \leqslant t \leqslant 2$. Find (a) the velocity v and acceleration a at time t and (b) the time at which the object is farthest from the origin and the maximum distance.
   Solution.
   (a) Velocity and acceleration are given by $v(t) = 5 - 12t$ and $a(t) = -12$.
   (b) The object is at $(0, 0)$ at $t = 0$ and at $t = 5/6$. The velocity is zero at $t = 5/12$ when s is a relative maximum; $s(5/12) = 25/24$ ft. up the incline. But, $|s(2)| = |10 - 24| = 14$ ft. down the incline, a maximum numerical distance from $(0, 0)$.

3. Suppose the height s(ft.) at time t (sec.) of a projectile fired vertically upward from the ground is given by $s = 3200t - 16t^2$. Find (a) the projectile's initial velocity, (b) its acceleration at time t, (c) the time when its height is maximum, (d) its maximum height, (e) the time when it strikes the ground, and (f) its terminal velocity.
   Solution.
   (a) $v = s' = 3200 - 32t$, $v(0) = 3200$ ft./sec.
   (b) $a = v' = s'' = -32$ ft./sec^2. (constant)
   (c) Since $a < 0$, the maximum s occurs when $3200 - 32t = 0$, i.e., when $t = 100$ sec.
   (d) $s(100) = 320,000 - 160,000$
   $$= 160,000 \text{ ft.}$$
   (e) $s = 0$ when the projectile strikes the ground: $3200t - 16t^2 = 0$ therefore $t = 0$ when it leaves the ground, and $t = 200$ sec. when it strikes the ground.
   (f) $v(200) = 3200 - 6400 = -3200$ ft./sec.

4. With fuel spent, a rocket is 50,000 ft. above ground and traveling straight up. Its height s (ft.) at time t (sec.) thereafter is given by

$s = 50,000 + 8800t - 16t^2$. What is the maximum height reached?
Solution. $v = 8800 - 32t$. At maximum height, $v = 0$ and $a = -32 <$ 0, therefore, $t = 8800/32 = 275$ sec. Since $v' = -32 < 0$, s is a maximum when $t = 275$, hence

$$s(275) = 50,000 + 8800 (275) - 16(275)^2$$

$$= 1,260,000 \text{ ft. } (\cong 239 \text{ mi.})$$

5. An accelerator shoots particles counterclockwise around a circle 1.3×10^4 cm. in radius at a path speed of $|v| = 3.0 \times 10^{10}$ cm./sec. Find the x-component of velocity and the position of a particle whose y-component of velocity is 10 cm./sec.
Solution. First, $v^2 = v_x^2 + v_y^2$ and $9.0 \times 10^{20} = v_x^2 + 10^2$, or $v_x = \pm 10\sqrt{9.0 \times 10^{18} - 1}$. Further, the equation of the path is $x^2 + y^2 = \pi (1.3)^2 \times 10^8$, from which we get, by differentiating with respect to time t, $xv_x + yv_y = 0$. Hence, $\pm 10x\sqrt{9.0 \times 10^{18} - 1} + 10y = 0$ and $x^2 + x^2 (9.0 \times 10^{18} - 1) = \pi (1.3)^2 \times 10^8$, from which

$$x = \frac{1.3\pi^{1/2}}{3.0 \times 10^5}$$

Since $v_y > 0$ only in quadrants I and IV, x must be positive. We find

$$y = \pm \frac{1.3\pi^{1/2}}{3.0 \times 10^5} \sqrt{9.0 \times 10^{18} - 1}$$

Thus, we have determined the coordinates of the points where the particle is located.

6. A man-made satellite with a path speed of 5 mi./sec. travels around the earth in a circular orbit represented by the parametric equations $x = 4500 \cos \theta$, $y = 4500 \sin \theta$. Find the tangential and normal components of acceleration.
Solution. We first note that the satellite is some 500 miles up and its angular velocity is 1/900 rad./sec.

$$v_x = -4500 \sin \theta \frac{d\theta}{dt} \qquad v_y = 4500 \cos \theta \frac{d\theta}{dt}$$

$$= -4500 \sin \theta \left(\frac{1}{900}\right) \qquad = 4500 \cos \theta \left(\frac{1}{900}\right)$$

$$= -5 \sin \theta \qquad\qquad = 5 \cos \theta$$

$$a_x = -5 \cos \theta \, \frac{d\theta}{dt} \qquad a_y = -5 \sin \theta \, \frac{d\theta}{dt}$$

$$= -\frac{\cos \theta}{180} \qquad\qquad = -\frac{\sin \theta}{180}$$

$$a_T = \frac{v_x a_x + v_y a_y}{|v|} = \frac{\frac{1}{36} \sin \theta \cos \theta - \frac{1}{36} \sin \theta \cos \theta}{5} = 0$$

$$a_N = \frac{v_x a_y - v_y a_x}{|v|} = \frac{\frac{1}{36} \sin^2 \theta + \frac{1}{36} \cos^2 \theta}{5} = \frac{1}{180} \text{ mi./sec.}$$

7. Find the angular velocity and angular acceleration of a particle that moves along the curve $y = x^{1/2}$ in such a way that $v_x = k$ (constant).

Solution. Now, $\theta = \tan^{-1} \dfrac{y}{x} = \tan^{-1} \dfrac{1}{x^{1/2}}$

$$\omega = \frac{d\theta}{dt} = \frac{1}{1 + \dfrac{1}{x}} \left(-\frac{1}{2} x^{-3/2} \right) \frac{dx}{dt}$$

$$= -\frac{k}{2(1 + x)\sqrt{x}}$$

$$\alpha = \frac{d\omega}{dt} = -\frac{k}{2} \, \frac{(1 + x)\sqrt{x}\,(0) - \left[(1 + x)\dfrac{1}{2\sqrt{x}} + \sqrt{x} \right]}{(1 + x)^2 x} \frac{dx}{dt}$$

$$= \frac{k}{4} \frac{1 + 3x}{(1 + x)^2 \, x^{3/2}}$$

PROBLEMS WITH ANSWERS

8. A particle is constrained to move along the X-axis. If its distance x from the origin $(0, 0)$ at time $t > 0$ is given, find its (a) velocity and acceleration at time t, (b) initial velocity, (c) maximum distance to the right of the origin, and (d) the time when it passes through the origin, for the following conditions.

a. $x = 1600 - 16t^2$

> *Ans.* (a) $v = -32t;\ a = -32$
> (b) 0
> (c) 1600
> (d) 10

  b. $x = -t^2 + t + 6$

$$Ans. \quad (a) \ v = -2t + 1; a = -2$$
$$(b) \ 1$$
$$(c) \ 6\frac{1}{4}$$
$$(d) \ 3$$

  c. $x = -16t^2 + 1200t + 40,000$

$$Ans. \quad (a) \ v = -32t + 1200; a = -32$$
$$(b) \ 1200$$
$$(c) \ 40,000$$
$$(d) \ 100$$

  d. $x = t - t^3$

$$Ans. \quad (a) \ v = 1 - 3t^2; a = -6t$$
$$(b) \ 1$$
$$(c) \ \frac{2}{9}\sqrt{3}$$
$$(d) \ 1$$

9. A particle with given initial velocity is in vertical motion under the influence of gravity. Find its height y ft. at time t sec. from the given conditions (-32 ft./sec.^{2} = acceleration due to gravity), if:

   a. The particle is dropped from 1200 ft. elevation.

$$Ans. \quad y = -16t^2 + 1200$$

   b. The particle is projected upward from ground at 800 ft./sec. initial velocity.

$$Ans. \quad y = -16t^2 + 800t$$

   c. The particle is projected upward from space platform 500,000 ft. above the ground with initial velocity of 2000 ft./sec.

$$Ans. \quad y = -16t^2 + 2000t + 500,000$$

   d. The particle is projected downward from a balloon 5000 ft. above the ground with initial velocity of 100 ft./sec.

$$Ans. \quad y = -16t^2 - 100t + 5000$$

10. A particle moves around the ellipse $x^2 + 2y^2 = 1$. Where will the two components of velocity be equal?

$$Ans. \quad v_x = v_y \text{ at } \left(\frac{\sqrt{6}}{3}, -\frac{\sqrt{6}}{6}\right) \text{and at} \left(-\frac{\sqrt{6}}{3}, \frac{\sqrt{6}}{6}\right)$$

11. A projectile, when acted on only by the force of gravity, has the following path: $x = v_0 t \cos \alpha$, and $y = v_0 t \sin \alpha - \frac{1}{2} gt^2$, where $v_x = v_0 \cos \alpha$, $v_y = v_0 \sin \alpha - gt$, and α is the angle the direction of fire makes with the horizontal.

a. How far will the projectile travel in a horizontal direction?

b. Show that this range is a maximum for $\alpha = 45°$.

$$Ans. \quad a. \quad \text{Horizontal range} = \frac{v_0^2}{g} \sin 2\alpha$$

12. If a particle moves along a curve with constant speed, show that the tangential component of acceleration $a_T = 0$.

13. Show that for motion along the cycloid $x = b(t - \sin t)$, $y = b(1 - \cos t)$, the magnitude of the acceleration $|a|$ is constant.

14. A particle moves according to the parametric equations $x = 3t$, $y = t^2$. Show that $v_x = 3$, $v_y = 2t$, $|v| = \sqrt{9 + 4t^2}$, $a_x = 0$, $a_y = 2$, and $|a| = 2$.

15. A particle moves so that its coordinates are given by $x = t$ and $y = t^2$. Show that $\dfrac{dx}{dt} = 1$, $\dfrac{dy}{dt} = 2t$, $\theta = \tan^{-1} 2t$, $\dfrac{d^2 x}{dt^2} = 0$, $\dfrac{d^2 y}{dt^2} = 2$, $\phi = 90°$, $|v| = \sqrt{4t^2 + 1}$, and $|a| = 2$.

16. A particle's position in the plane at time t is given by $x = \log t$ and $y = \frac{1}{2}\left(t + \dfrac{1}{t}\right)$. Show that at $t = 1$, $v_x = 1$, $v_y = 0$, $|v| = 1$, the direction of $v = 0$, $a_x = -1$, $a_y = 1$, $|a| = \sqrt{2}$, the direction of $a = 135°$, $a_T = -1$, and $a_N = 1$.

8.8 Related Rates

Suppose that a relation exists between two variables u and v each of which depends upon time t. Then $\dfrac{du}{dt}$ and $\dfrac{dv}{dt}$ represent, respectively, the rates of change of u and v with respect to time. Since u and v are related, so will $\dfrac{du}{dt}$ and $\dfrac{dv}{dt}$ be related, and these relations may be used to solve a variety of problems.

Example 1. A boat is being hauled toward a pier which is 20 ft. above the water. The rope is pulled in at the rate of 6 ft./sec. How fast is the boat approaching the base of the pier when 25 ft. of rope remains to be pulled in (Fig. 8.13)?

Solution. At any time t we have

$$20^2 + x^2 = z^2$$

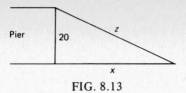

FIG. 8.13

Hence,

$$x \frac{dx}{dt} = z \frac{dz}{dt}$$

When $z = 25, x = 15$; therefore,

$$15 \frac{dx}{dt} = 25(-6), \text{ and}$$

$$\frac{dx}{dt} = -10 \text{ ft./sec.}$$

The boat approaches base at 10 ft./sec.

Example 2. A conical filter is 3 in. in radius and 4 in. deep. Liquid escapes through the filter at the rate of 2 cc/sec. How fast is the level of the liquid falling when the depth of the liquid is 3 in. (Fig. 8.14)?

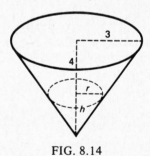

FIG. 8.14

Solution. Let V be the volume, h be the depth of the liquid at any time t, and r be the corresponding radius. Then, at any time t,

$$\frac{r}{h} = \frac{3}{4} \text{ and } r = \frac{3}{4}h$$

Since we seek $\frac{dh}{dt}$, it is best to express V as a function of h.

Thus,

$$V = \tfrac{1}{3}\pi r^2 h = \tfrac{3}{16}\pi h^3$$

$$\frac{dV}{dt} = \frac{3}{16}\pi h^2 \frac{dh}{dt}$$

Using 1 in. = 2.54 cm., we get

$$-2 = \frac{9}{16}\pi \left[3(2.54)\right]^2 \frac{dh}{dt} \text{ or } \frac{dh}{dt} = -0.02 \text{ cm./sec.}$$

The liquid level falls at the rate of 0.02 cm./sec.

Example 3. A man 6 ft. tall walks at 4 mi./hr. directly away from a light that hangs 18 ft. above the ground. (a) How fast does his shadow lengthen? (b) At what rate is the head of the shadow moving away from the base of the light (Fig. 8.15)?

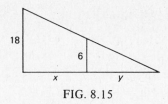

FIG. 8.15

Solution.

(a)
$$\frac{x+y}{18} = \frac{y}{6}$$

$$x = 2y$$

$$\frac{dx}{dt} = 2\frac{dy}{dt} \text{ or } \frac{dy}{dt} = \frac{1}{2}\frac{dx}{dt}$$

Thus, the shadow lengthens at a rate equal to one half that of the walker and is independent of position. Therefore, at any time,

$$\frac{dy}{dt} = 2 \text{ mi./hr.}$$

(b) The head of the shadow moves at a rate equal to the sum of the rates of the walker and of the lengthening of the shadow. Call $L = x + y$, then

$$\frac{dL}{dt} = \frac{dx}{dt} + \frac{dy}{dt}$$

$$= 6 \text{ mi./hr.}$$

This also is independent of position.

Example 4. A dive bomber loses altitude at the rate of 1600 mi./hr. How fast is the visible surface of the earth decreasing when the bomber is 1 mi. high?

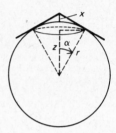

FIG. 8.16

Solution. First call the radius of the earth r and the height of the bomber x; let z be, as in Fig. 8.16, the distance from the center to the base of the visible spherical cap. We have

$$\frac{r}{r+x} = \frac{z}{r}$$

since each equals cos α.

Whence, $$z = \frac{r^2}{r+x}$$

Now the area of the spherical cap cut off by a plane z units from the center is

$$A = 2\pi r(r-z)$$

$$= 2\pi r^2 - \frac{2\pi r^3}{r+x}$$

$$\frac{dA}{dt} = \frac{2\pi r^3}{(r+x)^2} \frac{dx}{dt}$$

Setting $r = 4000$, $x = 1$, and $\frac{dx}{dt} = -1600$, we can evaluate $\frac{dA}{dt}$ exactly.

However, since $r + x$ is approximately equal to r, we can get an approximation for $\dfrac{dA}{dt}$ with little work.

$$\frac{dA}{dt} \cong 2\pi r \frac{dx}{dt} = 2\pi \times 4000 \times (-1600)$$

$$\cong -128\pi \times 10^5 \text{ sq. mi./hr.}$$

$$\cong -\frac{128 \times 22 \times 10^5}{7 \times 60 \times 60} \text{ sq. mi./sec.}$$

$$\cong -11{,}200 \text{ sq. mi./sec.}$$

The visible area decreases at the rate of approximately 11,200 sq. mi./sec.

PROBLEMS WITH SOLUTIONS

1. A pebble dropped into a pool creates a circular ripple whose radius increases at 3 ft./sec. How fast are the circle's area A and circumference C increasing when the radius r of the circle is 7 ft.?

   Solution.

   $$A = \pi r^2 \text{ and } \frac{dA}{dt} = 2\pi r \frac{dr}{dt} = 2\pi (7)(3) = 42\pi \text{ sq. ft./sec.}$$

   $$C = 2\pi r \text{ and } \frac{dC}{dt} = 2\pi \frac{dr}{dt} = 2\pi(3) = 6\pi \text{ ft./sec.}$$

2. Chips from a sawmill conveyer drop onto the ground forming a conical pile whose angle of repose is $45°$. (The radius r of the cone is equal to its height h). If the volume V grows at the rate of 4 cu. ft./min., how fast is the height of the pile increasing when it is 30 ft.?

   Solution. $V = \dfrac{1}{3} \pi r^2 h = \dfrac{1}{3} \pi h^3$ and $\dfrac{dV}{dt} = \pi h^2 \dfrac{dh}{dt}$, thus $4 = \pi(30)^2 \dfrac{dh}{dt}$, and $\dfrac{dh}{dt} = \dfrac{1}{225\pi}$ ft./min.

3. A pulley P is 10 ft. above the ground from point A. The end B of a taut rope moves horizontally away from point A and is 15 ft. from A. The rope is 25 ft. long, and a weight W is attached to the end of it. If B moves away from A at 5 ft./sec., how fast is W rising?

Solution. $\sqrt{x^2 + 100} = 25 - (10 - h) = h + 15$, and $x^2 + 100 = h^2 +$

$30h + 225$, thus $2x \dfrac{dx}{dT} = (2h + 30) \dfrac{dh}{dT}$. When $x = 15$, $h = 3$, and

$\dfrac{dx}{dT} = 5$, then $\dfrac{dh}{dT} = 4 \dfrac{1}{6}$ ft./sec.

4. If the length L of a rectangle increases at the rate of 2 ft./sec., how must the width W change so that the area A remains constant?

Solution. $A = LW$ and $\dfrac{dA}{dt} = L \dfrac{dW}{dt} + W \dfrac{dL}{dt}$, thus, $0 = L \dfrac{dW}{dt} + 2W$

and $\dfrac{dW}{dt} = - \dfrac{2W}{L}$ ft./sec.

5. Gas is being pumped into a spherical balloon at the rate of 3 cu. ft./sec. How fast are the radius r and the surface area A_s increasing when the radius of the balloon is 10 ft.?

Solution. $V = \dfrac{4}{3} \pi r^3$, $\dfrac{dV}{dt} = 4 \pi r^2 \dfrac{dr}{dt}$, and $3 = 4 \pi (10)^2 \dfrac{dr}{dt}$

Therefore, $\dfrac{dr}{dt} = \dfrac{3}{400 \pi}$ ft./sec. Further, $A_s = 4 \pi r^2$, $\dfrac{dA_s}{dt} = 8 \pi \dfrac{dr}{dt}$ and

$\dfrac{dA_s}{dt} = 8 \pi (10) \dfrac{3}{400 \pi} = \dfrac{3}{5}$ sq. ft./sec.

6. A cylinder has a radius r of 1/2 ft. The volume V of steam in it is increasing at 20 cu. ft./sec. How fast is the piston moving?

Solution. $V = \pi r^2 h$ and $\dfrac{dV}{dT} = \pi r^2 \dfrac{dh}{dT}$

With $r = 1/2$ and $\dfrac{dV}{dT} = 20$, $\dfrac{dh}{dT} = \dfrac{80}{\pi}$ ft./sec.

7. Water is being poured into a conical tank of 10 ft. radius and 12 ft. height at 40 cu. ft./min. How fast is the level of the water rising when it reaches the 6 ft. level?

Solution. $r = \dfrac{5}{6} h$

$$V = \dfrac{1}{3} \pi r^2 h = \dfrac{25}{108} \pi h^3$$

$$\dfrac{dV}{dT} = \dfrac{25}{36} \pi h^2 \dfrac{dh}{dT}$$

When $h = 6$, and $\dfrac{dV}{dT} = 40$, $\dfrac{dh}{dT} = 0.51$ ft./min.

PROBLEMS WITH ANSWERS

8. If the edge of a cube is changing at the rate of -2 in./sec., find the rate of change of (a) the length of a diagonal of the cube, (b) the length of a face diagonal, (c) the surface area when an edge is 1 in., and (d) the volume when an edge is 5 in.

$$Ans. \quad (a) \quad -2\sqrt{3} \text{ in./sec.}$$
$$(b) \quad -2\sqrt{2} \text{ in./sec.}$$
$$(c) \quad -24 \text{ sq. in./sec.}$$
$$(d) \quad -150 \text{ cu. in./sec.}$$

9. A baseball diamond is a square with 90 ft. sides. If a runner on first base starts to steal second at the rate of 30 ft./sec., how fast is his distance from home plate changing when he is halfway to second base?

$$Ans. \quad 6\sqrt{5} \text{ ft./sec.}$$

10. An airplane flying due west at 300 mi./hr. at an elevation of 5 mi. passes directly over a ship traveling due north at 36 mi./hr. How fast are the airplane and the ship separating 10 min. later?

$$Ans. \quad 300.67 \text{ mi./hr.}$$

11. An engine cylinder has a diameter of 10 in. At what rate is the piston moving when the steam is entering the cylinder at the rate of 16 cu. ft./sec.?

$$Ans. \quad 2304/25\pi \text{ ft./sec.}$$

12. A revolving light is 1/2 mi. from a straight beach and makes one revolution per minute. A ray from the light illuminates a spot on the beach. How fast is the spot moving along the beach when the ray makes a 45° angle with the shore line?

$$Ans. \quad 2\pi \text{ mi./min.}$$

13. A grade crossing has a double gate which has two arms that rotate upward about the same axis. The arm over the road is 12 ft. long and the one over the sidewalk is 5 ft. long. Each arm rotates at the rate of 4 rad./min. At what rate is the distance between their extremities changing when they make (a) a 45° angle with the horizontal and (b) a 60° angle with the horizontal?

$$Ans. \quad (a) \quad \frac{4\pi}{13} \text{ ft./sec.}$$
$$(b) \quad \frac{2\sqrt{3}\,\pi}{\sqrt{89}} \text{ ft./sec.}$$

14. A rod 10 ft. long moves so that its ends A and B remain constantly on the X- and Y-axes, respectively. If A is 8 ft. from the origin $(0,0)$ and is moving away from it at 2 ft./sec., (a) at what rate is B descending, (b) at what rate is the area of the triangle AOB changing, and (c) at what rate is the distance from $(0,0)$ to the center of the rod changing?

> *Ans.* (a) $2\dfrac{2}{3}$ ft./sec.
>
> (b) -2 sq. ft./sec.
>
> (c) Does not change.

15. An airplane, flying due north at an elevation of 1 mile and with a speed of 300 mi./hr., passes directly over a ship traveling due west at 30 mi./hr. How fast are they separating 2 minutes later?

> *Ans.* $\dfrac{3030}{\sqrt{102}}$ mi./hr.

8.9   Polar Coordinates

Whereas in rectangular coordinates $\dfrac{dy}{dx}$ represents the slope of the curve $y = f(x)$, in polar coordinates $\dfrac{d\rho}{d\theta}$ does *not* represent the slope of the curve $\rho = f(\theta)$. It merely represents the rate of change of the radius vector ρ with respect to the angle θ. In order to determine slope in polar coordinates, we make use of the relations between rectangular coordinates (x,y) and polar coordinates (ρ,θ); where x and y are functions of ρ *and* θ, these are:

$$x = \rho \cos \theta \text{ and } y = \rho \sin \theta$$

Hence,
$$\frac{dx}{d\theta} = \frac{d\rho}{d\theta} \cos \theta - \rho \sin \theta$$

$$\frac{dy}{d\theta} = \frac{d\rho}{d\theta} \sin \theta + \rho \cos \theta$$

$$\text{slope} = \frac{dy}{dx} = \frac{\dfrac{dy}{d\theta}}{\dfrac{dx}{d\theta}} = \frac{\dfrac{d\rho}{d\theta} \sin \theta + \rho \cos \theta}{\dfrac{d\rho}{d\theta} \cos \theta - \rho \sin \theta}, \text{ provided } \frac{dx}{d\theta} \neq 0$$

When numerator and denominator are both divided by $\dfrac{d\rho}{d\theta} \cos \theta$, this reduces to

$$\text{Slope} = \frac{\tan \theta + \dfrac{\rho}{\rho'}}{i - \tan \theta \cdot \dfrac{\rho}{\rho'}}, \quad \text{where } \rho' = \frac{d\rho}{d\theta} \qquad (1)$$

With this formula, we can readily find the slope of a curve whose equation is given in polar coordinates.

An important angle is the angle ψ between the radius vector and the

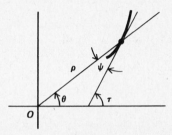

FIG. 8.17

tangent measured counterclockwise from the radius vector to the tangent. (Fig. 8.17.) Now,

$$\begin{aligned}
\text{Slope} &= \tan \tau \\
&= \tan (\theta + \psi) \\
&= \frac{\tan \theta + \tan \psi}{1 - \tan \theta \tan \psi} \qquad (2)
\end{aligned}$$

Comparing (1) and (2) we find

$$\tan \psi = \frac{\rho}{\rho'} \qquad (3)$$

The angle between two curves is given by (Fig. 8.18)

$$\theta_{12} = \psi_2 - \psi_1$$

where θ_{12} is the angle measured counterclockwise from curve 1 to curve 2; hence,

$$\begin{aligned}
\tan \theta_{12} &= \tan (\psi_2 - \psi_1) \\
&= \frac{\tan \psi_2 - \tan \psi_1}{1 + \tan \psi_1 \tan \psi_2} \qquad (4)
\end{aligned}$$

Formula (3) is used to evaluate $\tan \psi_1$ and $\tan \psi_2$.

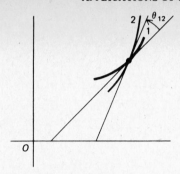

FIG. 8.18

Example 1. Find the slope of the curve $\rho = 2 - \cos\theta$ (a) at any point and (b) at $\theta = \dfrac{\pi}{4}$.

Solution.

(a)
$$\text{Slope} = \frac{\tan\theta + \dfrac{\rho}{\rho'}}{1 - \tan\theta\,\dfrac{\rho}{\rho'}}$$

$$= \frac{\tan\theta + \dfrac{2 - \cos\theta}{\sin\theta}}{1 - \tan\theta\,\dfrac{2 - \cos\theta}{\sin\theta}}$$

$$= \frac{\sin\theta\,\tan\theta + 2 - \cos\theta}{2(\sin\theta - \tan\theta)}$$

(b)
$$\text{Slope} = \frac{\dfrac{\sqrt{2}}{2} + 2 - \dfrac{\sqrt{2}}{2}}{\sqrt{2} - 2}$$

$$= \frac{2}{\sqrt{2} - 2}, \text{ at } \theta = \frac{\pi}{4}$$

Example 2. Find the angle of intersection of curve 1, $\rho = \sin\theta$, and curve 2, $\rho = 1 + \cos\theta$.

Solution. For curve 1, we have: $\tan\psi_1 = \dfrac{\rho}{\rho'} = \dfrac{\sin\theta}{\cos\theta} = \tan\theta$

For curve 2 we have: $\tan \psi_2 = \dfrac{\rho}{\rho'} = \dfrac{1 + \cos \theta}{- \sin \theta}$

Curve 1 is a circle and curve 2 is a cardioid. They intersect at two points, $\left(1, \dfrac{\pi}{2}\right)$ and $(0, \pi)$. For any common point, we have

$$\tan \theta_{12} = \frac{\tan \psi_2 - \tan \psi_1}{1 + \tan \psi_1 \tan \psi_2}$$

$$= \frac{\dfrac{1 + \cos \theta}{- \sin \theta} - \tan \theta}{1 + \left(\dfrac{1 + \cos \theta}{- \sin \theta}\right) \tan \theta}$$

$$= \frac{1 + \cos \theta}{\sin \theta}$$

At the first point of intersection, this reduces to

$$\tan \theta_{12} = \frac{1 + \cos \dfrac{\pi}{2}}{\sin \dfrac{\pi}{2}} = 1$$

Therefore, $\theta_{12} = \pi/4$

At the second point, $\tan \theta_{12} = \dfrac{1 + \cos \pi}{\sin \pi} = \dfrac{0}{0}$, and indeterminate form;

but $\dfrac{1 + \cos \theta}{\sin \theta} = \cot \dfrac{\theta}{2}$ and $\cot \dfrac{\pi}{2} = 0$. Hence, $\theta_{12} = 0$ and the curves are tangent.

PROBLEMS WITH SOLUTIONS

1. Find the slope of the curve $\rho = \cos \theta$ (a) at any point and (b) at $\theta = \frac{1}{3} \pi$.
   Solution.
   (a) $\rho' = - \sin \theta$

$$\text{Slope} = \frac{\tan \theta + \dfrac{\cos \theta}{- \sin \theta}}{1 - \dfrac{\tan \theta \cos \theta}{- \sin \theta}} = \frac{\sin^2 \theta - \cos^2 \theta}{2 \sin \theta \cos \theta}$$

(b) When $\theta = \dfrac{1}{3}\pi$, the slope $= \dfrac{3/4 - 1/4}{\left(2\dfrac{\sqrt{3}}{2}\right)\left(\dfrac{1}{2}\right)} = \dfrac{1}{\sqrt{3}} = \dfrac{\sqrt{3}}{3}$.

2. Find the slope of the curve $\rho = 1 + \cos\theta$ (a) at any point and (b) at $\theta = \dfrac{\pi}{3}$.

   Solution.

   (a) Slope $= \dfrac{\tan\theta + \dfrac{1 + \cos\theta}{-\sin\theta}}{1 - \dfrac{\tan\theta\,(1 + \cos\theta)}{-\sin\theta}}$

   $= \dfrac{\sin^2\theta - \cos^2\theta - \cos\theta}{2\sin\theta\cos\theta + \sin\theta}$

   (b) When $\theta = \dfrac{1}{3}\pi$, the slope $= \dfrac{3/4 - 1/4 - 1/2}{2\left(\dfrac{\sqrt{3}}{2}\right)\left(\dfrac{1}{2}\right) + \dfrac{\sqrt{3}}{2}} = 0$.

3. Find the angle of intersection of $\rho = \cos\theta$ and $\rho = 1 - \cos\theta$.
   Solution. $2\cos\theta = 1$ and $\cos\theta = 1/2$; therefore, $\theta = 1/3\pi$. Since $\rho = \cos\theta$, $\rho' = -\sin\theta$, and $\tan\psi_1 = \dfrac{-\cos\theta}{\sin\theta}$, and $\rho = 1 - \cos\theta$, $\rho' = \sin\theta$, and $\tan\psi_2 = \dfrac{1 - \cos\theta}{\sin\theta}$, $\tan\theta_{12} = \dfrac{\tan\psi_2 - \tan\psi_1}{1 + \tan\psi_1\tan\psi_2} =$

   $\dfrac{\dfrac{1 - \cos\theta}{\sin\theta} + \dfrac{\cos\theta}{\sin\theta}}{1 + \dfrac{-\cos\theta}{\sin\theta}\dfrac{1 - \cos\theta}{\sin\theta}} = \dfrac{\sin\theta}{1 - \cos\theta}$. When $\theta = \dfrac{1}{3}\pi$, $\tan\theta_{12} = \sqrt{3}$,

   and $\theta_{12} = \dfrac{1}{3}\pi$.

4. Find the slope of the curve $\rho = 1 - \dfrac{1}{1 + \theta}$ at the origin.
   Solution. Here, the origin has coordinates $(0,0)$.

   $$\rho' = \dfrac{1}{(1 + \theta)^2} \text{ and slope} = \dfrac{0 + \dfrac{0}{1}}{1 - 0\cdot\dfrac{0}{1}} = 0$$

5. Find the slope of the curve $\rho^2 = \cos 2\theta$ at the origin.

Solution. Here the origin has two sets of coordinates $\left(0, \dfrac{\pi}{4}\right)$ and

$\left(0, \dfrac{3\pi}{4}\right)$; further, $\rho' = -\dfrac{\sin 2\theta}{\rho}$

At $(0, \pi/4)$, the slope $= \dfrac{\tan \theta - \rho^2 \csc 2\theta}{1 + \tan \theta \, \rho^2 \csc 2\theta} \Bigg|_{\theta = \pi/4} = 1$

In a like manner, the slope at $(0, 3\pi/4)$ is -1.

6. Find the angle of intersection of curve 1, $\rho = \cos \theta$, and curve 2, $\rho = \sin \theta$ (Fig. 8.19).

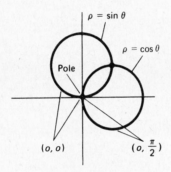

FIG. 8.19

Solution. For curve 1, $\tan \psi_1 = \dfrac{\rho}{\rho'} = -\cot \theta$; for curve 2, $\tan \psi_2 = \tan \theta$. Both curves intersect at $(\sqrt{2}/2, \pi/4)$ *and* at the origin (the pole). Note that for curve 1 the pole has coordinates $(0, \pi/2)$, while for curve 2 the pole has coordinates $(0, 0)$. The pole must, therefore, be handled specially. For any point with common coordinates,

$$\tan \theta_{12} = \frac{\tan \psi_2 - \tan \psi_1}{1 + \tan \psi_1 \tan \psi_2}$$

$$= \frac{\tan \theta + \cot \theta}{1 - \cot \theta \tan \theta}$$

Since the denominator is zero when $\theta = \pi/4$, $\tan \theta_{12}$ does not exist. This indicates that the curves intersect at right angles. Further, the denominator is identically zero; this indicates that the two tangents to the curves for a given vectorial angle are orthogonal (at right angles). At the pole, the line tangent to curve 1 is vertical and the line tangent to curve 2 is horizontal; they are, therefore, orthogonal.

7. Find the angle of intersection of curve 1, $\rho = \cos \theta$, and curve 2, $\rho = \cos \dfrac{\theta}{2}$, at the point of intersection $(1/2, \pi/3)$. These curves also intersect at other points.

Solution. For curve 1, $\tan \psi_1 = - \cot \theta$; for curve 2, $\tan \psi_2 = - 2 \cot \dfrac{\theta}{2}$.

At $\theta = \pi/3$, $\tan \theta_{12} = \dfrac{- 2\sqrt{3} + 1/\sqrt{3}}{1 + (- 1/\sqrt{3})(- 2\sqrt{3})}$

$$= - \frac{5}{9} \sqrt{3}$$

Therefore, the acute angle $\theta_{12} = \text{Tan}^{-1} \dfrac{5}{9} \sqrt{3}$.

PROBLEMS WITH ANSWERS

8. Find the slope of each curve at the point indicated.

a. $\rho = a \sin \theta + b \cos \theta, (\rho, \theta)$

Ans. $\dfrac{a \tan 2\theta + b}{a - b \tan 2\theta}$

b. $\rho = a \sec^2 \dfrac{\theta}{2}, (\rho, \theta)$

Ans. $\dfrac{\tan \theta \tan \dfrac{\theta}{2} + 1}{\tan \dfrac{\theta}{2} - \tan \theta}$

c. $\rho = 1 + \log \cos \theta, (\rho, \theta)$

Ans. $\dfrac{\tan \theta - (1 + \log \cos \theta) \cot \theta}{2 + \log \cos \theta}$

d. $\rho = a (\sec \theta - \cos \theta), \left(\dfrac{a}{2} \sqrt{2}, \dfrac{\pi}{4} \right)$

Ans. 2

9. Find the smaller angle of intersection between the following pairs of curves.

   a. $\rho = a(1 + \sin\theta), \rho = a(1 - \sin\theta)$

Ans. $\dfrac{\pi}{2}$

   b. $\rho = 4\cos\theta, \rho = 4(1 - \cos\theta)$

Ans. $\dfrac{\pi}{3}$

   c. $\rho = 2(1 + \cos\theta), \rho = 6\cos\theta$

Ans. $\dfrac{\pi}{6}$

   d. $\rho = 4\sec^2\dfrac{\theta}{2}, \rho\sin^2\dfrac{\theta}{2} = 8$

Ans. $\dfrac{\pi}{2}$

   e. $\rho = \sin\theta + \cos\theta, \rho = 2\sin\theta$

Ans. $\dfrac{\pi}{4}$

10. Find the slope of the curve $\rho = a\sin 2\theta$ at the point $\left(a, \dfrac{\pi}{4}\right)$.

Ans. -1

11. Find the angle between the radius vector and the tangent to the curve $\rho = a\theta$ at the point $\left(\dfrac{a\pi}{2}, \dfrac{\pi}{2}\right)$.

Ans. $\tan^{-1}\dfrac{\pi}{2}$

12. Find the angle of intersection of the curves $\rho = a(1 - \sin\theta)$ and $\rho = a(1 + \sin\theta)$ at the point (a, π).

Ans. $\dfrac{\pi}{2}$

8.10 Differentials

Up to this point, we have considered $\dfrac{dy}{dx}$ as a single symbol and not as a fraction. But it is clear, however, that since $\dfrac{dy}{dx}$ is the tangent of a certain angle, it could be represented as a fraction if we so desired. Further, we could arbitrarily choose any value for the denominator of this fraction; then, of course, the numerator would be fixed so that the ratio would be equal to $\dfrac{dy}{dx}$. The symbol itself suggests that we define quantities dy and dx, and then take the ratio of these for the derivative of y with respect to x. Let us recall that

$$\frac{dy}{dx} = \lim_{\Delta x \to 0} \frac{\Delta y}{\Delta x} = f'(x)$$

Now, $\dfrac{\Delta y}{\Delta x} = \tan \alpha,$ Δy and Δx being increments of the dependent and independent variables, respectively (Fig. 8.20). Hence, $\dfrac{\Delta y}{\Delta x}$ is the ratio

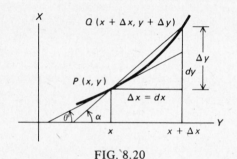

FIG. 8.20

of two increments. But $\dfrac{dy}{dx} = \tan \theta$ could also be called a ratio. If we set $\Delta x = dx$ and call this the **differential** of the independent variable x, then $\dfrac{dy}{dx} = f'(x),$ or $dy = f'(x)\, dx = \dfrac{dy}{dx}\, dx$. We define dy as the differential of the function. *The differential of a function equals the derivative of the function multiplied by the differential of the independent variable.*

PROBLEMS WITH ANSWERS

Find the differential of the following functions.

1. $y = x^3 + 7.$ *Ans.* $dy = 3x^2\, dx$

2. $y = 4 \sin 3x.$ *Ans.* $dy = 12 \cos 3x\, dx$

3. $y = e^{x^2} - \log x.$ *Ans.* $dy = \left(2xe^{x^2} - \dfrac{1}{x}\right) dx$

4. $y = xe^x$.   Ans. $dy = e^x(1 + x)\,dx$

5. $y + \log y + \tan 2x = 5$.   Ans. $dy + \dfrac{1}{y}\,dy + 2\sec^2 2x\,dx = 0$

8.11 Approximations by Means of Differentials

Since the limit of $\dfrac{\Delta y}{\Delta x}$, as $\Delta x \to 0$, equals $\dfrac{dy}{dx}$, then, by the very notion of a limit, it follows that, for sufficiently small values of Δx, we have approximately (the symbol $\cong$ stands for "approximately equal to") $dy \cong \Delta y$. This enables us to use differentials as approximations to increments. Whereas it may be somewhat troublesome to compute the exact value of an increment Δy of a function for a given increment Δx of the independent variable, it may be a relatively simple matter to compute the differential dy. Further, where actual measurements are being used, it would usually be foolish to compute Δy exactly since the measurements themselves are nothing but approximations. If Δx is small enough, the difference between dy and Δy will be small.

Where errors of measurement are involved, it is customary to call Δy the error in y, $\dfrac{\Delta y}{y}$ the relative error, and $100\,\dfrac{\Delta y}{y}$ the percentage error. Approximations to these are:

$$dy \cong \text{the error in } y$$

$$\frac{dy}{y} \cong \text{the relative error}$$

$$100\,\frac{dy}{y} \cong \text{the percentage error}$$

Note that if $\log y$ is first computed then the relative error is automatically introduced upon taking differentials since the differential of $\log y$ is $\dfrac{1}{y}\,dy$.

Example 1. A spherical ball bearing when new measures 3.00 in. in radius. What is the approximate volume of metal lost after it wears down to $r = 2.98$ in.?
Solution.

$$V = \tfrac{4}{3}\pi r^3$$

$$dV = 4\pi r^2 \, dr$$
$$= 4\pi(3^2)(0.02) = 0.72\,\pi$$
$$= 2.26 \text{ cu. in.}$$

If we assume that the figures $r = 3.00$ in. and $r = 2.98$ in. are exact, then the exact answer would be

$$\triangle V = \tfrac{4}{3}\pi\,[3^3 - (2.98)^2]$$
$$= (0.71521066 \cdots)\,\pi$$
$$= 2.26 \text{ cu. in. (to two decimal places)}$$

Example 2. Find (a) the error and (b) the percentage error made in computing the volume of a cube, if a 1% error is made in computing the length of an edge.
Solution.

(a) $V = x^3$ and $dV = 3x^2 \, dx$. Now, we are given $100\,\dfrac{dx}{x} = 1$, i.e., $\dfrac{dx}{x} =$ 0.01. Therefore, we must modify dV to read

$$dV = 3x^3\,\frac{dx}{x} \text{ by multiplying and dividing by } x.$$

$$dV = 3x^3\,(0.01)$$
$$= 0.03\,x^3 \text{ cu. units, which depends, necessarily, upon } x.$$

(b) $\dfrac{dV}{V} = \dfrac{3x^2}{x^3}\,dx = 3\dfrac{dx}{x} = 0.03$

Hence, a 3% error in the calculation of the volume will be made, if a 1% error is made in the measurement of an edge.

Example 3. Kinetic energy is given by $E = \tfrac{1}{2}\,mv^2$. If E is known only to within 2% for a given mass m, find (a) the relative error and (b) the percentage error made in estimating the velocity v from this equation.
Solution.

(a)
$$E = \tfrac{1}{2}\,mv^2$$
$$\log E = \log \tfrac{1}{2}\,m + 2\log v$$

$$\frac{dE}{E} = 2\,\frac{dv}{v}$$

$$0.02 = 2\frac{dv}{v}$$

$$\frac{dv}{v} = 0.01$$

(b) $\qquad\qquad 100\frac{dv}{v} = 1\%$

PROBLEMS WITH SOLUTIONS

1. The side of a unit square is measured in error to be 0.98, and from this value the area is computed. Find (a) the exact error and (b) the approximate error.
   Solution.
   (a) The exact area is 1 sq. unit; the computed area is $(0.98)^2 =$ 0.9604 sq. unit. Thus, the exact error $\triangle A$ is 0.0396 sq. unit.
   (b) The area A of a square of side x is $A = x^2$. Now, $dA \cong \triangle A$ and $dA = 2x\,dx$. Further, $x = 1$ and $dx = 0.02$; therefore, $dA = 2(0.02) = 0.04$ sq. unit.

2. The edge of a cube is 10 ft., but it is measured to be 9.97 ft., and from this value the volume is computed. Find (a) the exact error and (b) the approximate error.
   Solution.
   (a) The exact volume is 1000 cu. ft.; the computed volume is $(9.97)^3 = 991.026973$ cu. ft.; thus, the exact error $\triangle V$ is 8.973027 cu. ft.
   (b) The volume of a cube of edge x is $V = x^3$. Now, $dV \cong \triangle V$ and $dV = 3x^2\,dx$. Further, $x = 10$ and $dx = 0.03$; therefore, $dV = 300\,(0.03) = 9$ cu. ft.

3. Find (a) the error and (b) the percentage error made in computing the volume of a sphere if a 1% error is made in computing the radius.
   Solution.
   (a) $V = \dfrac{4}{3}\pi r^3$ and $dV = 4\pi r^2\,dr.$

   The percentage error given is $100\,\dfrac{dr}{r} = 1$; hence, $\dfrac{dr}{r} = 0.01$; therefore

$$dV = 4\pi r^3 \frac{dr}{r}$$

$$= 4\pi r^3 (0.01) = \frac{4\pi r^3}{100} \text{ cu. units}$$

(b) $\dfrac{dV}{V} = \dfrac{4\pi r^2\, dr}{\frac{4}{3}\pi r^3} = 3\,\dfrac{dr}{r} = 0.03$

Therefore, if a 1% error is made in measuring the radius, a 3% error will result in computing the volume.

4. Find (a) the error and (b) the percentage error made in computing the area A of a circle if a 1% error is made in computing the radius r.

Solution.

(a) $A = \pi r^2$ and

$$dA = 2\pi r\, dr$$

$$= 2\pi r^2 \frac{dr}{r}$$

$$= 2\pi r^2 (0.01)$$

$$= \frac{2\pi r^2}{100} \text{ sq. units}$$

(b) $\dfrac{dA}{A} = \dfrac{2\pi r\, dr}{\pi r^2} = 2\,\dfrac{dr}{r} = 0.02$

Therefore, if a 1% error is made in measuring the radius, a 2% error will result in computing the area.

5. Find (a) the error and (b) the percentage error made in computing the volume V of a right circular cone of fixed height h, if a 1% error is made in computing the radius r.

Solution.

(a) $V = \dfrac{1}{3}\pi r^2 h$ and

$$dV = \frac{2}{3}\pi r h\, dr$$

$$= \frac{2\pi h r^2}{3}\frac{dr}{r}$$

$$= \frac{2\pi h r^2}{300} \text{ cu. units}$$

(b) $\dfrac{dV}{V} = 2\dfrac{dr}{r} = 0.02$

Therefore, if a 1% error is made in measuring the radius, a 2% error will result in computing the volume.

PROBLEMS WITH ANSWERS

6. For each function f defined below, write the differential of f at x.

(a) $f(x) = x^n$

$Ans.\quad df = nx^{n-1}\,dx$

(b) $f(x) = e^{ax}$

$Ans.\quad df = ae^{ax}\,dx$

(c) $f(x) = A \sin (bx + \pi/3)$

$Ans.\quad df = bA \cos (bx + \pi/3)\,dx$

(d) $f(x) = \sqrt{a^2 + x^2}$

$Ans.\quad df = \dfrac{x\,dx}{\sqrt{a^2 + x^2}}$

(e) $f(x) = x^n \cos mx$

$Ans.\quad df = (nx^{n-1} \cos mx - mx^n \sin mx)\,dx$

(f) $y = x \log x$

$Ans.\quad dy = (1 + \log x)\,dx$

(g) $y = \mathrm{Tan}^{-1} x$

$Ans.\quad dy = \dfrac{dx}{1 + x^2}$

(h) $y = e^{-x} \sec 2x$

$Ans.\quad dy = e^{-x} \sec 2x\,(2 \tan 2x - 1)\,dx$

(i) $y = \dfrac{x - 1}{x + 2}$

$Ans.\quad dy = \dfrac{3\,dx}{(x + 2)^2}$

(j) $y = \mathrm{Cos}^{-1} (1 - x^2)$

$Ans.\quad dy = \dfrac{2x\,dx}{|x|\,\sqrt{2 - x^2}}$

7. Find, by using differentials, an approximate value for $\sqrt[3]{509}$. (Hint: Let $y = \sqrt[3]{x}$ and find dy.)

$Ans.\quad$ 7.984 correct to 3 decimals

8. What is the allowable approximate percentage error in the measurement of the diameter of a sphere, if the volume is not to be in error by more than 3%?

$Ans.\quad$ 1%

9. A painter in making up his bid for painting an observatory's hemispherical dome estimates the radius of the dome to be 28 ft. when it is actually 29 ft. If the cost of painting is 2¢ per sq. ft., approximately how much extra will it cost to give the dome 2 coats?

Ans. $14.08

10. The period t (time in seconds for one complete oscillation) of a pendulum is $t = 2\pi \sqrt{\dfrac{l}{g}}$, where $g = 32.2$ ft./sec². and l = length of the pendulum in feet. (a) If a pendulum takes 4 sec. for a complete oscillation (2 sec. for "tick" 2 sec. for "tock"), find its length. (b) If, because of extremely cold weather, the pendulum of such a great-grandfather clock should shrink to 13 ft., how much time would it gain or lose in a day?

Ans. (a) 13.05 ft.

(b) Gain about 2.7 min./day

11. If $f = uv$, then $df = u\,dv + v\,du$. Show that

$$\frac{df}{f} = \frac{du}{u} + \frac{dv}{v}$$

(that is, the relative error of a product is the sum of the relative errors of the factors).

12. If $f = u/v$, then $df = \dfrac{u\,dv - v\,du}{v^2}$. Show that

$$\frac{df}{f} = \frac{du}{u} - \frac{dv}{v}$$

(that is, the relative error of the quotient is the relative error of the numerator minus the relative error of the denominator).

13. How exactly must the diameter of a circle be measured so that (a) the area will be correct to within 1%, and (b) the perimeter will be correct to within 1%?

Ans. (a) error $\leqslant \frac{1}{2}$%

(b) error $\leqslant 1$ %

14. How exactly must the diameter of a sphere be measured so that (a) the volume will be correct to within 1%, and (b) the surface area will be correct to within 1%?

Ans. (a) error $\leqslant \frac{1}{3}$%

(b) error $\leqslant \frac{1}{2}$%

15. How exactly must the edge of a cube be measured so that (a) the volume will be correct to within 1%, and (b) the surface area will be correct to within 1%?

$$\textit{Ans.} \quad \text{(a) error} \leqslant \tfrac{1}{3}\%$$
$$\text{(b) error} \leqslant \tfrac{1}{2}\%$$

8.12 Differentiation of Arc Length

We seek the instantaneous rate of change of arc length per unit change in x. This will evidently be given by

$$\frac{ds}{dx} = \lim_{\Delta x \to 0} \frac{\Delta s}{\Delta x}$$

where s represents the length of arc measured from some fixed point P

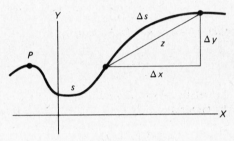

FIG. 8.21

on the curve (Fig. 8.21). In order to determine $\dfrac{ds}{dx}$, we note that, $z^2 = (\Delta x)^2 + (\Delta y)^2$; thus,

$$\frac{\Delta s}{\Delta x} = \frac{\Delta s}{z} \cdot \frac{z}{\Delta x}$$

$$= \frac{\Delta s}{z} \frac{\sqrt{(\Delta x)^2 + (\Delta y)^2}}{\Delta x}$$

$$= \frac{\Delta s}{z} \sqrt{1 + \left(\frac{\Delta y}{\Delta x}\right)^2}$$

Although we shall not prove it here, it is true that $\lim\limits_{\Delta x \to 0} \dfrac{\Delta s}{z} =$

$\lim\limits_{\text{arc} \to 0} \dfrac{\text{arc}}{\text{chord}} = 1$; hence,

$$\lim_{\Delta x \to 0} \frac{\Delta s}{\Delta x} = \lim_{\Delta x \to 0} \sqrt{1 + \left(\frac{\Delta y}{\Delta x}\right)^2}$$

$$\frac{ds}{dx} = \sqrt{1 + \left(\frac{dy}{dx}\right)^2} \tag{1}$$

This could be written, in terms of differentials:

$$ds = \sqrt{1 + y'^2} \, dx \tag{2}$$

or again

$$(ds)^2 = (dx)^2 + (dy)^2 \tag{3}$$

These are important forms, and the student should realize that essentially they say that ds, dx, and dy play the role of the sides of a right triangle. Intuitively this seems very reasonable, since the arc approaches the chord as the arc tends to zero. A little reflection will make it clear why the situation is quite different in polar coordinates where the expression corresponding to (3) is

$$(ds)^2 = (d\rho)^2 + \rho^2 (d\theta)^2 \tag{4}$$

This follows from (3) immediately upon transforming from rectangular to polar coordinates by means of the relations $x = \rho \cos \theta$ and $y = \rho \sin \theta$. The student should become thoroughly familiar with these formulas since considerable use will be made of them later.

8.13   Curvature

In the preceding article, we developed the formula for $\dfrac{ds}{dx}$, the rate of change of arc length with respect to unit change in x. A more useful notion, however, is that of the rate of change of arc length per unit change in inclination, where the inclination $= \theta$ and the slope $=$ $\tan \theta$ (Fig. 8.22). The reciprocal of this, namely, the rate of change of inclination per unit change in arc length, is

$$K = \frac{d\theta}{ds} = \lim_{\Delta s \to 0} \frac{\Delta \theta}{\Delta s}$$

where K is the **curvature** (at a point P).

We shall now derive a formula for K in terms of the more familiar quantities y' and y''. To do this, we write

$$\theta = \tan^{-1} y'$$

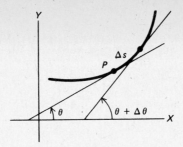

FIG. 8.22

and differentiate this with respect to x getting

$$\frac{d\theta}{dx} = \frac{1}{1 + y'^2} y''$$

Then $\dfrac{d\theta}{ds} = \dfrac{d\theta}{dx} \cdot \dfrac{dx}{ds},$

$$= \frac{y''}{1 + y'^2} \left(\frac{1}{1 + y'^2}\right)^{1/2}$$

by (1) in Section 8.12. Finally the curvature K is given by

$$K = \frac{y''}{(1 + y'^2)^{3/2}} \tag{1}$$

In general, the sign of K is chosen as that of y'', although many authors treat K as always being positive. The curvature measures the rate at which the tangent line turns per unit distance moved along the curve, or, simply, it measures the rate of change of direction of the curve.

Example 1. Find the curvature of a circle of radius r (Fig. 8.23).
Solution 1. Let the equation of the circle be

$$x^2 + y^2 = r^2$$

Then, $\qquad 2x + 2yy' = 0$

$$y' = -\frac{x}{y}$$

$$y'' = -\frac{y - xy'}{y^2} = -\frac{y - x\left(-\dfrac{x}{y}\right)}{y^2}$$

$$= -\frac{r^2}{y^2}$$

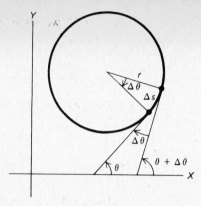

FIG. 8.23

$$K = \dfrac{-\dfrac{r^2}{y^3}}{\left(1 + \dfrac{x^2}{y^2}\right)^{3/2}}$$

$$= -\frac{1}{r}, \text{ where } y'' < 0, \text{ i.e., for } y > 0.$$

$$= \frac{1}{r}, \text{ where } y'' > 0, \text{ i.e., for } y < 0.$$

Normally, however, the curvature of a circle is considered positive everywhere since this is the result that would be obtained, if we were to compute curvature directly from the fundamental definition. This is done in Solution 2.

Solution 2.

$$K = \lim_{\Delta s \to 0} \frac{\Delta \theta}{\Delta s}$$

Recalling that the arc length $s = r\theta$ for a circle, we have

$$K = \lim_{\Delta s \to 0} \frac{\Delta s/r}{\Delta s} = \frac{1}{r}$$

Thus, the curvature is a *constant* for a circle and is equal to the reciprocal of the radius.

By making use of (4) in Section 8.12 we arrive at the following formula for curvature in polar coordinates

$$K = \frac{\rho^2 + 2\rho'^2 - \rho\rho''}{(\rho^2 + \rho'^2)^{3/2}} \tag{2}$$

Example 2. Find the curvature of the spiral $\rho = a\theta$.
Solution.

$$\rho = a\theta$$
$$\rho' = a$$
$$\rho'' = 0$$
$$K = \frac{a^2\theta^2 + 2a^2}{(a^2\theta^2 + a^2)^{3/2}}$$
$$= \frac{1}{a}\frac{\theta^2 + 2}{(\theta^2 + 1)^{3/2}}$$

Example 3. Find the curvature of the cardioid $\rho = a(1 - \cos\theta)$ at the point $\left(a, \frac{\pi}{2}\right)$.

Solution. $\quad \rho = a(1 - \cos\theta)$

$$\rho' = a\sin\theta$$
$$\rho'' = a\cos\theta$$
$$K = \frac{a^2(1 - \cos\theta)^2 + 2a^2\sin^2\theta - a(1 - \cos\theta)(a\cos\theta)}{[a^2(1 - \cos\theta)^2 + a^2\sin^2\theta]^{3/2}}$$
$$= \frac{3 - 3\cos\theta}{2\sqrt{2a}(1 - \cos\theta)^{3/2}} = \frac{3}{2\sqrt{2a\rho}}$$
$$K = \frac{3\sqrt{2}}{4a} \quad \text{at the point } \left(a, \frac{\pi}{2}\right)$$

8.14 Circle of Curvature

In Example 1, we found that the curvature of a circle is the reciprocal of the radius of the circle. Similarly, for any curve, we define *the radius of curvature as the absolute value of the reciprocal of curvature*; that is, for any point on the curve

$$R = \frac{1}{|K|} \tag{1}$$

$$R = \frac{(1 + y'^2)^{3/2}}{y''} \quad \text{(rectangular coordinates)} \tag{2}$$

$$R = \frac{(\rho^2 + \rho'^2)^{3/2}}{\rho^2 + 2\rho'^2 - \rho\rho''} \quad \text{(polar coordinates)} \tag{3}$$

The student should memorize definition (1), but not necessarily (2) and (3); if the formulas for curvature are known, it is a simple matter to compute the radius of curvature from $R = \frac{1}{|K|}$.

A circle drawn with this radius R and with center on the normal to the curve on the concave side will have the same curvature as the curve itself. This circle is called the **circle of curvature**; it is an aid to the understanding of the geometric significance of curvature.

8.15 Center of Curvature

In order to find the coordinates of the center of the circle of curvature, referred to briefly as **center of curvature**, proceed as follows: Let the equation of the circle be

$$(x - \alpha)^2 + (y - \beta)^2 = R^2 \tag{1}$$

where $R^2 = \frac{1}{K^2}$. Upon differentiating (1) twice and simplifying, we get

$$y' = -\frac{x - \alpha}{y - \beta} \tag{2}$$

$$y'' = -\frac{R^2}{(y - \beta)^3} \tag{3}$$

Solving (2) and (3) simultaneously for α and β, making use of the value of R, we get

$$\alpha = x - \frac{y'(1 + y'^2)}{y''} \tag{4}$$

$$\beta = y + \frac{1 + y'^2}{y''} \tag{5}$$

Thus, the coordinates of the circle of curvature are given in terms of x, y, y', and y''.

Example. Find the coordinates of the center of curvature of the cubical parabola $y = x^3$ at the point $(1, 1)$ (Fig. 8.24).

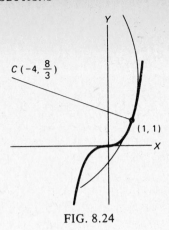

FIG. 8.24

Solution. $y = x^3$

$$y' = 3x^2$$

$$y'' = 6x$$

$$\alpha = x - \frac{3x^2(1 + 9x^4)}{6x}$$

$$\beta = y + \frac{1 + 9x^4}{6x}$$

$$\alpha = 1 - \frac{3(1 + 9)}{6} = -4$$

$$\beta = 1 + \frac{10}{6} = \frac{8}{3}$$

PROBLEMS WITH SOLUTIONS

1. Find the curvature K for $y = \log \sin x$.
   Solution. Now $y' = \cot x$, $y'' = -\csc^2 x$ and

   $$K = \frac{y''}{(1 + y'^2)^{3/2}} = \frac{-\csc^2 x}{(1 + \cot^2 x)^{3/2}}$$

   $$= \frac{\csc^2 x}{(\csc^2 x)^{3/2}} = -\sin x$$

2. Find the curvature K for $y = e^x \sin x$ at $x = \frac{\pi}{4}$.

   Solution. $K(x) = \dfrac{2e^x \cos x}{[1 + e^{2x}(1 + 2 \sin x \cos x)]^{3/2}}$

$$K\left(\frac{\pi}{4}\right) = \frac{\sqrt{2}\, e^{\pi/4}}{(1 + 2e^{\pi/2})^{3/2}}$$

3. Find the curvature K for $\rho = \dfrac{a}{\theta}$.

   Solution. $K = \dfrac{\rho^2 + 2\rho'^2 - \rho\rho''}{(\rho^2 + \rho'^2)^{3/2}}$

$$= \frac{\dfrac{a^2}{\theta^2} + \dfrac{2a^2}{\theta^4} - \dfrac{2a^2}{\theta^4}}{\left(\dfrac{a^2}{\theta^2} + \dfrac{a^2}{\theta^4}\right)^{3/2}}$$

$$= \frac{a^2/\theta^2}{\left(\dfrac{a^2}{\theta^2}\right)^{3/2} \left(1 + \dfrac{1}{\theta^2}\right)^{3/2}}$$

$$= \frac{\theta}{a(1 + \theta^2)^{3/2}}$$

4. Find the curvature K for

$$x = a(\cos\theta + \theta\sin\theta)$$
$$y = a(\sin\theta - \theta\cos\theta)$$

   Solution. $dx = a\theta\cos\theta\, d\theta$, $dy = a\theta\sin\theta\, d\theta$

$$y' = \tan\theta, \quad \text{and} \quad y'' = \frac{1}{a\theta}\sec^3\theta\,; \text{ thus,}$$

$$K = \frac{\dfrac{1}{a\theta}\sec^3\theta}{(1 + \tan^2\theta)^{3/2}} = \frac{1}{a\theta}$$

5. Find the radius of curvature R for $y = x^3$ at $(2, 8)$.

   Solution. $K(x) = \dfrac{6x}{(1 + 9x^4)^{3/2}}$

$$K(2) = \frac{12}{145^{3/2}}$$

$$R(2) = \frac{1}{|K(2)|} = \frac{145^{3/2}}{12}$$

6. Find the point on $y = e^{-x}$ where the curvature is a maximum.

Solution. $K = \dfrac{e^{-x}}{(1 + e^{-2x})^{3/2}}$

$\dfrac{dK}{dx} = \dfrac{-e^{-x}(1 + e^{-2x})^{1/2}(1 - 2e^{-2x})}{(1 + e^{-2x})^3}$

$1 - 2e^{-2x} = 0$

$x = 0.3466$

7. Find the coordinates of the center of curvature of $y = \sin x$.
Solution. Here $y' = \cos x$ and $y'' = -\sin x$; therefore,

$$\alpha = x + \frac{\cos x (1 + \cos^2 x)}{\sin x}$$

$$\beta = \sin x - \frac{1 + \cos^2 x}{\sin x}$$

$$= \frac{2 \cos^2 x}{\sin x}$$

PROBLEMS WITH ANSWERS

We list in the following exercises some of the standard curves along with the radius of curvature R and the coordinates α and β of the center of curvature. Note that in some cases α and β involve only the parameter x, but that in other cases y is also present. Allowing y to be present rather than eliminating it does no harm, since y is a function of x and hence is determined for each value of x; it merely simplifies matters to retain y in some instances.

In Problems 8 through 17, find R, α, and β.

8. Parabola
$y^2 = 4px$

Ans. $R = \dfrac{2(p + x)^{3/2}}{\sqrt{p}}$

$\alpha = 3x + 2p, \beta = -\dfrac{y^3}{4p^2}$

9. Semicubical parabola
$y^2 = x^3$

Ans. $R = \dfrac{\sqrt{x}}{6}(4 + 9x)^{3/2}$

$\alpha = -\dfrac{x}{2}(9x + 2), \beta = \dfrac{4}{3}\sqrt{x}(3x + 1)$

10. Catenary

$$y = \frac{a}{2} \left(e^{x/a} + e^{-x/a} \right) \qquad\qquad Ans. \quad R = \frac{y^2}{a}$$

$$\alpha = x - \frac{y}{a} \sqrt{y^2 - a^2}, \beta = 2$$

11. Exponential
$y = e^x$

$$Ans. \quad R = \frac{(1 + e^{2x})^{3/2}}{e^x}$$

$$\alpha = x - 1 - e^{2x}, \beta = 2e^x + e^{-x}$$

12. Ellipse
$b^2 x^2 + a^2 y^2 = a^2 b^2$

$$Ans. \quad R = \frac{(a^2 - e^2 x^2)^{3/2}}{ab} \quad (e = \text{eccentricity})$$

$$\alpha = \frac{a^2 - b^2}{a^4} x^3, \beta = -\frac{a^2 - b^2}{b^4} y^3$$

13. Hyperbola
$b^2 x^2 - a^2 y^2 = a^2 b^2$

$$Ans. \quad R = \frac{(e^2 x^2 - a^2)^{3/2}}{ab} \quad (e = \text{eccentricity})$$

$$\alpha = \frac{a^2 + b^2}{a^4} x^3, \beta = -\frac{a^2 + b^2}{b^4} y^3$$

14. Four cusped hypocycloid
$x^{2/3} + y^{2/3} = a^{2/3}$

$$Ans. \quad R = 3(axy)^{1/3}$$
$$\alpha = x + 3x^{1/3} y^{2/3}, \beta = y + 3x^{2/3} y^{1/3}$$

15. Lemniscate
$x = a \cos \theta \sqrt{\cos 2\theta}$
$y = a \sin \theta \sqrt{\cos 2\theta}$

$$Ans. \quad R = \frac{a}{3 \sqrt{\cos 2\theta}}$$

$$x = \frac{2a \cos^3 \theta}{3 \sqrt{\cos 2\theta}}, \beta = -\frac{2a \sin^3 \theta}{3 \sqrt{\cos 2\theta}}$$

16. Cycloid
$x = a(\theta - \sin \theta)$
$y = a(1 - \cos \theta)$

$$Ans. \quad R = 2\sqrt{2ay} = 4a \sin \frac{\theta}{2}$$

$$\alpha = a(\theta + \sin \theta), \beta = -a(1 - \cos \theta)$$

17. Logarithmic curve

$$y = \log x \qquad\qquad Ans. \quad R = \frac{(1 + x^2)^{3/2}}{x}$$

$$\alpha = 2x + \frac{1}{x}, \beta = y - 1 - x^2$$

8.16 Extended Law of the Mean

We have already met the "mean value theorem for derivatives" (also called "the law of the mean") as Theorem IV, Chapter 6. For the next section, we need an extension of this theorem so as to include two functions. Its statement is as follows:

THEOREM I. (Generalized mean value theorem for derivatives, or "extended law of the mean"). *Let f and g be continuous on the closed interval* [a,b] *and differentiable in the open interval* (a,b). *If* $g'(x) = 0$ *for no point x in* (a,b), *then there exists some point* x_1 *in* (a,b) *such that*

$$\frac{f(b) - f(a)}{g(b) - g(a)} = \frac{f'(x_1)}{g'(x_1)} , a < x_1 < b \tag{1}$$

This theorem is cf value in the evaluation of indeterminate forms.

8.17 Indeterminate Forms

In Chapter 3, it was pointed out that $\lim\limits_{x \to 0} \dfrac{\sin x}{x} = 1$ and $\lim\limits_{x \to 0} \dfrac{1 - \cos x}{x} = 0$. We shall now develop a general method whereby indeterminate forms of the type $\dfrac{0}{0}$ can be evaluated. To do this, we make use of the **extended law of the mean.** Suppose that $f(a) = 0 = g(a)$ and write x for b. Then (1) of Section 8.16 becomes

$$\frac{f(x)}{g(x)} = \frac{f'(x_1)}{g'(x_1)} , a < x_1 < x \tag{1}$$

Now $\lim\limits_{x \to a} \dfrac{f(x)}{g(x)} = \dfrac{0}{0}$, since $f(a) = 0 = g(a)$. But, clearly $x_1 \longrightarrow a$ as $x \longrightarrow a$. Therefore, taking limits of both sides of (1) we get

$$\lim_{x \to a} \frac{f(x)}{g(x)} = \lim_{x \to a} \frac{f'(x)}{g'(x)} \tag{2}$$

If this last limit exists, relation (2) is known as **L'Hospital's rule,** and it holds whether a is finite or infinite. It should be emphasized that in order to determine $\lim\limits_{x \to a} \dfrac{f(x)}{g(x)}$, we actually take the derivative of the numerator and the derivative of the denominator separately, and then evaluate the quotient of the derivatives; i.e., compute $\lim\limits_{x \to a} \dfrac{f'(x)}{g'(x)}$, provided this is not again indeterminate. The student should be warned not to confuse this operation with that of differentiating a quotient.

If $\lim\limits_{x \to a} \dfrac{f'(x)}{g'(x)}$ is itself an indeterminate form, then we may start over again and apply L'Hospital's rule to this getting $\lim\limits_{x \to a} \dfrac{f'(x)}{g'(x)} = \lim\limits_{x \to a} \dfrac{f''(x)}{g''(x)}$, etc. Most of the problems that the student will ordinarily meet will yield to a finite number of repeated applications of this rule.

L'Hospital's rule also applies to other indeterminate forms besides $0/0$; namely, ∞/∞, 1^{∞}, 0^{0}, ∞^{0}, and $\infty - \infty$. Thus, if $\lim\limits_{x \to a} f(x)/g(x)$ yields one of these indeterminate forms, a finite limit may exist after repeated applications of L'Hospital's rule.

Example 1. $\operatorname*{Lim}\limits_{x \to a} \dfrac{f(x)}{g(x)} = \dfrac{\infty}{\infty}$, where $\lim\limits_{x \to a} f(x) = \lim\limits_{x \to a} g(x) = \infty$.

Solution. $\operatorname*{Lim}\limits_{x \to a} \dfrac{f(x)}{g(x)} = \lim\limits_{x \to a} \dfrac{1/g(x)}{1/f(x)} = \dfrac{0}{0}$; and so L'Hospital's rule may now be applied.

Example 2. $\operatorname*{Lim}\limits_{x \to a} f(x)\, g(x) = \infty \cdot 0$, where $\lim\limits_{x \to a} f(x) = \infty$ and $\lim\limits_{x \to a} g(x) = 0$.

Solution. $\operatorname*{Lim}\limits_{x \to a} f(x)g(x) = \lim\limits_{x \to a} \dfrac{f(x)}{1/g(x)} = \dfrac{\infty}{\infty}$, which is suitable for the application of L'Hospital's rule.

Example 3. $\operatorname*{Lim}\limits_{x \to a} f(x)^{g(x)} = 1^{\infty}$, where $\lim\limits_{x \to a} f(x) = 1$ and $\lim\limits_{x \to a} g(x) = \infty$.

Solution. Taking the logarithm of both sides of this equation, we have:

$$\log \lim_{x \to a} f(x)^{g(x)} = \lim_{x \to a} \log f(x)^{g(x)} = \lim_{x \to a} g(x) \log f(x) = \infty \cdot 0$$

which, by Example 2, yields ∞/∞; i.e., it can be reduced to $\lim\limits_{x \to a} \dfrac{f(x)}{1/g(x)} = \dfrac{\infty}{\infty}$. If $\lim\limits_{x \to a} \dfrac{\log f(x)}{1/g(x)} = b$, then $\lim\limits_{x \to a} f(x)^{g(x)} = e^{b}$.

The method used in Example 3 also supplies to the other exponential indeterminate forms 0^{0} and ∞^{0}.

Example 4. $\text{Lim}_{x \to a} [f(x) - g(x)] = \infty - \infty$, where $\lim_{x \to a} f(x) = \lim_{x \to a} g(x) = \infty$.

Solution.

$$\text{Lim}_{x \to a} [f(x) - g(x)] = \lim_{x \to a} \frac{1}{1/f(x)} - \frac{1}{1/g(x)} = \lim_{x \to a} \frac{1/g(x) - 1/f(x)}{1/f(x) \cdot 1/g(x)} = \frac{0}{0}.$$

SUMMARY

Type of indeterminate form   *Apply L'Hospital's rule to:*

1. $\lim \dfrac{f}{g} = \dfrac{0}{0}$ $\lim \dfrac{f}{g} = \lim \dfrac{f'}{g'} = \lim \dfrac{f''}{g''} = \lim \dfrac{f'''}{g'''} = \cdots$

2. $\lim \dfrac{f}{g} = \dfrac{\infty}{\infty}$ The first one of these which is not indeterminate will give the answer.

3. $\lim fg = \infty \cdot 0$ $\left\{ \lim \dfrac{f}{1/g} \text{ or } \lim \dfrac{g}{1/f} \right.$

4. $\lim f^g = 1^\infty$ $\left\{ \lim \dfrac{\log f}{1/g} \text{ or } \lim \dfrac{g}{1/\log f} \right.$

5. $\lim f^g = 0^0$

6. $\lim f^g = \infty^0$ If L'Hospital's rule applied to one of these yields b for the limit, then the answer to the original problem is e^b.

7. $\lim (f - g) = \infty - \infty$ $\lim \dfrac{1/g - 1/f}{(1/g)(1/f)}$

The answer will be given directly by L'Hospital's rule.

Example 5. Find $\lim\limits_{x \to 0} \dfrac{\sin 2x + \tan x}{3x}$.

Solution. This limit is of type $\dfrac{0}{0}$. Applying L'Hospital's rule, we obtain

$$\lim_{x \to 0} \frac{2 \cos 2x + \sec^2 x}{3} = 1$$

Example 6. Find $\lim\limits_{x \to 0} \dfrac{\sin 2x \tan x}{3x}$.

Solution. This limit by L'Hospital's rule equals

$$\lim_{x \to 0} \frac{\sin 2x \sec^2 x + 2 \cos 2x \tan x}{3} = 0$$

Example 7. Find $\lim\limits_{x\to 0} \dfrac{\sin 2x + \tan x}{3x^2}$.

Solution. This limit equals $\lim\limits_{x\to 0} \dfrac{2\cos 2x + \sec^2 x}{6x} = \dfrac{3}{0} = \infty$

Example 8. Find $\lim\limits_{x\to\infty} \dfrac{x^n}{e^x}$.

Solution. This is of type $\dfrac{\infty}{\infty}$. We get

$$\lim_{x\to\infty} \frac{x^n}{e^x} = \lim_{x\to\infty} \frac{nx^{n-1}}{e^\alpha} = \lim_{x\to\infty} \frac{n(n-1)x^{n-2}}{e^x} = \cdots$$

$$= \lim_{x\to\infty} \frac{n!}{e^x} = 0$$

Example 9. Find $\lim\limits_{x\to\infty} (1+x)^{1/x}$.

Solution. This is of type ∞^0 . We write

$$\lim_{x\to\infty} \frac{\log (1+x)}{x} = \lim_{x\to\infty} \frac{\dfrac{1}{1+x}}{1} = 0$$

Therefore,

$$\lim_{x\to\infty} (1+x)^{1/x} = e^0 = 1$$

Example 10. Show that $\lim f^g = 0^\infty$ is not indeterminate, but equals 0.

Solution. Write $A = \lim f^g$

Then $\log A = \lim g \log f$

$$= \lim (\infty)(-\infty)$$

$$= -\infty$$

Therefore, $A = e^{-\infty} = 0$

Example 11. Evaluate $\lim\limits_{x\to 0} \dfrac{e^x - e^{-x} - 2x}{x - \sin x}$.

Solution. This equals $\dfrac{0}{0}$ type.

$$\lim_{x\to 0} \frac{e^x + e^{-x} - 2}{1 - \cos x} = \frac{0}{0}$$

$$= \lim_{x\to 0} \frac{e^x - e^{-x}}{\sin x} = \frac{0}{0} = \lim_{x\to 0} \frac{e^x + e^{-x}}{\cos x} = 2$$

Example 12. Evaluate $\lim\limits_{x \to \frac{\pi}{2}} (\tan x - \sec x)$.

Solution. This is of type $\infty - \infty$. We write

$$\lim_{x \to \frac{\pi}{2}} (\tan x - \sec x)$$

$$= \lim_{x \to \frac{\pi}{2}} \left(\frac{\sin x}{\cos x} - \frac{1}{\cos x} \right)$$

$$= \lim_{x \to \frac{\pi}{2}} \left(\frac{\sin x - 1}{\cos x} \right) = \frac{0}{0}$$

$$= \lim_{x \to \frac{\pi}{2}} \frac{\cos x}{- \sin x} = 0$$

PROBLEMS WITH SOLUTIONS

1. Find $\lim\limits_{x \to 0} \dfrac{\sin x}{x}$.

   Solution. $\lim\limits_{x \to 0} \dfrac{\sin x}{x} = \lim\limits_{x \to 0} \dfrac{\cos x}{1} = 1$

2. Find $\lim\limits_{x \to 1} \dfrac{x^2 - 3x + 2}{x^2 - 4x + 3}$.

   Solution. $\lim\limits_{x \to 1} \dfrac{x^2 - 3x + 2}{x^2 - 4x + 3} = \lim\limits_{x \to 1} \dfrac{2x - 3}{2x - 4} = \dfrac{1}{2}$

3. Find $\lim\limits_{x \to 0} \dfrac{\sin x + \tan x}{2x}$.

   Solution. $\lim\limits_{x \to 0} \dfrac{\sin x + \tan x}{2x} = \lim\limits_{x \to 0} \dfrac{\cos x + \sec^2 x}{2} = 1$

4. Find $\lim\limits_{x \to 0} \dfrac{xe^x - x}{1 - \cos x}$.

   Solution. This equals $\lim\limits_{x \to 0} \dfrac{xe^x + e^x - 1}{\sin x} = \dfrac{0}{0}$. By applying L'Hospital's rule again, we get

   $$\lim_{x \to 0} \frac{xe^x + 2e^x}{\cos x} = 2$$

5. Find $\lim\limits_{x \to 7} \dfrac{2 - \sqrt{x-3}}{x^2 - 49}$.

   Solution. This equals $\lim\limits_{x \to 7} \dfrac{-1/2(x-3)^{-1/2}}{2x} = -\dfrac{1}{56}$.

6. Find $\lim\limits_{x \to 1} \dfrac{\log x - \sin(x-1)}{(x-1)^2}$.

   Solution. This equals $\lim\limits_{x \to 1} \dfrac{1/x - \cos(x-1)}{2(x-1)} = \dfrac{0}{0}$.

   By applying L'Hospital's rule again we get

   $$\lim_{x \to 1} \frac{-1/x^2 + \sin(x-1)}{2} = -\frac{1}{2}$$

7. Find $\lim\limits_{x \to \infty} \dfrac{x^2}{e^x}$.

   Solution. This equals $\lim\limits_{x \to \infty} \dfrac{2x}{e^x} = \lim\limits_{x \to \infty} \dfrac{2}{e^x} = 0$.

8. Find $\lim\limits_{x \to 0} \dfrac{\log \sin 2x}{\log \sin x}$.

   Solution. This equals

   $$\lim_{x \to 0} \frac{2 \cos 2x \sin x}{\sin 2x \cos x} = \lim_{x \to 0} \frac{-2 \sin x \sin 2x + 2 \cos x \cos 2x}{-\sin x \sin 2x + 2 \cos x \cos 2x} = 1$$

9. Find $\lim\limits_{x \to \infty} \dfrac{\log x}{e^x}$.

   Solution. This equals $\lim\limits_{x \to \infty} \dfrac{1/x}{e^x} = 0$.

10. Find $\lim\limits_{x \to 0} \sin x \csc x$.

    Solution. This equals $\lim\limits_{x \to 0} \dfrac{\sin x}{\sin x} = \lim\limits_{x \to 0} \dfrac{\cos x}{\cos x} = 1$.

11. Find $\lim\limits_{x \to 0} \sin x \cot x$.

    Solution. This equals $\lim\limits_{x \to 0} \dfrac{\sin x}{\tan x} = \lim\limits_{x \to 0} \dfrac{\cos x}{\sec^2 x} = 1$.

12. Find $\lim\limits_{x \to 0} x \log x$.

Solution. This equals $\lim\limits_{x\to 0} \dfrac{\log x}{1/x} = \lim\limits_{x\to 0} \dfrac{1/x}{-1/x^2} = 0.$

PROBLEMS WITH ANSWERS

13. $\lim\limits_{x\to 1} \dfrac{x^3 - x}{x^4 - 3x^2 + 1}.$ *Ans.* 0

14. $\lim\limits_{x\to 2} \dfrac{x^3 - 6x + 4}{x^2 + x - 6}.$ *Ans.* $\dfrac{6}{5}$

15. $\lim\limits_{\theta\to 0} \dfrac{1 - \cos \theta}{\theta^2}.$ *Ans.* $\dfrac{1}{2}$

16. $\lim\limits_{y\to 0} \dfrac{\log \sin y}{\log y}.$ *Ans.* 1

17. $\lim\limits_{x\to \infty} \dfrac{x^n}{n^x}.$ *Ans.* 0

18. $\lim\limits_{z\to \infty} z^{1/z}.$ *Ans.* 1

19. $\lim\limits_{\theta\to 0} \dfrac{\theta}{\sin^{-1} \theta}.$ *Ans.* 1

20. $\lim\limits_{x\to \infty} \dfrac{a^x}{x^b}.$ *Ans.* ∞

21. $\lim\limits_{x\to \infty} \dfrac{\log x}{x}.$ *Ans.* 0

22. $\lim\limits_{x\to \frac{1}{2}} (1 - 2x) \tan \pi x.$ *Ans.* $\dfrac{2}{\pi}$

CHAPTER 9
APPLICATIONS OF INTEGRATION

9.1 Definite Integral

Many of the applications of integration involve definite integrals which are of the form $\int_a^b f(x)\,dx$ and which are evaluated by means of Theorems VI and VIII, Chapter 6. First, a function G is determined such that $G'(x) = f(x)$. Then

$$\int_a^b f(x)\,dx = G(b) - G(a)$$

This is usually called *the definite integral of f(x) from a to b,* where a and b are the limits of integration; a is the **lower limit** and b is the **upper limit**.

Example 1. Find $\int_0^1 (x^2 + e^x)\,dx$.

Solution. $\int_0^1 (x^2 + e^x)\,dx = \dfrac{x^3}{3} + e^x \Big|_0^1$

$$= (\tfrac{1}{3} + e) - (1)$$
$$= e - \tfrac{2}{3}$$

Example 2. Find $\int_0^\pi \sin x\,dx$.

Solution. $\int_0^\pi \sin x\,dx = -\cos x \Big|_0^\pi = 1 + 1 = 2$

PROBLEMS WITH SOLUTIONS

1. $\displaystyle\int_{-1}^1 (x^2 - 1)\,dx = \frac{1}{3}x^3 - x \Big|_{-1}^1 = \left(-\frac{2}{3}\right) - \left(\frac{2}{3}\right) = -\frac{4}{3}$

2. $\int_0^{2\pi} \sin x \, dx = -\cos x \Big|_0^{2\pi} = (-1) - (-1) = 0$

3. $\int_3^{10} (x^2 + x + 1) \, dx = \dfrac{x^3}{3} + \dfrac{x^2}{2} + x \Big|_3^{10}$

$$= \dfrac{1000}{3} + \dfrac{100}{2} + 10 - \left(9 + \dfrac{9}{2} + 3\right)$$

$$= 376\dfrac{5}{6}$$

4. $\int_0^4 x^{1/2} \, dx = \dfrac{2}{3} x^{3/2} \Big|_0^4 = \dfrac{16}{3}$

5. $\int_1^e \dfrac{dx}{x} = \log x \Big|_1^e = \log e - \log 1 = 1$

6. $\int_0^1 \sqrt{1 - x^2} \, dx = \dfrac{x}{2} \sqrt{1 - x^2} + \dfrac{1}{2} \operatorname{Sin}^{-1} x \Big|_0^1 = \dfrac{\pi}{4}$

7. $\int_{-\pi/2}^{\pi/2} \cos x \, dx = \sin x \Big|_{-\pi/2}^{\pi/2} = 2$

8. $\int_0^{\pi/4} \sec x \tan x \, dx = \sec x \Big|_0^{\pi/4} = \sqrt{2} - 1$

9. $\int_1^e x \log x \, dx = \dfrac{x^2}{2} \left(\log x - \dfrac{1}{2}\right) \Big|_1^e = \dfrac{1}{4} (e^2 + 1)$

10. $\int_0^1 x e^x \, dx = e^x (x - 1) \Big|_0^1 = 1$

11. $\int_0^{1/2} \dfrac{dx}{\sqrt{1 - x^2}} = \operatorname{Sin}^{-1} x \Big|_0^{1/2} = \dfrac{\pi}{6}$

12. $\displaystyle\int_0^1 \frac{dx}{1 + x^2} = \left.\text{Tan}^{-1} \, x \, \right|_0^1 = \frac{\pi}{4}$

13. $\displaystyle\int_{\sqrt{2}}^2 \frac{dx}{x\sqrt{x^2 - 1}} = \left.\text{Sec}^{-1} \, x \, \right|_{\sqrt{2}}^2 = \frac{\pi}{3} - \frac{\pi}{4} = \frac{\pi}{12}$

14. $\displaystyle\int_0^1 \frac{dx}{\sqrt{x^2 + 1}} = \left.\log\left(x + \sqrt{x^2 + 1}\right)\right|_0^1 = \log\left(1 + \sqrt{2}\right)$

PROBLEMS WITH ANSWERS

15. $\displaystyle\int_1^2 (1 - x + x^2)\, dx.$ *Ans.* 11/6

16. $\displaystyle\int_0^1 \cos 2\pi x \, dx.$ *Ans.* 0

17. $\displaystyle\int_0^3 \sqrt{9 - x^2} \, dx.$ *Ans.* $9\pi/4$

18. $\displaystyle\int_a^b \frac{dx}{x^2 - 9}.$ *Ans.* $\dfrac{1}{6} \log \dfrac{(b - 3)\,(a + 3)}{(b + 3)\,(a - 3)}$

19. $\displaystyle\int_0^{\pi/3} \tan x \, dx.$ *Ans.* $\log 2$

20. $\displaystyle\int_0^{\pi/4} \sec x \tan x \, dx.$ *Ans.* $\sqrt{2} - 1$

21. $\displaystyle\int_{\sqrt{\pi/2}}^{\sqrt{2\pi/2}} 2x \csc x^2 \, dx.$ *Ans.* $-\log\left(\sqrt{2} - 1\right)$

22. $\int_{1}^{2} \frac{4}{4x^2 - 1} dx.$ *Ans.* log 9/5

23. $\int_{0}^{1} x\sqrt{1 - x^2} \, dx.$ *Ans.* 1/3

24. $\int_{1}^{2} 2xe^{x^2} \, dx.$ *Ans.* $e^4 - e$

25. $\int_{0}^{2} 7^x \, dx.$ *Ans.* $\dfrac{48}{\log 7}$

26. $\int_{\pi/2}^{\pi} \sin 2x \, dx.$ *Ans.* -1

27. $\int_{0}^{4\pi} \cos 3x \, dx.$ *Ans.* 0

28. $\int_{-\pi/4}^{0} \sec^2 x \, dx.$ *Ans.* 1

9.2 Improper Integrals

The question arises as to what meaning is to be attached to a definite integral when one (or both) of the limits of integration is infinite or when the integrand itself becomes infinite at or between the limits of integration. When any of these cases occur, the integral is called an **improper integral**.

Case I. Infinite limits of integration

Consider $\int_{a}^{\infty} f(x) \, dx.$ The value of this integral is defined to be $\lim_{b \to \infty} \int_{a}^{b} f(x) \, dx$ provided this limit exists.

Example 1.   Find $\int_0^\infty e^{-3x}\, dx$.

Solution. $\lim\limits_{b\to\infty} \int_0^b e^{-3x}\, dx = \lim\limits_{b\to\infty} -\dfrac{1}{3}e^{-3x}\Big|_0^b$

$$= \lim_{b\to\infty}\left(-\frac{1}{3}e^{-3b} + \frac{1}{3}\right) = \frac{1}{3}$$

Therefore, $\int_0^\infty e^{-3x}\, dx = \dfrac{1}{3}$

Example 2.   Find $\int_1^\infty \dfrac{dx}{x}$.

Solution. $\lim\limits_{b\to\infty} \int_1^b \dfrac{dx}{x} = \lim\limits_{b\to\infty} (\log b) = \infty$

Hence, we may say that $\int_1^\infty \dfrac{dx}{x}$ has no value or is infinite; thus,

$$\int_1^\infty \frac{dx}{x} = \infty$$

Example 3.   Find $\int_0^\infty \cos x\, dx$.

Solution. $\lim\limits_{b\to\infty} \int_0^b \cos x\, dx = \lim\limits_{b\to\infty} (\sin b)$

Since $\sin b$ approaches no limit at all as $b \longrightarrow \infty$, the original problem has no answer. $\int_0^\infty \cos x\, dx$ does not exist, and has no meaning.

   Case II. Integrand discontinuous
Let $f(x)$ be continuous in (a, b) except at x_1, and consider

$$\int_a^b f(x)\, dx, \quad f(x_1) = \infty, \quad a \leqslant x_1 \leqslant b$$

The value of this integral depends upon the behavior of $f(x)$ in the neighborhood of x_1 and is defined to be

$$\int_a^b f(x)\,dx = \lim_{\epsilon_1 \to 0} \int_a^{x_1 - \epsilon_1} f(x)\,dx + \lim_{\epsilon_2 \to 0} \int_{x_1 + \epsilon_2}^b f(x)\,dx$$

provided the right-hand member exists, and is itself not an indeterminate form of the type $\infty - \infty$. If $x_1 = b$, say, there would be the obvious modification

$$\int_a^b f(x)\,dx = \lim_{\epsilon \to 0} \int_a^{b-\epsilon} f(x)\,dx$$

Example 4. Find $\displaystyle\int_{-1}^1 \frac{dx}{x^{2/3}}$.

Solution. The function becomes infinite at the origin; hence,

$$\int_{-1}^1 \frac{dx}{x^{2/3}} = \lim_{\epsilon_1 \to 0} \int_{-1}^{-\epsilon_1} \frac{dx}{x^{2/3}} + \lim_{\epsilon_2 \to 0} \int_{\epsilon_2}^1 \frac{dx}{x^{2/3}}$$

$$= \lim_{\epsilon_1 \to 0} \left. 3x^{1/3} \right|_{-1}^{-\epsilon_1} + \lim_{\epsilon_2 \to 0} \left. 3x^{1/3} \right|_{\epsilon_2}^1$$

$$= \lim_{\epsilon_1 \to 0} 3[(-\epsilon_1)^{1/3} - (-1)^{1/3}] + \lim_{\epsilon_2 \to 0} 3(1 - \epsilon_2^{1/3})$$

$$= 3 + 3 = 6$$

Example 5. Find $\displaystyle\int_1^2 \frac{dx}{(x-1)^2}$.

Solution. The integrand becomes infinite at $x = 1$. Here, we define the value of the integral to be

$$\lim_{\epsilon \to 0} \int_{1+\epsilon}^2 \frac{dx}{(x-1)^2} = \lim_{\epsilon \to 0} \left. -\frac{1}{(x-1)} \right|_{1+\epsilon}^2$$

$$= \lim_{\epsilon \to 0} \left(-1 + \frac{1}{\epsilon} \right) = \infty$$

Example 6. Find $\int_{-\pi/2}^{\pi/2} \tan x \, dx$.

Solution. Here, the integrand becomes infinite at both limits of integration; hence,

$$\int_{-\pi/2}^{\pi/2} \tan x \, dx = \lim_{\substack{\epsilon_1 \to 0 \\ \epsilon_2 \to 0}} \int_{-\pi/2+\epsilon_1}^{\pi/2-\epsilon_2} \tan x \, dx$$

(These two limits are to be taken quite independently of each other.)

$$= \lim_{\substack{\epsilon_1 \to 0 \\ \epsilon_2 \to 0}} \left[\log \sec \left(\frac{\pi}{2} - \epsilon_2 \right) - \log \sec \left(-\frac{\pi}{2} + \epsilon_1 \right) \right]$$

$$= \infty - \infty$$

This is indeterminate, and the original integral has no meaning.

PROBLEMS WITH SOLUTIONS

1. $\int_1^\infty \frac{1}{x^2} \, dx = \lim_{b \to \infty} \int_1^b \frac{1}{x^2} \, dx = \lim_{b \to \infty} -\frac{1}{x} \Big|_1^b$

$$= \lim_{b \to \infty} \left(1 - \frac{1}{b} \right) = 1$$

2. $\int_{-\infty}^0 e^x \, dx = \lim_{a \to -\infty} \int_a^0 e^x \, dx = \lim_{a \to -\infty} e^x \Big|_a^0$

$$= \lim_{a \to -\infty} (1 - e^a) = 1$$

3. $\int_{-\infty}^\infty xe^{-x^2} \, dx = \lim_{a \to -\infty} \int_a^0 xe^{-x^2} \, dx + \lim_{b \to \infty} \int_0^b xe^{-x^2} \, dx$

$$= \lim_{a \to -\infty} -\frac{1}{2} e^{-x^2} \Big|_a^0 + \lim_{b \to \infty} -\frac{1}{2} e^{-x^2} \Big|_0^b$$

$$= \lim_{a \to -\infty} \frac{1}{2} (e^{-a^2} - 1) + \lim_{b \to \infty} \frac{1}{2} (1 - e^{-b^2})$$

$$= -\frac{1}{2} + \frac{1}{2} = 0$$

4. $\displaystyle\int_0^1 \frac{1}{\sqrt{x}}\, dx = \lim_{\epsilon \to 0} \int_\epsilon^1 \frac{1}{\sqrt{x}}\, dx = \lim_{\epsilon \to 0} 2\sqrt{x}\,\Big|_\epsilon^1$

$$= \lim_{\epsilon \to 0} (2 - 2\sqrt{3}\,) = 2$$

5. $\displaystyle\int_{-1}^1 \frac{1}{x^{1/3}}\, dx = \lim_{\epsilon_1 \to 0} \int_{-1}^{-\epsilon_1} \frac{1}{x^{1/3}}\, dx + \lim_{\epsilon_2 \to 0} \int_{\epsilon_2}^1 \frac{1}{x^{1/3}}\, dx$

$$= \lim_{\epsilon \to 0} \frac{3}{2} x^{2/3}\,\Big|_{-1}^{-\epsilon_1} + \lim_{\epsilon_2 \to 0} \frac{3}{2} x^{2/3}\,\Big|_{\epsilon_2}^{1}$$

$$= \lim_{\epsilon_1 \to 0} \left[\frac{3}{2}(-\epsilon_1)^{2/3} - \frac{3}{2}\right] + \lim_{\epsilon_2 \to 0} \left[\frac{3}{2} - \frac{3}{2}(\epsilon_2)^{2/3}\right]$$

$$= -\frac{3}{2} + \frac{3}{2} = 0$$

6. $\displaystyle\int_{-1}^1 \frac{dx}{x^3} = \lim_{\epsilon_1 \to 0} \int_{-1}^{-\epsilon_1} \frac{1}{x^3}\, dx + \lim_{\epsilon_2 \to 0} \int_{\epsilon_2}^1 \frac{1}{x^3}\, dx$

$$= \lim_{\epsilon_1 \to 0} -\frac{1}{2}\frac{1}{x^2}\,\Big|_{-1}^{-\epsilon_1} + \lim_{\epsilon_2 \to 0} -\frac{1}{2}\frac{1}{x^2}\,\Big|_{\epsilon_2}^{1}$$

$$= \lim_{\epsilon_1 \to 0} \frac{1}{2}\left(1 - \frac{1}{\epsilon_1^2}\right) + \lim_{\epsilon_2 \to 0} \frac{1}{2}\left(\frac{1}{\epsilon_2^2} - 1\right) = \infty - \infty$$

However, this does not exist. Note that when $\epsilon_1 = \epsilon_2$ the integrals yield the **principal value.** In this case, the principal value is 0.

PROBLEMS WITH ANSWERS

7. $\displaystyle\int_{-\infty}^{-1} \frac{dx}{x\sqrt{x^2-1}}.$ *Ans.* $\dfrac{\pi}{2}$

8. $\displaystyle\int_1^\infty \frac{dx}{x^3}.$ *Ans.* $\dfrac{1}{2}$

9. $\displaystyle\int_1^\infty \frac{\log x}{x}\, dx.$ *Ans.* Does not exist.

10. $\displaystyle\int_0^{\pi/2} \frac{\cos x}{\sqrt{\sin x}}\, dx.$ *Ans.* 2

11. $\displaystyle\int_0^3 \frac{2x\,dx}{(x^2-1)^{2/3}}\, dx.$ *Ans.* 9

12. $\displaystyle\int_0^a \frac{dx}{\sqrt{a-x}}.$ *Ans.* $2\sqrt{a}$

13. $\displaystyle\int_0^1 \log x\,dx.$ *Ans.* -1

14. $\displaystyle\int_0^1 \frac{dx}{x^{1/2}}\, dx.$ *Ans.* $2\sqrt{x}\ \Big|_0^1 = 2$

9.3 Setting Up Problems

In all of the applications which follow, we make essential use of:

1. Integration by summation in the formulation of a given problem (Chapter 4).
2. Antidifferentiation and the fundamental theorem of calculus, first form (Theorem VIII, Chapter 6).

Therefore, initially, we *think* in terms of the sum and limit, setting up in our minds what is in effect an expression of some such form as

$$\lim_{\substack{n\to\infty \\ \Delta x_i \to 0}} \sum_{i=1}^n f(x_i)\,\Delta x_i$$

But we do not *write* this. Instead, we write the equivalent integral $\int_a^b f(x)\,dx$, and then proceed to evaluate this by antidifferentiation methods and Theorem VIII (Chapter 6).

9.4 Plane Area

Although area has been discussed in earlier chapters, we sketch these two examples of how we set the problem up by means of these ideas. The area problem is relatively simple.

Example 1. Let f be continuous in the open interval (a, b), and let $f(x) \geqslant 0$ in (a, b). Find the area enclosed by $y = f(x)$, $y = 0$, $x = a$, and $x = b$.

Solution. The summation procedure is to divide (a, b) into n subintervals by the points whose x-coordinates are $a = x_0,\ x_1,\ x_2, \ldots, x_i, \ldots,$ $x_n = b$, and set $\Delta x_i = |x_i - x_{i-1}|$, $i = 1, 2, 3, \ldots$. The area will then be given by

$$A = \lim_{\substack{n \to \infty \\ \Delta x_i \to 0}} \sum_{i=1}^{n} f(x_i)\, \Delta x_i = \int_a^b f(x)\, dx$$

We shorten the work by *thinking* that the area of the ith rectangle is $f(x_i)\, \Delta x_i$, but for this, we *write* $f(x)\, dx$. The total area will be $\lim \Sigma f(x_i)\, \Delta x_i$, but, for this, we write $\int_a^b f(x)\, dx$. Now, we evaluate this integral for the answer.

Example 2. Find the area enclosed by $y = x(x + 1)\,(x - 2)$, $y = 0$, $x = 2$, and $x = 3$ (Fig. 9.1).

Solution. The area of a small rectangle of width dx and height y is $y\, dx$, or $x(x + 1)\,(x - 2)\, dx$; hence, the required area is the *limit of the sum of the* $y_i\, \Delta x_i$ *as* $\Delta x_i \longrightarrow 0$ $(i = 1, 2, \ldots, n)$, or

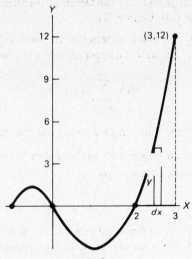

FIG. 9.1

$$\int_2^3 x(x+1)(x-2)\,dx = \int_2^3 (x^3 - x^2 - 2x)\,dx$$

$$= \frac{x^4}{4} - \frac{x^3}{3} - x^2 \Big|_2^3 = \frac{59}{12} \text{ sq. units}$$

If, for all x in $(a, b), f(x)$ is continuous and $f(x) \geqslant 0$, then $\int_a^b f(x)\,dx$ is a nonnegative number. If $f(x) \leqslant 0$, then $\int_a^b f(x)\,dx$ is a nonpositive number and the area is to be considered negative or zero. If $f(x)$ is sometimes positive and sometimes negative, then $\int_a^b f(x)\,dx$ will give the area adding the positive areas above the X-axis and the negative areas below the X-axis algebraically.

Example 3. Find the area between the X-axis and the curve $y = x^2 - x - 12$ bounded by the lines $x = 0$ and $x = 4$.

Solution. The curve lies below the X-axis between $x = 0$ and $x = 4$; thus,

$$A = \int_0^4 (x^2 - x - 12)\,dx$$

$$= \left(\frac{x^3}{3} - \frac{x^2}{2} - 12x\right)\Big|_0^4 = -\frac{104}{3} \text{ sq. units}$$

Although the area turns out to be negative in the problem of Example 3, because it lies below the X-axis, yet, in most cases, we shall not be interested in the sign of the area. Hence, we would take the absolute value of the area and give 104/3 sq. units as the answer. This procedure will cause no trouble unless, in a given problem, some of the area is above the X-axis and some of it is below the X-axis. In this case, the definite integral representing the area, when evaluated over the whole interval, will add the two areas algebraically. This will not give the sum of the absolute values of the areas; therefore, such an integral must be broken up into separate integrals for the positive and negative areas.

Example 4. Find the total area, regardless of sign, contained between the X-axis and the curve $y = x(x+1)(x-2)$ (Fig. 9.2).

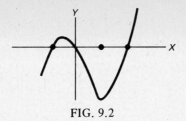

FIG. 9.2

Solution. The curve crosses the X-axis at $x = -1$, 0, and 2. The total area, regardless of sign is, therefore,

$$A = \int_{-1}^{0} (x^3 - x^2 - 2x)\, dx - \int_{0}^{2} (x^3 - x^2 - 2x)\, dx$$

$$= \left(\frac{x^4}{4} - \frac{x^3}{3} - x^2 \right) \Bigg|_{-1}^{0} - \left(\frac{x^4}{4} - \frac{x^3}{3} - x^2 \right) \Bigg|_{0}^{2}$$

$$= [-(\tfrac{1}{4} + \tfrac{1}{3} - 1)] - (4 - \tfrac{8}{3} - 4)$$

$$= \tfrac{5}{12} - (-\tfrac{8}{3}) = \tfrac{37}{12} \text{ sq. units}$$

Conditions of symmetry should be made use of, where they exist, to shorten computations. If the area is symmetric with respect to the X-axis, set up the integral for the part above the X-axis and multiply by two. The student should be sure that symmetry is present before applying this principle.

Example 5. Find the area of a circle of radius r.

Solution. Let the equation of the circle be

$$x^2 + y^2 = r^2$$

$$A = 4 \int_{0}^{r} \sqrt{r^2 - x^2}\, dx$$

$$= 4 \left(\frac{x}{2} \sqrt{r^2 - x^2} + \frac{r^2}{2} \operatorname{Sin}^{-1} \frac{x}{r} \right) \Bigg|_{0}^{r} = \pi r^2 \text{ sq. units}$$

Example 6. Find the area common to the two curves $y = x^2$ and $y^2 = x$.

Solution. These parabolas intersect in the points $(0, 0)$ and $(1, 1)$;

hence, the area included by them is the difference of the areas under the curves, that is,

$$A = \int_0^1 \sqrt{x}\, dx - \int_0^1 x^2\, dx$$

$$= \left(\frac{2}{3} x^{3/2} - \frac{x^3}{3} \right) \Big|_0^1$$

$$= \tfrac{1}{3} \text{ sq. unit}$$

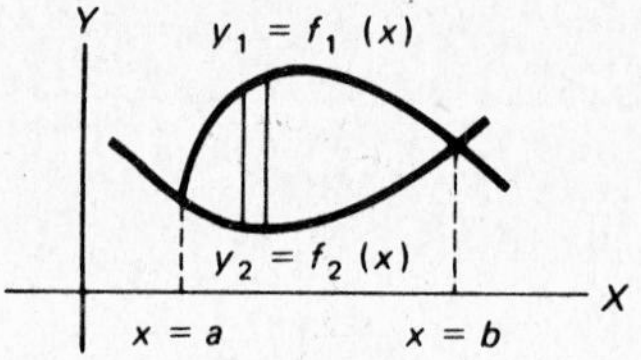

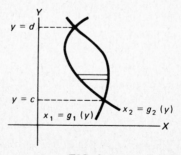

FIG. 9.3

Generalizing Example 5 (Fig. 9.3), we get the area common to two curves:

$$A = \int_a^b y_1 dx - \int_a^b y_2 dx$$

$$= \int_a^b (y_1 - y_2)\, dx \tag{1}$$

If the curves are situated as in Fig. 9.4, then it is not convenient to perform the integration with respect to x. Instead, we transform the

FIG. 9.4

integrals (1) as follows:

$$A = \int_c^d x_1 \, dy - \int_c^d x_2 \, dy$$

Here, $x_1 = g_1(y)$ and $x_2 = g_2(y)$ result from solving $y_1 = f_1(x)$ and $y_2 = f_2(x)$ for x in terms of y.

Example 7. Find the area common to the curves $2(y - 1)^2 = x$ and $(y - 1)^2 = x - 1$ (Fig. 9.5).

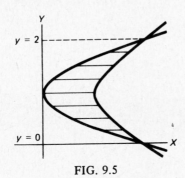

FIG. 9.5

Solution. These curves intersect in the points $(2, 0)$ and $(2, 2)$; the area is given by

$$A = \int_0^2 [1 + (y - 1)^2] \, dy - \int_0^2 2(y - 1)^2 \, dy$$

$$= \int_0^2 [1 - (y - 1)^2] \, dy$$

$$= \left(y - \tfrac{1}{3}(y - 1)^3 \right) \Big|_0^2$$

$$= (2 - \tfrac{1}{3}) - (\tfrac{1}{3})$$

$$= \tfrac{4}{3} \text{ sq. units}$$

If the equation of the curve is given in parametric form $x = f(t)$, $y = g(t)$, then the area becomes

$$A = \int_a^b y \, dx = \int_c^d g(t) f'(t) \, dt$$

where $x = a$ when $t = c$, $x = b$ when $t = d$.

In polar coordinates the area formula is

$$A = \frac{1}{2} \int_{\theta_0}^{\theta_1} \rho^2 \, d\theta \tag{2}$$

We reason as follows (see Fig. 9.6): the differential (limit of increment) area is no longer a strip but approximately the sector of a circle of radius ρ and central angle $d\theta$. The area of such a sector is $\frac{1}{2}\rho^2 \, d\theta$. The sum of all such sectors $\int \frac{1}{2}\rho^2 \, d\theta$ gives the total area sought.

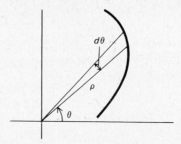

FIG. 9.6

Example 8. Find the area under one arch of the cycloid $x = a(t - \sin t)$, $y = a(1 - \cos t)$.

Solution.

$$A = \int y \, dx$$

$$= \int a^2 (1 - \cos t)(1 - \cos t) \, dt$$

For one arch t runs from 0 to 2π; hence,

$$A = a^2 \int_0^{2\pi} (1 - \cos t)^2 \, dt$$

$$= a^2 \int_0^{2\pi} (1 - 2 \cos t + \cos^2 t) \, dt$$

$$= a^2 \left(t - 2 \sin t + \tfrac{1}{2}t + \tfrac{1}{4} \sin 2t \right) \Big|_0^{2\pi}$$

$$= 3\pi a^2 \text{ sq. units}$$

This is three times the area of the generating circle.

Example 9. Find the area enclosed by the cardioid $\rho = a(1 - \cos \theta)$.

Solution.
$$A = \tfrac{1}{2} \int \rho^2 \, d\theta$$

$$= \frac{1}{2} \int_0^{2\pi} a^2 (1 - \cos \theta)^2 \, d\theta$$

By the methods of Example 8, this equals $\frac{3}{2} \pi a^2$ sq. units.

PROBLEMS WITH ANSWERS

1. Find the area of the ellipse $b^2 x^2 + a^2 y^2 = a^2 b^2$.　　*Ans.* πab

2. Find the area under one arch of $y = \sin x$.　　*Ans.* 2

3. Find the area cut off from the semicubical parabola $y^2 = x^3$ by the line $x = 4$.　　*Ans.* $25 \frac{3}{5}$

4. Find the area bounded by the parabola $x^{1/2} + y^{1/2} = a^{1/2}$ and the coordinate axes.

   Ans. $\dfrac{a^2}{6}$

5. Find the area of one loop of the lemniscate $x = a \cos \theta \sqrt{\cos 2\theta}$, $y = a \sin \theta \sqrt{\cos 2\theta}$.

   Ans. $\dfrac{a^2}{2}$

6. Find the area of one loop of $\rho = a \sin 2\theta$.　　*Ans.* $\dfrac{\pi a^2}{8}$

7. Find the area inside the circle $\rho = a \cos \theta$ and outside the cardioid $\rho = a(1 - \cos \theta)$.

   Ans. $\dfrac{a^2}{3}(3\sqrt{3} - \pi)$

9.5 Length of a Curve

In rectangular coordinates, the differential of arc length is given by

$$ds = \sqrt{dx^2 + dy^2} \tag{1}$$

$$= \sqrt{1 + \left(\frac{dy}{dx}\right)^2} \, dx \tag{2}$$

$$= \sqrt{1 + \left(\frac{dx}{dy}\right)^2} \, dy \tag{3}$$

The length of an element of an arc is Δs, but we write ds and the total length of the curve is the (limit of the) sum of all such elements, namely, $\int ds$. Therefore, using (2) or (3) the length of a curve is given by

$$s = \int ds = \int_a^b \sqrt{1 + y'^2}\, dx$$

$$= \int_c^d \sqrt{1 + x'^2}\, dy \tag{4}$$

Example 1. Find the total length of the circumference of a circle of radius r.

Solution. Let $x^2 + y^2 = r^2$, then

$$x\, dx + y\, dy = 0$$

or

$$\frac{dy}{dx} = y' = -\frac{x}{y}$$

$$y'^2 = \frac{x^2}{y^2}$$

$$1 + y'^2 = 1 + \frac{x^2}{y^2} = \frac{x^2 + y^2}{y^2} = \frac{r^2}{y^2}$$

$$\sqrt{1 + y'^2} = \frac{r}{y} = \frac{r}{\sqrt{r^2 - x^2}}$$

The circumference C is therefore given by

$$C = 4 \int_0^r \frac{r\, dx}{\sqrt{r^2 - x^2}}$$

$$= 4r \left[\mathrm{Sin}^{-1} \frac{x}{r} \right]_0^r = 2\pi r$$

In polar coordinates the differential of arc length is (see Section 8.12 (4)).

$$ds = \sqrt{d\rho^2 + \rho^2 \, d\theta^2} \tag{5}$$

$$= \sqrt{\rho^2 + \left(\frac{d\rho}{d\theta}\right)^2} \, d\theta \tag{6}$$

$$= \sqrt{1 + \rho^2 \left(\frac{d\theta}{d\rho}\right)^2} \, d\rho \tag{7}$$

Length of curve is, therefore,

$$s = \int ds = \int_{\theta_1}^{\theta_2} \sqrt{\rho^2 + \rho'^2} \, d\theta \tag{8}$$

$$= \int_{\rho_1}^{\rho_2} \sqrt{1 + \rho^2 \theta'^2} \, d\rho$$

Example 2. Find the length of the cardioid $\rho = a(1 - \cos \theta)$.

Solution.
$$\rho' = a \sin \theta$$
$$\rho'^2 = a^2 \sin^2 \theta$$
$$\rho^2 + \rho'^2 = a^2(1 - \cos \theta)^2 + a^2 \sin^2 \theta$$
$$= 2a^2(1 - \cos \theta)$$

$$s = \int_0^{2\pi} \sqrt{2a^2(1 - \cos \theta)} \, d\theta$$

$$= 2a \int_0^{2\pi} \sqrt{\frac{1 - \cos \theta}{2}} \, d\theta$$

$$= 2a \int_0^{2\pi} \sin \frac{1}{2}\theta \, d\theta$$

$$= 2a(-2 \cos \tfrac{1}{2}\theta) \Big|_0^{2\pi} = 8a$$

If the parametric equations of the curve are $x = f(t)$ and $y = g(t)$, then

$$s = \int ds = \int_{t_1}^{t_2} \sqrt{\left(\frac{dx}{dt}\right)^2 + \left(\frac{dy}{dt}\right)^2} \, dt$$

Example 3. Find the length of the curve $x = 2 - t$, $y = \frac{1}{2}t^2$ from $t = 0$ to $t = 1$.

Solution. $s = \int_0^1 \sqrt{(-1)^2 + (t)^2}\ dt = \int_0^1 \sqrt{1 + t^2}\ dt$

$$= \frac{t}{2}\sqrt{1 + t^2} + \frac{1}{2}\log(t + \sqrt{1 + t^2})\ \Big|_0^1$$

$$= \frac{1}{2}\sqrt{2} + \frac{1}{2}\log(1 + \sqrt{2})$$

PROBLEMS WITH SOLUTIONS

1. Find the length of the parabola $y = x^2$ from 0 to 1.
    Solution.

    The arc length $s = \int_0^1 \sqrt{1 + y'^2}\ dx = \int_0^1 \sqrt{1 + 4x^2}\ dx$

    $$= 2\int_0^1 \sqrt{\frac{1}{4} + x^2}\ dx$$

    $$= 2\left[\frac{x}{2}\sqrt{\frac{1}{4} + x^2} + \frac{1}{8}\log\left(x + \sqrt{\frac{1}{4} + x^2}\right)\right]_0^1$$

    $$= \frac{\sqrt{5}}{2} + \frac{1}{4}\log(2 + \sqrt{5})$$

2. Find the length of the spiral $\rho = e^{2\theta}$ from $\rho = 0$ to $\rho = 1$.
    Solution.

    The arc length $s = \int_0^1 \sqrt{1 + \left(\rho\frac{d\theta}{d\rho}\right)^2}\ d\rho = \int_0^1 \sqrt{1 + \frac{\rho^2}{4\rho^2}}\ d\rho$

    $$= \int_0^1 \sqrt{1 + \frac{1}{4}}\ d\rho = \frac{\sqrt{5}}{2}\int_0^1 d\rho = \frac{\sqrt{5}\rho}{2}\Big|_0^1 = \frac{\sqrt{5}}{2}$$

    (since $\theta = \frac{1}{2}\log\rho$)

3. Find the length of the spiral of Archimedes $\rho = a\theta$ from $\theta = 0$ to $\theta = 2\pi$.

Solution.

$$s = \int_0^{2\pi} \sqrt{\rho^2 + \left(\frac{d\rho}{d\theta}\right)^2}\; d\theta = \int_0^{2\pi} \sqrt{a^2\theta^2 + a^2}\; d\theta$$

$$= \int_0^{2\pi} a\sqrt{1+\theta^2}\; d\theta = a\left[\frac{\theta}{2}\sqrt{1+\theta^2} + \tfrac{1}{2}\log\left(\theta + \sqrt{1+\theta^2}\right)\right]_0^{2\pi}$$

$$\cong 21a$$

4. Find the length of the logarithmic spiral $\rho = e^\theta$ from $\theta = 0$ to $\theta = \pi/2$.

Solution. $\quad s = \int_0^{\pi/2} \sqrt{\left(\frac{d\rho}{d\theta}\right)^2 + \rho^2}\; d\theta = \int_0^{\pi/2} e^\theta \sqrt{2}\; d\theta$

$$= \sqrt{2}\,e^\theta \;\Big|_0^{\pi/2} = \sqrt{2}\,(e^{\pi/2} - 1)$$

PROBLEMS WITH ANSWERS

5. Find the total length of the hypocycloid of four cusps $x^{2/3} + y^{2/3} = a^{2/3}$ *Ans.* $6a$

6. Find the length of one arch of the cycloid $x = a(\theta - \sin\theta)$, $y = a(1 - \cos\theta)$. *Ans.* $8a$

7. Find the length of the curve $x = 3t^2$, $y = 2t^3$ from 0 to 3.
Ans. $20\sqrt{10} - 2$

8. Find the length of the catenary $\tfrac{1}{2}(e^x + e^{-x})$ from -1 to 1.
Ans. $e - \dfrac{1}{e}$

9.6 Solids of Revolution

Let the area under the curve $y = f(x)$, namely, $\int_a^b f(x)\,dx$, be revolved about the X-axis thus generating a volume (Fig. 9.7). The area of a cross section of this solid by a plane perpendicular to the X-axis is πy^2; hence, the volume of a little slice dx thick is $\pi y^2\, dx$. The total volume

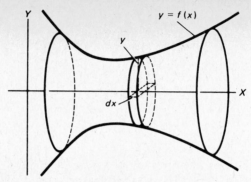

FIG. 9.7

of the solid of revolution between two parallel planes $x = a$ and $x = b$ is, therefore, the sum of all such slices, or

$$V = \pi \int_a^b y^2 \, dx \qquad (1)$$

Example 1. The area under the curve $y = x^3$ from $x = 0$ to $x = 1$ is resolved about the X-axis. Find the volume thus generated.

Solution.
$$V = \pi \int_0^1 y^2 \, dx = \pi \int_0^1 x^6 \, dx$$

$$= \frac{\pi x^7}{7} \bigg|_0^1 = \frac{\pi}{7} \text{ cu. unit}$$

If the area $\int_c^d x \, dy$ is resolved about the Y-axis, the volume becomes

$$V = \pi \int_c^d x^2 \, dy \qquad (2)$$

Example 2. The area between the Y-axis, the curve $y = x^2$, and the lines $y = 1$ and $y = 2$, is revolved about the Y-axis. Find the volume generated.

Solution.
$$V = \pi \int_1^2 x^2 \, dy = \pi \int_1^2 y \, dy$$

$$= \frac{\pi y^2}{2} \bigg|_1^2 = \frac{3}{2}\pi \text{ cu. units}$$

FIG. 9.8

Example 3. The area common to the two parabolas $y = x^2$ and $y^2 = x$ is revolved about the Y-axis. Find the volume generated (Fig. 9.8).

Solution. At any height y a slice will be a disc with a hole in it, like a washer. The area of this disc is $\pi(x_1^2 - x_2^2)$, where x_1 refers to the x in $y = x^2$, and x_2 refers to the x in $y^2 = x$. Hence, the volume of this elementary disc is $\pi(x_1^2 - x_2^2)\, dy$, and the volume sought is

$$V = \pi \int_0^1 (x_1^2 - x_2^2)\, dy$$

$$= \pi \int_0^1 (y - y^4)\, dy$$

$$= \pi\left(\frac{y^2}{2} - \frac{y^5}{5}\right)\Bigg|_0^1 = \frac{3}{10}\pi \text{ cu. unit}$$

We could make use of a cylindrical element of volume in examples such as that of Example 3. When an elementary area strip (as in Fig. 9.9) is revolved about the Y-axis, the volume element generated is a thin, cylindrical shell with area $2\pi x(y_2 - y_1)$. The volume element is $2\pi x(y_2 - y_1)\, dx$; summing these, we have for the volume

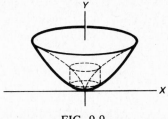

FIG. 9.9

$$V = 2\pi \int_a^b x(y_2 - y_1)\, dx \qquad (3)$$

The student should not try to memorize these formulas, because there are many others similar to them, but he should master the technique of setting them up.

Example 4. Solve the problem in Example 3 by the use of cylindrical elements.

Solution. We have already set this up in (3) above; thus,

$$V = 2\pi \int_0^1 x(\sqrt{x} - x^2)\, dx$$

$$= 2\pi\left(\frac{2}{5}x^{5/2} - \frac{x^4}{4}\right)\Bigg|_0^1 = \frac{3}{10}\pi \text{ cu. unit}$$

PROBLEMS WITH SOLUTIONS

Find the volume generated by revolving the area given in Problems 1 through 5.

1. The area enclosed by the ellipse $\dfrac{x^2}{a^2} + \dfrac{y^2}{b^2} = 1$ about the X-axis.

    Solution. $V = 2\pi \displaystyle\int_0^a \frac{b^2}{a^2}(a^2 - x^2)\, dx = \frac{4}{3}\pi ab^2$

2. The area enclosed by $x^{1/2} + y^{1/2} = a^{1/2}$, $x = 0$, $y = 0$, about the X-axis.

    Solution. $V = \pi \displaystyle\int_0^a (a^{1/2} - x_1^{1/2})^4\, dx$

    $$= \pi \int_0^a (a^2 - 4a^{3/2}x^{1/2} + 6ax - 4a^{1/2}x^{3/2} + x^2)\, dx$$

    $$= \frac{\pi a^3}{15}$$

3. The area bounded by $y = \dfrac{8a^3}{x^2 + 4a^2}$ and the X-axis about the X-axis.

Solution. $V = 2\pi \displaystyle\int_0^\infty \dfrac{64a^6}{(x^2 + 4a^2)^2}\, dx$

$\qquad = 128\pi a^6 \displaystyle\int_0^\infty \dfrac{dx}{(x^2 + 4a^2)^2}$

$\qquad = 128\pi a^6 \, \dfrac{1}{8a^2}\left(\dfrac{x}{x^2 + 4a^2} + \dfrac{1}{2a^2}\, \mathrm{Tan}^{-1} \dfrac{x}{2a}\right)\Big|_0^\infty$

$\qquad = 16\pi a^4 \left(0 + \dfrac{\pi}{4a}\right) = 4\pi^2 a^3$

4. The area bounded by $y = \dfrac{1}{x}$, $x = 1$, and $y = 0$, about the X-axis.

Solution. $V = \pi \displaystyle\int_0^\infty \dfrac{1}{x^2}\, dx = -\dfrac{\pi}{x}\Big|_0^\infty$

Compare this finite volume with the infinite area under the same curve (see Example 2, Section 9.2).

5. The area of a circle about a tangent (Fig. 9.10).

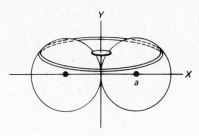

FIG. 9.10

Solution. Let the circle be $(x - r)^2 + y^2 = r^2$, and revolve it about the Y-axis. V_1 refers to the volume of the smaller disc (shaded) and V_2 refers to the volume of the larger disc which includes the smaller disc.

$$V_2 = 2 \int_0^r \pi \left(r + \sqrt{r^2 - y^2} \right)^2 dy$$

$$= 2\pi \int_0^r \left(r^2 + 2r\sqrt{r^2 - y^2} + r^2 - y^2 \right) dy$$

$$= 2\pi \int_0^r \left(2r^2 + 2r\sqrt{r^2 - y^2} - y^2 \right) dy$$

$$= 2\pi \left[2r^2 y + 2r \left(\frac{y}{2}\sqrt{r^2 - y^2} + \frac{r^2}{2} \operatorname{Sin}^{-1} \frac{y}{r} \right) - \frac{y^3}{3} \right]_0^r$$

$$= \frac{(10 + 3\pi)}{3} \pi r^3$$

$$V_1 = 2 \int_0^r \pi \left(r - \sqrt{r^2 - y^2} \right)^2 dy$$

$$= 2\pi \int_0^r r^2 - 2r(\sqrt{r^2 - y^2} + r^2 - y^2) \, dy$$

$$= 2\pi \left[2r^2 y - 2r \left(\frac{y}{2}\sqrt{r^2 - y^2} + \frac{r^2}{2} \operatorname{Sin}^{-1} \frac{y}{r} \right) - \frac{y^3}{3} \right]_0^r$$

$$= \frac{(10 - 3\pi)}{3} \pi r^3$$

$$V = V_2 - V_1 = \left[\frac{10 + 3\pi}{3} - \frac{10 - 3\pi}{3} \right] \pi r^3$$

$$= 2\pi^2 r^3 = 2\pi r \cdot \pi r^2$$

PROBLEMS WITH ANSWERS

Find the volume formed by revolving the areas given in Problems 6 through 10.

6. The area of a semicircle of radius r about a diameter.

Ans. $\frac{4}{3}\pi r^3$ (volume of a sphere)

7. The area of the triangle formed by the lines $x = 0$, $y = h$, and $y = \frac{h}{r} x$ about the Y-axis.

Ans. $\frac{1}{3}\pi r^2 h$ (volume of a cone)

8. The area under one arch of $y = \sin x$ about the X-axis.

$$Ans. \quad \frac{\pi^2}{2}$$

9. The area in the second quadrant under $y = e^x$ about the X-axis.

$$Ans. \quad \pi \int_{-\infty}^{0} e^{2x} \, dx = \frac{\pi}{2}$$

10. The area under one arch of $y = \sin x$ about the Y-axis.

$$Ans. \quad 2\pi \int_{0}^{\pi} x \sin x \, dx = 2\pi^2$$

9.7 Volumes of Known Cross Section

Consider the sections of a solid made by parallel planes. For purposes of illustration, let these planes be perpendicular to the X-axis. If it is possible to write down the area of each section in terms of its distance from some fixed point on the X-axis, say the origin O, then the volume of the solid can be determined. For, as in Fig. 9.11, the area of the

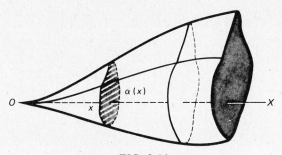

FIG. 9.11

cross section at distance x is known to be a function of x, say $\alpha(x)$, and the volume element then is $\alpha(x) \, dx$. Hence, the volume from $x = a$ to $x = b$ would be

$$V = \int_{a}^{b} \alpha(x) \, dx \tag{1}$$

Solids of revolution are a special case of this.

Example 1. A solid has a circular base (radius r), and every section

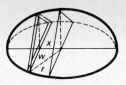

FIG. 9.12

perpendicular to a diameter is an equilateral triangle. Find the volume of the solid (Fig. 9.12).

Solution. At a distance x from the center along the diameter shown the cross section is an equilateral triangle whose base is, say, $2w$ and whose altitude is then $w\sqrt{3}$. The area of this triangle is $w^2\sqrt{3}$. But this can be expressed in terms of x since $w^2 + x^2 = r^2$. The area is then $\sqrt{3}(r^2 - x^2)$. The volume element becomes $\sqrt{3}(r^2 - x^2)\,dx$. The solid is symmetric, and so we write

$$V = 2\int_0^r \sqrt{3}(r^2 - x^2)\,dx$$

$$= 2\sqrt{3}\left(r^2 x - \frac{x^3}{3}\right)\Bigg|_0^r = \frac{4}{3}\sqrt{3}\,r^3 \text{ cu. units}$$

Example 2. A newel post cap has a square base 6 in. on one side. Each section parallel to the base is a square whose side is proportional to $(4 - x)$, where x is the distance (in inches) above the base. Find the volume of the cap (Fig. 9.13).

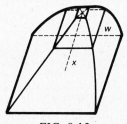

FIG. 9.13

Solution. $w = k(4 - x)$

$w = 6$ when $x = 0; \therefore k = \frac{3}{2}$

$w = \frac{3}{2}(4 - x)$

A slice dx thick parallel to the base and x in. above it will, therefore, have a volume element $w^2 dx$ or $\frac{9}{4}(4 - x)^2\,dx$. The volume of the cap is

$$V = \frac{9}{4} \int_0^4 (4 - x)^2 \, dx$$

$$= \frac{9}{4} \left(-\frac{1}{3}\right)(4 - x)^3 \Bigg|_0^4 = 48 \text{ cu. in.}$$

PROBLEMS WITH SOLUTIONS

1. Find the volume of a solid with a circular base of radius r, if every section perpendicular to a diameter is an isosceles triangle with altitude h (Fig. 9.14).

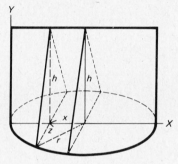

FIG. 9.14

Solution. $z = \sqrt{r^2 - x^2}$

$$V = 2 \int_0^r h \sqrt{r^2 - x^2} \, dx = \frac{\pi r^2 h}{2}$$

2. Find the volume of a solid with a circular base of radius r, if every section perpendicular to a diameter is a square (Fig. 9.15).

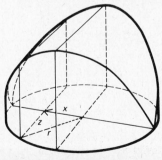

FIG. 9.15

Solution. $z = \sqrt{r^2 - x^2}$

$$V = 8 \int_0^r (r^2 - x^2)\, dx = \frac{16}{3} r^3$$

3. A special ice-cream cone is shown in Fig. 9.16. It has a 2 in. square top and is 3 in. high. Find the liquid volume it will hold.

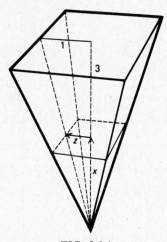

FIG. 9.16

Solution. $\dfrac{x}{z} = \dfrac{3}{1},\ z = \dfrac{1}{3} x$

$$V = \int_0^3 \left(\frac{2}{3} x \right)^2 dx$$

$$= \frac{4}{9} \cdot \frac{x^3}{3} \Big|_0^3 = 4 \text{ cu. in.}$$

4. Find the volume of the ellipsoid $\dfrac{x^2}{a^2} + \dfrac{y^2}{b^2} + \dfrac{z^2}{c^2} = 1$.

Solution. At a distance x from the origin along the X-axis,

$$\frac{y^2}{b^2} + \frac{z^2}{c^2} = 1 - \frac{x^2}{a^2} = \frac{a^2 - x^2}{a^2}$$

$$\text{and } \frac{y^2}{\dfrac{b^2}{a^2}(a^2 - x^2)} + \frac{z^2}{\dfrac{c^2}{a^2}(a^2 - x^2)} = 1$$

Therefore, the area at distance x is

$$= \frac{\pi bc}{a^2} (a^2 - x^2)$$

Hence, $V = 2 \int_0^a \frac{\pi bc}{a^2} (a^2 - x^2) \, dx$

$$= \frac{2\pi bc}{a^2} \left(a^2 x - \frac{x^3}{3} \right) \Big|_0^a$$

$$= \frac{2\pi bc}{a^2} \left(\frac{2}{3} a^3 \right) = \frac{4}{3} \pi abc$$

5. Find the volume of a wedge cut from a cylinder of radius R by a plane which passes through a diameter at the base and makes an angle $\theta = \text{Tan}^{-1} \dfrac{H}{R}$ with the base (Fig. 9.17).

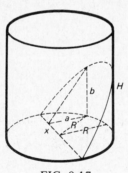

FIG. 9.17

Solution 1. $a = \sqrt{R^2 - x^2}, \dfrac{b}{a} = \dfrac{H}{R}$, and $b = \dfrac{H}{R}\sqrt{R^2 - x^2}$; therefore,

$$V = 2 \int_0^R \frac{1}{2}\sqrt{R^2 - x^2} \, \frac{H}{R}\sqrt{R^2 - x^2} \, dx$$

$$= \frac{H}{R} \int_0^R (R^2 - x^2) \, dx$$

$$= \frac{H}{R} \left(R^2 x - \frac{x^3}{3} \right) \Big|_0^R = \frac{2}{3} R^2 H$$

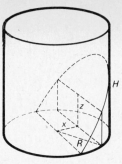

FIG. 9.18

Solution 2. (Fig. 9.18)

$$\frac{z}{x} = \frac{H}{R}, \text{ hence } z = \frac{H}{R}x, \text{ therefore,}$$

$$V = 2\frac{H}{R}\int_0^R x\sqrt{R^2 - x^2}\, dx$$

$$= -\frac{2}{3}\frac{H}{R}(R^2 - x^2)^{3/2}\ \Big|_0^R = \frac{2}{3}R^2 H$$

PROBLEMS WITH ANSWERS

6. Find the volume common to two cylinders, each of radius r, whose axes intersect at right angles.

$$Ans. \quad V = 8\int_0^r (r^2 - x^2)\, dx = \frac{16}{3}r^3$$

7. Find the volume of a solid whose base is a segment of the parabola $y^2 = x$ cut off by the chord $x = 5$ and whose section by a plane perpendicular to the axis of the parabola at a distance x from the vertex is a rectangle with height equal to $\frac{1}{2}(5 - x)$.

$$Ans. \quad \frac{4}{3}\sqrt{125} \text{ cu. units}$$

9.8 Area of Surfaces of Revolution

When the curve $y = f(x)$ is revolved about the X-axis, a surface is generated (Fig. 9.19). To find the area of this surface, consider the area generated by an element of arc ds. This area is roughly that of a cylinder of radius y, and we write $dS = 2\pi y\, ds$. Summing all such elements of

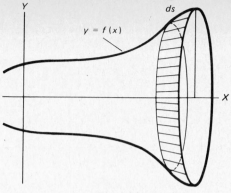

FIG. 9.19

surface area, we have

$$S = 2\pi \int y \, ds$$

$$= 2\pi \int_a^b y \sqrt{1 + y'^2} \, dx \qquad (1)$$

Appropriate modifications of this formula will be necessary if the curve is revolved about some other line or if polar coordinates are used.

Example 1. Find the surface area of a sphere of radius r.
Solution. This area could be generated by revolving the upper half of the circle $x^2 + y^2 = r^2$ about the X-axis; thus,

$$S = 2\pi \int_{-r}^{r} y \sqrt{1 + y'^2} \, dx$$

$$= 4\pi \int_0^r \left(\sqrt{r^2 - x^2} \right) \left(\frac{r}{y} \right) dx$$

$$= 4\pi r \int_0^r dx$$

$$= 4\pi r x \bigg|_0^r = 4\pi r^2$$

Example 2. Revolve that arch of the cycloid $x = a(\theta - \sin \theta), y = a(1 - \cos \theta)$ which passes through the origin about the Y-axis, and compute the surface area generated.

Solution.

$$S = 2\pi \int x \, ds$$

$$= 2\pi \int x \sqrt{1 + y'^2} \, dx$$

Now,

$$dx = a(1 - \cos \theta) \, d\theta$$

$$dy = a \sin \theta \, d\theta$$

$$y' = \frac{\sin \theta}{1 - \cos \theta}$$

$$y'^2 = \frac{\sin^2 \theta}{(1 - \cos \theta)^2}$$

$$1 + y'^2 = \frac{2}{1 - \cos \theta} = \frac{1}{\sin^2 \frac{1}{2} \theta}$$

$$\sqrt{1 + y'^2} = \frac{1}{\sin \frac{1}{2} \theta}$$

$$S = 2\pi a^2 \int_0^{2\pi} (\theta - \sin \theta) \frac{1}{\sin \frac{1}{2} \theta} (1 - \cos \theta) \, d\theta$$

$$= 4\pi a^2 \int_0^{2\pi} (\theta - \sin \theta) \sin \frac{1}{2} \theta \, d\theta$$

$$= 4\pi a^2 \int_0^{2\pi} \left(\theta \sin \frac{1}{2} \theta - \sin \theta \sin \frac{1}{2} \theta \right) d\theta$$

$$= 4\pi a^2 \int_0^{2\pi} \left(\theta \sin \frac{1}{2} \theta - 2 \sin^2 \frac{1}{2} \theta \cos \frac{1}{2} \right) d\theta$$

Integrating the first of these by parts we have

$$4\pi a^2 \int_0^{2\pi} \theta \sin \frac{1}{2} \theta \, d\theta = 4\pi a^2 \left(-2\theta \cos \frac{1}{2} \theta \right) \Big|_0^{2\pi} + 2 \int_0^{2\pi} \cos \frac{1}{2} \theta \, d\theta$$

$$= 4\pi a^2 \left(4\pi + 4 \sin \frac{1}{2} \theta \right) \Big|_0^{2\pi} = 16\pi^2 a^2$$

This is the answer, since

$$4\pi a^2 \int_0^{2\pi} \left(-2\sin^2\frac{1}{2}\theta\,\cos\frac{1}{2}\theta\right)d\theta = 4\pi a^2 \left(-\frac{4}{3}\sin^3\frac{1}{2}\theta\right)\Big|_0^{2\pi} = 0$$

Example 3. Find the surface area generated by revolving the upper half of the cardioid $\rho = a(1 - \cos\theta)$ about the polar axis.

Solution. The radius of the circle through which the element ds swings is $\rho \sin\theta$. The surface element is therefore $2\pi\rho \sin\theta\,ds$; thus,

$$S = 2\pi \int_0^\pi \rho \sin\theta \sqrt{\rho^2 + \rho'^2}\,d\theta$$

$$= 2\pi a \int_0^\pi (1 - \cos\theta)\sin\theta \sqrt{2a^2(1 - \cos\theta)}\,d\theta$$

$$= 4\pi a^2 \int_0^\pi (1 - \cos\theta)\sin\theta\,\sin\frac{1}{2}\theta\,d\theta$$

$$= 8\pi a^2 \int_0^\pi \sin^3\frac{1}{2}\theta\,\sin\theta\,d\theta$$

$$= 16\pi a^2 \int_0^\pi \sin^4\frac{1}{2}\theta\,\cos\frac{1}{2}\theta\,d\theta$$

$$= 16\pi a^2 \left(\frac{2}{5}\sin^5\frac{1}{2}\theta\right)\Big|_0^\pi = \frac{32}{5}\pi a^2$$

PROBLEMS WITH SOLUTIONS

1. Find the surface area generated by revolving the ellipse $x = a\cos\theta$, $y = b\sin\theta$, $a \geqslant b$, about the X-axis.

   Solution. $dx = -a\sin\theta\,d\theta$ and $dy = b\cos\theta$; thus,

   $$ds = \sqrt{a^2\sin^2\theta + b^2\cos^2\theta}\,d\theta = a\sqrt{1 - e^2\cos^2\theta}\,d\theta,$$

   $$\text{where } e = \frac{\sqrt{a^2 - b^2}}{a} \neq 0$$

   Hence, $\quad S = 4\pi \int_0^{\pi/2} ab\sin\theta \sqrt{1 - e^2\cos^2\theta}\,d\theta$

Let $y = e \cos \theta$, then

$$S = \frac{4\pi ab}{a} \int_0^e \sqrt{1 - y^2} \, dy$$

$$= \frac{4\pi ab}{e} \left(\frac{y}{2} \sqrt{1 - y^2} + \frac{1}{2} \operatorname{Sin}^{-1} y \right)\Big|_0^e = \frac{2\pi ab}{e} \left(e\sqrt{1 - e^2} + \operatorname{Sin}^{-1} e \right)$$

$$= 2\pi ab \left(\sqrt{1 - e^2} + \frac{1}{e} \operatorname{Sin}^{-1} e \right)$$

2. Find the lateral surface area of a cone of radius r and altitude h. (Let l denote the slant height.)

   Solution. $S = 2\pi \int_0^h y \sqrt{1 + \frac{r^2}{h^2}} \, dx$

   $$= 2\pi \int_0^h \frac{r}{h} x \sqrt{1 + \frac{r^2}{h^2}} \, dx$$

   $$= 2\pi \int_0^h \frac{rl}{h^2} x \, dx = \frac{2\pi rl}{h^2} \cdot \frac{h^2}{2} = \pi rl$$

3. Find the surface area generated by revolving the lemniscate $\rho_2 = a^2 \cos 2\theta$ about the polar axis.

   Solution. $\rho^2 = a^2 \cos 2\theta$, $\rho = a\sqrt{\cos 2\theta}$,

   $$\rho' = -\frac{a \sin 2\theta}{\sqrt{\cos 2\theta}}, \text{ and } \quad \rho'^2 = \frac{a^2 \sin^2 2\theta}{\cos 2\theta}$$

   $$ds = \sqrt{a^2 \cos 2\theta + \frac{a^2 \sin^2 2\theta}{\cos 2\theta}} \, d\theta$$

   $$= \sqrt{\frac{a^2 (\cos^2 2\theta + \sin^2 2\theta)}{\cos 2\theta}} \, d\theta = \frac{a}{\sqrt{\cos 2\theta}} \, d\theta$$

   $$S = 2\pi \int_0^{\pi/4} 2a\sqrt{\cos 2\theta} \sin \theta \cdot \frac{a}{\sqrt{\cos 2\theta}} \, d\theta$$

   $$= 4\pi a^2 \int_0^{\pi/4} \sin \theta \, d\theta = -4\pi a^2 \cos \theta \Big|_0^{\pi/4}$$

   $$= -4\pi a^2 \left(\frac{\sqrt{2}}{2} - 1 \right) = 2\pi a^2 \left(2 - \sqrt{2} \right)$$

PROBLEMS WITH ANSWERS

4. Find the surface area of a zone of altitude h of a sphere of radius r. (A zone is the portion of a sphere included between two parallel planes; the altitude of a zone is the distance between the parallel planes.)

$$Ans.\quad 2\pi rh$$

5. Find the surface area generated by revolving one arch of $y = \cos x$ about the X-axis.

$$Ans.\quad 2\pi \left[\sqrt{2} + \log\ 1 + \sqrt{2} \right]$$

6. Find the surface area generated by revolving one arch of the cycloid $x = a(\theta - \sin \theta)$, $y = a(1 - \cos \theta)$ about the X-axis.

$$Ans.\quad \frac{64}{3}\pi a^2$$

7. Find the surface area generated when the hypocycloid $x^{2/3} + y^{2/3} = a^{2/3}$ is revolved about the X-axis.

$$Ans.\quad S = 4\pi \int_0^a (a^{2/3} - x^{2/3})^{3/2}\ \left(1 + \frac{y^{2/3}}{x^{2/3}}\right)^{1/2}\ dx = \frac{12\pi a^2}{5}$$

9.9 Work

Let a body move along a straight line under the application of a constant force F lbs. acting in the direction of motion. If the particle is displaced x ft., the work done is $W = Fx$ ft.-lbs. Thus, the work done in lifting a body weighing 100 lbs. a vertical distance of 2 ft. against the force of gravity is 200 ft.-lbs.

To generalize this notion of work, consider a variable force F acting in the direction of motion which takes place along a curve. The distance (or displacement) element along the curve is $ds;$ if the force F is a function of s, then the element of work done will be $dW = F(s)\,ds$. Summing all such elements, we have the total work done, or

$$W = \int_{s_1}^{s_2} F(s)\,ds \tag{1}$$

In the applications which will concern us, the motion will take place along a line which may be taken as one of the axes. For example, if the motion is along the X-axis (1) becomes

$$W = \int_{x_1}^{x_2} F(x)\,dx \tag{2}$$

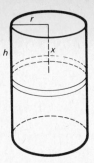

FIG. 9.20

Example 1. A vertical cylindrical tank of radius r ft. and height h ft. is full of water. Find the work done in emptying the tank by pumping the water out over the top rim (Fig. 9.20).

Solution. Consider an element of water (a disc) at a depth x from the top. The volume of this slice is $\pi r^2 \, dx$. Its weight is therefore $w\pi r^2 \, dx$, where the weight of 1 cu. ft. of the liquid is w (density); here $w = 62.5$ lbs., since the liquid is water. In general, w will depend upon the liquid considered. The work done in lifting this weight (force) x feet is, therefore, $dW = 62.5\pi r^2 x \, dx$, and the total work is

$$W = \int_0^h 62.5\pi r^2 x \, dx$$

$$= \frac{62.5\pi r^2 h^2}{2} \text{ ft.-lbs.}$$

$$= \frac{w}{2}\pi r^2 h^2 \text{ (in the case of a liquid of density } w)$$

Example 2. An anchor chain of a ship weighs 50 lbs. per linear foot, while the anchor itself weighs 2000 lbs. What is the work done in pulling up anchor, if 100 ft. of chain are out, assuming that the lift is vertical?

Solution. Let x be the number of feet of anchor chain out at any time, and consider an element dx. This element weighs $50 \, dx$ lbs., and it must be lifted x ft. The work required is, therefore,

$$W = \int_0^{100} 50x \, dx + 2000 \times 100$$

where 2000×100 represents the work of lifting the anchor itself; thus,

$$W = 25x^2 \Big|_0^{100} + 200,000$$

$$= 250,000 + 200,000$$

$$= 450,000 \text{ ft.-lbs.}$$

Example 3. The force required to stretch a certain spring is proportional to the elongation. If a force of one pound stretches the spring $\frac{1}{2}$ in., what is the work done in stretching the spring 2 in.?

Solution. Call L the natural length of the spring and x the elongation. Then $F = kx$, $1 = k \cdot \frac{1}{2}$, and $k = 2$. Hence, $F = 2x$ and the work is

$$W = \int_0^2 2x \, dx$$

$$= x^2 \Big|_0^2 = 4 \text{ in.-lbs.}$$

PROBLEMS WITH SOLUTIONS

1. A horizontal cylindrical oil tank is 10 ft. long and has a radius of 4 ft. If it is filled with oil of density w, find the work done in pumping the oil out over the top (Fig. 9.21).

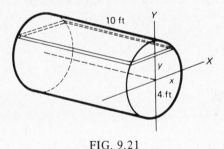

FIG. 9.21

Soltuion. $W = w \displaystyle\int_{-4}^{4} (10) \, (2x) \, (4 - y) \, dy$ and $x^2 + y^2 = 16$; hence,

$$W = 20w \int_{-4}^{4} (4 - y) \sqrt{16 - y^2} \, dy$$

$$= 20w \left[4 \left\{ \frac{y}{2} \sqrt{16 - y^2} + \frac{16}{2} \operatorname{Sin}^{-1} \frac{y}{4} \right\} - \frac{1}{3} (16 - 2)^{3/2} \right]_{-4}^{4}$$

$$= 20w \, [4(8\pi)] = 640\pi w \text{ ft.-lb.}$$

2. Newton's law of universal gravitation states that *the force of attraction between two bodies is directly proportional to the product of the masses and inversely proportional to the square of the distance between them*, i.e., $F = \dfrac{Gmm'}{s^2}$, where G is a constant, m and m' are the masses, and s is the distance. Find the work required to separate two bodies from a distance s to a distance ks for $k > 0$.

Solution. $W = \displaystyle\int_{s}^{ks} \dfrac{Gmm'}{s^2}\,ds = \dfrac{Gmm'}{s} \cdot \dfrac{k-1}{k}$

3. A bar is stretched from its original length l by a gradually increasing load. If s is the amount of stretch for a given force F, then by Hooke's law $s = \dfrac{Fl}{eA}$ (the so-called "flea" formula), where e is the coefficient of expansion characteristic of the metal and A is the cross-sectional area of the bar. The work done in stretching the bar is, therefore,

$$W = \frac{eA}{l} \int_{0}^{s} s\,ds = \frac{eAs^2}{2l}$$

Find the work done in stretching a wrought iron bar ($e = 30{,}000{,}000$, $A = 2$ sq. in.) from an original length of 100 in. to 100.5 in.

Solution. $W = \dfrac{eA}{l} \displaystyle\int_{0}^{s} s\,ds = \dfrac{eAs^2}{2l} = \dfrac{30{,}000{,}000 \times 2 \times \left(\frac{1}{2}\right)^2}{2 \times 100}$

$$= 75{,}000 \text{ in.-lbs.}$$

4. Let p denote the pressure (force F per unit area) which a gas exerts on a piston of area A; then $F = pA$, and $F\Delta s = pA\,\Delta s$, where Δs is the increment in distance over which the piston moves. Now, $A\Delta s = dV$, where V is the volume; therefore,

$$W = \int_{V_1}^{V_2} p\,dV$$

   a. Air expands according to Boyle's law (no change in temperature, i.e., $pV = p_0 V_0 = $ constant). Find the work done in expanding from V_0 to V_1.

b. Air expands adiabatically if $pV^k = p_0 V_0^k$ = constant, where $k \cong 1.41$. Find the work done in expanding adiabatically from V_0 to V_1.

Solution.

a.
$$W = \int_{V_0}^{V_1} p\, dV = p_0 V_0 \int_{V_0}^{V_1} \frac{dV}{V} = p_0 V_0 \log \frac{V_1}{V_0}$$

b.
$$W = \int_{V_0}^{V_1} p_0 V_0^k\, V^{-k}\, dV$$

$$= p_0 V_0^k \frac{V^{1-k}}{1-k} \Big|_{V_0}^{V_1} = \frac{p_0 V_0 - p_1 V_1}{k-1}$$

5. Two negative charges are separated by a distance s and the charges have magnitudes q and q'. Coulomb's law states that *the force of repulsion between these two charges is directly proportional to the product of the charges and inversely proportional to the square of the distance between them.* (If the proper units are chosen, the proportionality constant is unity.) Find the work required to move one charge a distance of $s/2$ from the other.

Solution. $F = \dfrac{qq'}{s^2}$

$$W = \int_{s}^{s/2} \left(-\frac{qq'}{s^2} \right) ds = \frac{qq'}{s} \Big|_s^{s/2} = \frac{qq'}{s}$$

PROBLEMS WITH ANSWERS

6. A hemispherical tank of radius r is full of gasoline of density w. What is the work done in pumping the gasoline out over the rim of the tank?

Ans. $\frac{1}{4}\pi w r^4$

7. The natural length of a spring is 4 in. and the force required to compress it is $F = 3x$, where x is the amount of compression in inches. Find the work done in compressing the spring until it is only 2.8 in. long.

Ans. 2.16 in.-lbs.

8. A conical tank full of water is 6 ft. deep (vertex down), and the top has a radius of 2 ft. Find the work required to empty the tank by pumping the water to a point 3 ft. above the top of the tank.

Ans. $12\pi w + 24\pi w = 36\pi w$ ft.-lbs.

9. A 500 lb. sandbag is being lifted vertically. If the sand leaks out at the rate of 1 lb. per foot of lift, how much work is done in a 20 ft. lift?

$$Ans. \quad W = \int_0^{20} \left(500 - \frac{1}{2}s\right) ds = 9900 \text{ ft.-lb.}$$

9.10 Pressure

When an area is submerged in a liquid, there is a pressure (force per unit area) on it due to the weight of the liquid above it. It is a fundamental principle of hydrostatics that the pressure $p = wh$, where w is the weight per unit volume of the liquid and h is the depth of submersion. It is also fundamental that the pressure is uniform in all directions. These principles and the calculus enable us to find the total force F due to liquid pressure on a submerged area of variable depth.

Think of a vertical area as being submerged, and consider a horizontal strip of length l and depth h (Fig. 9.22). Let this element be

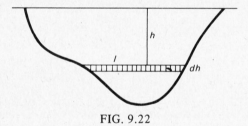

FIG. 9.22

of width dh. Then the area of the strip is $l\,dh$ and the pressure on it is wh. The force on this element is pressure × area, or $dF = whl\,dh$. Therefore, the total force F will be given by

$$F = \int_a^b whl\,dh \tag{1}$$

This integral may be evaluated as soon as l is expressed as a function of h.

Example 1. A plate in the form of an equilateral triangle of side $2a$ is submerged vertically in water until one edge is just in the surface of the water. Find the total force on one side of such a plate (Fig. 9.23).

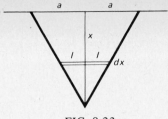

FIG. 9.23

Solution. If $2l$ is the length of an element submerged to a depth x

then $\dfrac{l}{a\sqrt{3}-x} = \dfrac{a}{a\sqrt{3}}$.

Then, $l = a - \dfrac{\sqrt{3}}{3}x$

The total force is then given by

$$F = \int_0^{a\sqrt{3}} wx\left(a - \frac{\sqrt{3}}{3}x\right)dx$$

$$= w\left[\frac{ax^2}{2} - \frac{\sqrt{3}\,x^3}{9}\right]_0^{a\sqrt{3}} = \frac{w}{2}a^3 \text{ lbs.}$$

Example 2. The center of a circular floodgate of radius 2 ft. in a reservoir is at a depth of 6 ft. Find the total force on the gate (Fig. 9.24).

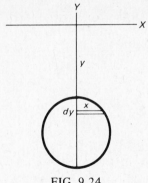

FIG. 9.24

Solution. Taking axes as shown, the equation of the circle is

$$x^2 + (y + 6)^2 = 4$$

At a depth y the width of an element is dy and the length is

$$2x = 2\sqrt{4 - (y + 6)^2}$$

The total force is then

$$F = 2w \int_{-4}^{-8} y \sqrt{4 - (y + 6)^2} \, dy$$

Set $y + 6 = z; dy = dz$

$$F = 2w \int_{2}^{-2} (z - 6)\sqrt{4 - z^2} \, dz$$

$$= 2w \left[-\frac{1}{3}(4 - z^2)^{3/2} - 3z\sqrt{4 - z^2} - 12 \, \mathrm{Sin}^{-1} \frac{z}{2} \right]_{2}^{-2} = 24\pi w \text{ lbs.}$$

Example 3. A plate in the form of the parabola $y = x^2$ is lowered vertically into water to a depth of one foot, vertex downward. Find the total force on one side of the plate (Fig. 9.25).

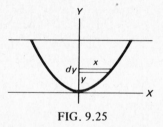

FIG. 9.25

Solution. The length of an element submerged $(1 - y)$ ft. is $2x$. The total force is then

$$F = 2w \int_{0}^{1} (1 - y)x \, dy$$

$$= 2w \int_{0}^{1} (1 - y)\sqrt{y} \, dy$$

$$= 2w \left[\frac{2}{3} y^{3/2} - \frac{2}{5} y^{5/2} \right]_{0}^{1} = \frac{8}{15} w \text{ lbs.}$$

PROBLEMS WITH SOLUTIONS

1. The cross section of a dam is shown in Fig. 9.26. Find the total force on the dam if it is 300 ft. wide.

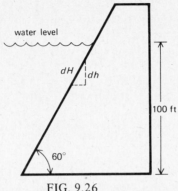

water level

dH dh

100 ft

60°

FIG. 9.26

Solution. Consider a strip of slant height dH, where $\dfrac{dH}{dh} = \dfrac{2\sqrt{3}}{3}$.

Then $F \displaystyle\int_0^{-100} \left(300\, wh\, \dfrac{2\sqrt{3}}{3} \right) dh = w\sqrt{3}\,(10^6)$ lbs.

2. The triangular plate shown in Fig. 9.27 has a base of 2 ft. and an altitude of 1 ft. The plate is immersed in water so that its base lies on the water's surface. Let the density of water be w. Find the total force exerted on one side of the plate.

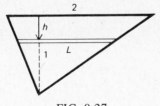

2

h

1   L

FIG. 9.27

Solution. Consider a thin slice of the plate at depth h and of thickness dh. Its length $L = 2(1 - h) = 2 - 2h$. Therefore,

$$F = w \int_0^1 h(2 - 2h)\, dh = 2w \int_0^1 (h - h^2)\, dh =$$

$$w \left(h^2 - \tfrac{2}{3} h^3 \right) \Big|_0^1 = \frac{1}{3} w \text{ lbs.}$$

Note that the answer is independent of the kind of triangle (isosceles, acute, or obtuse).

3. A triangular plate (Fig. 9.28) of base 2 ft. and altitude 1 ft. has its vertex on the surface of water with density w and its base parallel to this surface. Find the force on one side of the plate.

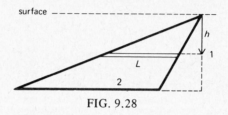

FIG. 9.28

Solution. $L = 2h$; therefore,

$$F = w \int_0^1 2h^2 \, dh = \frac{2}{3} w$$

Compare this answer with that of Problem 2. Note that the answer is also independent of the kind of triangle.

4. A cylindrical tank of radius r ft. is placed horizontally and filled with a liquid of density w. Find the force on one end of the tank (Fig. 9.29).

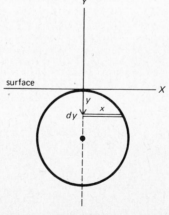

FIG. 9.29

Solution. $F = 2w \int xy \, dy$

$$= 2w \int_0^{-2r} y\sqrt{r^2 - (y+r)^2} \, dy$$

Let $y + r = z$, then $dy = dz$, and

$$F = 2w \int_r^{-r} (z - r) \sqrt{r^2 - z^2} \, dz$$

$$= 2w \left[-\frac{1}{3}(r^2 - z^2)^{3/2} - r\left(\frac{z}{2}\sqrt{r^2 - z^2} + \frac{r^2}{2}\operatorname{Sin}^{-1}\frac{z}{r}\right) \right]_r^{-r} = w\pi r^2$$

5. The cross section of an open gutter is a semicircle of radius r. The gutter of length L has closed ends; it is placed horizontally and is filled with water of density w. Find the total force on the gutter including ends (Fig. 9.30).

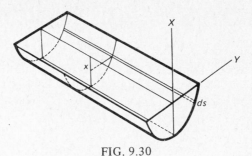

FIG. 9.30

Solution. $F_1 = \int xLw \, ds$ (on the curved surface)

$$= Lwr \int_0^r x(r^2 - x^2)^{-1/2} \, dx = Lwr^2$$

$$F_2 = 2w \int_0^r x2y \, dx \quad \text{(on the ends)}$$

$$= 4w \int_0^r x\sqrt{r^2 - x^2} \, dx = \frac{4}{3}wr^3$$

Total force on gutter is $F_1 + F_2 = Lwr^2 + \frac{4}{3}wr^3$.

PROBLEMS WITH ANSWERS

6. A trapezoid is submerged vertically in water with its upper edge 10 ft. below the surface and its lower edge 20 ft. below the surface. If the upper and lower bases of the trapezoid are, respectively, 8 ft. and 15 ft. long, find the total force on one face.

$$Ans. \quad F = \int_{10}^{20} wh \left(\frac{7h + 10}{10} \right) dh = 1783 \tfrac{1}{3} \ w$$

7. A 6 ft. × 8 ft. rectangular floodgate is placed vertically in water with the 6 ft. side in the surface of the water. Find the force on one side.

$$Ans. \quad 192 w \ \text{lbs.}$$

8. A cylindrical tank of radius 5 ft. is placed horizontally and is half full of gasoline that weighs w lbs./cu. ft. Find the force exerted on one end of the tank.

$$Ans. \quad \frac{250}{3} w \ \text{lbs.}$$

9. A hemispherical bowl 2 ft. in radius is filled with water. Find the total force exerted on the bowl.

$$Ans. \quad F = \int wx(2\pi \sqrt{4 - x^2}) \, ds = 8\pi w \ \text{lbs.}$$

9.11   Center of Mass

For a point mass m_1 lying at a distance r_1 from a line L, the **first moment** of the mass with respect to the line is defined in Fig. 9.31.

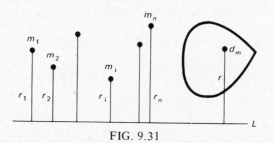

FIG. 9.31

$$\text{1st moment} = r_1 m_1 \qquad\qquad (1)$$

For n such particles, we have the sum

$$\text{1st moment} = r_1 m_1 + r_2 m_2 + \cdots + r_n m_n = \Sigma r_i m_i \qquad (2)$$

For a continuous mass distribution, this sum becomes an integral.

$$1\text{st moment} = \int r \, dm \tag{3}$$

where r represents the distance of the element of mass dm from the line L.

The **center of mass** measured from L is

$$\overline{r} = \text{center of mass} = \frac{\int r \, dm}{\int dm} \tag{4}$$

From (4), we can readily compute the center of mass for a given mass measured from a given line (or from a given plane in the case where the mass is three-dimensional). Sometimes the center of mass is called the **center of gravity**, or **c.g.**; for masses that are pure geometrical figures, the term **centroid** is often used. We shall use the abbreviation c.g. since its use rarely causes any confusion.

Case I. One-dimensional mass. Consider a wire in the shape of the curve $y = f(x)$. We modify (4) as follows, in order to compute the coordinates of the c.g. Now *mass equals density times volume*, but, in this case, the "volume" is the length of the curve; hence, mass = density × length, or

$$dm = \rho \, ds$$

The coordinates of the c.g. are

$$\overline{x} = \frac{\int \rho x \, ds}{\int \rho \, ds} \text{ and } \overline{y} = \frac{\int \rho y \, ds}{\int \rho \, ds} \tag{5}$$

Formulas (5) hold whether density ρ is a constant or not; in case ρ is a constant, it may be canceled.

Example 1. A wire of uniform density is in the form of a semicircle. Find its c.g. (Fig. 9.32).

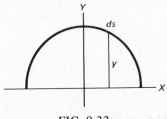

FIG. 9.32

Solution. Consider the wire as the upper half of the circle $x^2 + y^2 = r^2$. For reasons of symmetry $\bar{x} = 0$.

$$\bar{y} = \frac{\int \rho y \, ds}{\int \rho \, ds}$$

$$= \frac{2 \int_0^r y \sqrt{1 + y'^2} \, dx}{\pi r} = \frac{2r}{\pi}$$

Example 2. The density of a certain rod 1 ft. long varies directly as the square of the distance from one end. Find the c.g.

Solution. Place the rod on the X-axis, one end at the origin so that $\rho = kx^2$, then

$$\bar{x} = \frac{\int \rho y \, ds}{\int \rho \, ds}, \quad \bar{y} = 0$$

$$\bar{x} = \frac{\int_0^a kx^3 \, dx}{\int_0^a kx^2 \, dx} = \frac{3}{4} a$$

Case II. Two-dimensional mass. Consider a plate as an area with a given contour and the mass as density $\times$ area (Fig. 9.33). Thus,

$$dm = \rho \, dA = \rho (y_1 - y_2) \, dx = \rho (x_1 - x_2) \, dy$$

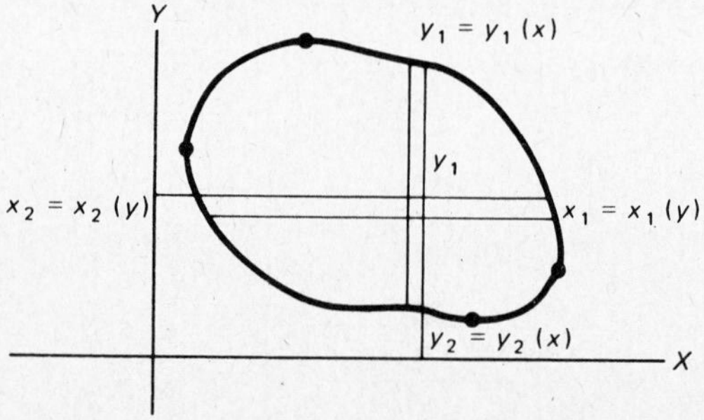

FIG. 9.33

and the coordinates of the c.g. are given by

$$\bar{x} = \frac{\int \rho x (y_1 - y_2)\, dx}{\int \rho (y_1 - y_2)\, dx} \quad \text{and} \quad \bar{y} = \frac{\int \rho y (x_1 - x_2)\, dy}{\int \rho (x_1 - x_2)\, dy} \tag{6}$$

Note that for $\bar{x}$, the area element is taken as $(y_1 - y_2)\, dx$, since all parts of this strip are at the same distance x from the Y-axis; similarly for $\bar{y}$ all parts of the strip $(x_1 - x_2)\, dy$ are at the same distance y from the X-axis.

In most problems commonly met, ρ is a constant and, in this case, it is possible to use the area element $(y_1 - y_2)\, dx$ in computing $\bar{y}$ by considering all of the mass of the strip as concentrated at the middle point, i.e., at $\frac{y_1 + y_2}{2}$. When this is done,

$$\bar{y} = \frac{\int \frac{1}{2} (y_1^2 - y_2^2)\, dx}{\int (y_1 - y_2)\, dx}$$

This may be used instead of the expression for $\bar{y}$ in (6).

Example 3. Find the c.g. of a semicircular plate of radius r and of uniform density (Fig. 9.34).

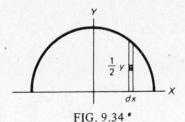

FIG. 9.34

Solution. Write $x^2 + y^2 = r^2$

$$\bar{x} = 0, \quad \bar{y} = \frac{\dfrac{1}{2} \displaystyle\int_{-r}^{r} y^2\, dx}{\displaystyle\int_{-r}^{r} y\, dx}$$

$$\bar{y} = \frac{1}{\pi r^2} \int_{-r}^{r} (r^2 - x^2)\, dx$$

$$= \frac{1}{\pi r^2} \left[r^2 x - \frac{x^3}{3} \right]_{-r}^{r} = \frac{4r}{3\pi}$$

Example 4. Find the c.g. of that area cut from the parabola $y^2 = 4px$ by the latus rectum (Fig. 9.35).

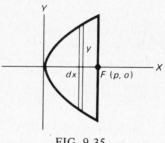

FIG. 9.35

Solution. Because of symmetry $\bar{y} = 0$.

$$\bar{x} = \frac{\displaystyle\int_0^p xy\, dx}{\displaystyle\int_0^p y\, dx} = \frac{\displaystyle\int_0^p x(2\sqrt{px})\, dx}{\displaystyle\int_0^p 2\sqrt{px}\, dx}$$

$$= \frac{\dfrac{2}{5} x^{5/2} \Big]_0^p}{\dfrac{2}{3} x^{3/2} \Big]_0^p} = \frac{3}{5} p$$

Case III. Three-Dimensional Mass. Here $dm = \rho\, dv$, but thus far we have discussed only solids of revolution and solids with known cross sections. But for a three-dimensional mass (volume) we take moments with respect to a plane and so define the first moment with respect to that plane. Suppose the area between the curve $y = f(x)$, $x = 0$, $y = c$, and $y = d$ is revolved about the Y-axis. Then taking slices perpendicular to the Y-axis as we did in computing the volume, we have $dm = \rho\pi x^2\, dy$. This mass is all at the same distance from the base

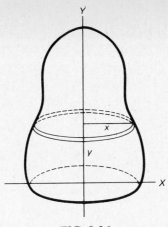

FIG. 9.36

plane perpendicular to the Y-axis. The first moment of this mass with respect to the base plane is (Fig. 9.36).

$$1\text{st moment} = \pi \int \rho y x^2 \, dy$$

The $\bar{y}$ of the c.g. is given by

$$\bar{y} = \frac{\int \rho y x^2 \, dy}{\int \rho x^2 \, dy} \tag{8}$$

Since the solid is one of revolution, this will completely locate the c.g. provided the solid is homogeneous or the density is a function of the distance from the axis of rotation.

Example 5. Find the c.g. of a homogeneous hemisphere of radius r (Fig. 9.37).
Solution. The c.g. lies on the diameter perpendicular to the base at a distance $\bar{y}$ above the base.

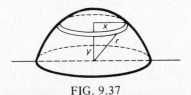

FIG. 9.37

$$\overline{y} = \frac{\int \pi y x^2 \, dy}{\int dv}$$

$$= \frac{\int_0^r \pi y (r^2 - y^2) \, dy}{\frac{2}{3} \pi r^3}$$

$$= \frac{3}{2r^3} \left[\frac{r^2 y^2}{2} - \frac{y^4}{4} \right]_0^r = \frac{3}{8} r$$

Example 6. Find the c.g. of a cone of radius r and altitude h, shown in Fig. 9.38. (Measure from the vertex.)

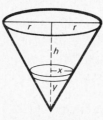

FIG. 9.38

Solution. $\overline{x} = 0$

$$\overline{y} = \frac{\int_0^h \pi y x^2 \, dy}{\int dv}$$

But $\dfrac{x}{y} = \dfrac{r}{h}$, or $x = \dfrac{r}{h} y$, and the volume of the cone is $\dfrac{1}{3} \pi r^2 h$; therefore,

$$\overline{y} = \frac{\int_0^h \pi y \left(\frac{r}{h} y \right)^2 \, dy}{\frac{1}{3} \pi r^2 h}$$

$$= \frac{3}{h^2} \frac{y^4}{4} \bigg|_0^h = \frac{3}{4} h$$

PROBLEMS WITH SOLUTIONS

1. The density of a rod L ft. long varies directly as the nth power of its distance from one end. Find the c.g.

Solution. $\rho = kx^n$

$$\bar{x} = \frac{k\displaystyle\int_0^L x^{n+1}\,dx}{k\displaystyle\int_0^L x^n\,dx} = \frac{\dfrac{k}{n+2}x^{n+2}\Big|_0^L}{\dfrac{k}{n+1}x^{n+1}\Big|_0^L} = \frac{n+1}{n+2}L$$

2. Find the c.g. of that portion of the area of the ellipse $\dfrac{x^2}{a^2} + \dfrac{y^2}{b^2} = 1$

which lies in the first quadrant.

Solution.

$$\bar{y} = \frac{\displaystyle\int_0^b y\left(\frac{a}{b}\sqrt{b^2-y^2}\,dy\right)}{\frac{1}{4}\pi ab} = \frac{4}{\pi b^2}\int_0^b y\sqrt{b^2-y^2}\,dy$$

$$= \frac{4}{\pi b^2}\left[-\frac{1}{3}(b^2-x^2)^{3/2}\right]_0^b = \frac{4b}{3\pi}$$

$$\bar{x} = \frac{\displaystyle\int_0^a x\left(\frac{b}{a}\sqrt{a^2-x^2}\right)dx}{\frac{1}{4}\pi ab} = \frac{4a}{3\pi}$$

3. Given a circular sector of radius r and angle 2θ with center at 0. Find the distance from 0 to the c.g. of the area of the sector (Fig. 9.39).

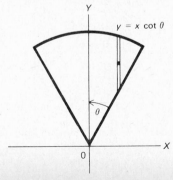

FIG. 9.39

Solution.

$$x^2 + y^2 = r^2 \quad \text{and} \quad y = x \cot \theta$$

$$\text{Area} = r^2 \theta$$

$$r^2 \theta \, \bar{y} = 2 \int_0^{r \sin \theta} \frac{(r^2 - x^2) - x^2 \cot^2 \theta}{2} \, dx$$

$$= r^2 x - \frac{x^3}{3}(1 + \cot^2 \theta) \Big|_0^{r \sin \theta}$$

$$= r^3 \sin \theta \left(1 - \frac{1}{3}\right)$$

$$= \frac{2}{3} r^3 \sin \theta$$

$$\bar{y} = \frac{2r \sin \theta}{3\theta}$$

4. Find the c.g. of a solid cylinder of radius r when the density varies directly as the distance from the bottom of the cylinder.
   Solution. $\bar{x} = 0$

$$\bar{y} = \frac{\displaystyle\int_0^h (ky) yx^2 \, dy}{\displaystyle\int_0^h (ky) x^2 \, dy}$$

$$= \frac{kr^2}{kr^2} \frac{\displaystyle\int_0^h y^2 \, dy}{\displaystyle\int_0^h y \, dy} = \frac{2h}{3}$$

5. Show that the c.g. of a straight wire of uniform density is at the center of the length.
   Solution. $ds = dx$

$$\bar{x} = \frac{\displaystyle\int_0^L x \, dx}{L} = \frac{x^2}{2L} \Big|_0^L = \frac{L}{2}$$

PROBLEMS WITH ANSWERS

In Problems 6 through 13, find the c.g. of the given masses.

6. A wire is bent in the form of a right triangle with legs a and b. Take the vertices at $(0,0)$, $(a,0)$, and $(0,b)$.

$$Ans. \quad x = \frac{\frac{1}{2}a^2 + \frac{1}{2}a\sqrt{a^2 + b^2}}{a + b + \sqrt{a^2 + b^2}}$$

$$y = \frac{\frac{1}{2}b^2 + \frac{1}{2}b\sqrt{a^2 + b^2}}{a + b + \sqrt{a^2 + b^2}}$$

7. A right-triangular area with legs a and b. Take the vertices at $(0,0)$, $(a,0)$, and $(0,b)$. (This is the point of intersection of the medians.)

$$Ans. \quad \overline{x} = \frac{1}{3}a, \overline{y} = \frac{1}{3}b$$

8. The first arch of the cycloid $x = a(\theta - \sin \theta)$, $y = a(1 - \cos \theta)$.

$$Ans. \quad \overline{x} = \pi a, \overline{y} = \frac{4}{3}a$$

9. The area under the first arch of the cycloid $x = a(\theta - \sin \theta)$, $y = a(1 - \cos \theta)$.

$$Ans. \quad \overline{x} = \pi a, \overline{y} = \frac{5}{6}a$$

10. The area under $y = \cos x$ between $x = -\dfrac{\pi}{2}$ and $x = \dfrac{\pi}{2}$.

$$Ans. \quad \overline{x} = 0, \overline{y} = \frac{\pi}{8}$$

11. A rectangle whether considered as an arc or an area.

$$Ans. \quad \text{c.g. at the geometric center}$$

12. The volume formed by rotating the area in the first quadrant under the parabola $y^2 = 4px$ between the vertex and the latus rectum about the Y-axis.

$$Ans. \quad \overline{x} = 0, \overline{y} = \frac{5}{6}p$$

13. A hemisphere of radius r, if density $\rho = ky$, where y is distance from the base plane.

$$Ans. \quad \overline{y} = \frac{8}{15}r$$

9.12 Moment of Inertia

Corresponding to (3), Section 9.11, for the **first moment** we define the **second moment** as

$$2\text{nd moment} = I = \int r^2 \, dm \tag{1}$$

This quantity plays an important part in the theory of rotating bodies and is called the **moment of inertia** of the body with respect to the line (called an axis) or to the plane.

Again, there are several cases according to the character of the mass considered. But these are treated in precisely the same way they were treated in computing the first moments, the only difference being that here the square of the distance is used instead of the first power of the distance.

Case I. One-dimensional mass. Let the curve $y = f(x)$, a wire, be given. To find the moment of inertia I of this wire with respect to the X-axis we write for (1),

$$I_x = \int \rho y^2 \, ds \tag{2}$$

Evaluate this integral in the usual way. I_y is defined similarly.

Example 1. Find the moment of inertia of a circumference of a circle about a diameter.

Solution. We consider $\rho = 1$ and write $x^2 + y^2 = r^2$. The moment of inertia of the whole circumference will be four times the moment of inertia of a quarter circumference, hence,

$$I_x = \int y^2 \, ds$$

$$= 4 \int_0^r y^2 \sqrt{1 + y'^2} \, dx$$

$$= 4 \int_0^r ry \, dx$$

$$= 4r \int_0^r \sqrt{r^2 - x^2} \, dx$$

$$= 4r \left[\frac{x}{2} \sqrt{r^2 - x^2} + \frac{r^2}{3} \sin^{-1} \frac{x}{r} \right]_0^r = \pi r^3$$

This is the moment of inertia of the mathematical circumference. If this arc is replaced by a wire of density ρ and mass $M = \rho s$, where $s = 2\pi r = $ length, then

$$I_x = \rho \pi r^3$$

But $\rho = \dfrac{M}{2\pi r}$; therefore, in terms of M we have

$$I_x = \frac{Mr^2}{2}$$

Case II. Two-dimensional mass. Here, the mass element is an area and the moments of inertia about the X- and Y-axes are, respectively,

$$I_x = \int \rho y^2 \, dA \tag{3}$$

$$I_y = \int \rho x^2 \, dA \tag{4}$$

where an area element is taken so that it all lies at the same distance from the given axis.

Example 2. Find I_x and I_y for the area common to the two curves $y = x^2$ and $y^2 = x$.

Solution. These two parabolas intersect in the points $(0,0)$ and $(1,1)$, and, because of conditions of symmetry, $I_x = I_y$.

$$I_x = \int_0^1 y^2 \left(\sqrt{y} - y^2\right) dy, \rho = 1$$

$$= \left[\frac{2}{7} y^{7/2} - \frac{y^5}{5}\right]_0^1 = \frac{3}{35}$$

For a plate of density $\rho = \dfrac{M}{1/3}$ (see Example 6, Section 9.4), this becomes

$$I_x = \frac{3}{35} \rho = \frac{9}{35} M$$

Example 3. Find the moment of inertia of a circular area about (a) a diameter and (b) a tangent.

Solution. Write $(x - r)^2 + y^2 = r^2$. Since this circle is tangent to the Y-axis at the origin, the answers are I_x and I_y.

(a) $I_x = \int \rho y^2 (x - r) \, dy, \rho = 1$

$$= 4 \int_0^r y^2 \sqrt{r^2 - y^2} \, dy$$

$$= 4 \left[\frac{y}{8} (2y^2 - r^2)\sqrt{r^2 - y^2} + \frac{r^4}{8} \mathrm{Sin}^{-1} \frac{y}{r} \right]_0^r = \frac{\pi r^4}{4} \text{ (area)}$$

(Formula 49, Appendix B)

$$= \frac{\rho \pi r^4}{4} = \frac{Mr^2}{4} \text{ (plate of mass } M)$$

(b) $I_y = \int x^2 y \, dx$

$$= 2 \int_0^{2r} x^2 \sqrt{r^2 - (x - r)^2} \, dx$$

Set $x - r = z$

$$I_y = 2 \int_{-r}^r (z + r)^2 \sqrt{r^2 - z^2} \, dz$$

$$= 2 \int_{-r}^r (z^2 + 2rz + r^2) \sqrt{r^2 - z^2} \, dz$$

$$= 2 \left[\frac{z}{8} (2z^2 - r^2) \sqrt{r^2 - z^2} + \frac{r^4}{8} \mathrm{Sin}^{-1} \frac{z}{r} \right.$$

$$\left. - \frac{2}{3} r(r^2 - z^2)^{3/2} + \frac{r^2 z}{2} \sqrt{r^2 - z^2} + \frac{r^4}{2} \mathrm{Sin}^{-1} \frac{z}{r} \right]_{-r}^r$$

(Formulas 44 and 49, Appendix B)

$$= \tfrac{5}{4} \pi r^4 \text{ (area)}$$

$$= \tfrac{5}{4} Mr^2 \text{ (plate)}$$

Note that the methods used in Example 3 have some involved integrations. It is possible to compute the moment of inertia about a diameter with comparative ease after introducing the idea of the moment of inertia of a mass (one- or two-dimensional) about an axis perpendicular to the plane of the mass. This is called the **polar moment of inertia** I_0 and is defined, as before, as the mass times the square of the distance, or

$$I_0 = \int (x^2 + y^2)\, dm \tag{5}$$

This can be broken up into the two parts:

$$I_0 = \int x^2\, dm + \int y^2\, dm \tag{6}$$

$$= I_y + I_x$$

Hence, when the axis is taken through the origin and perpendicular to the xy-plane, the moment of inertia about this axis is the sum of the regular moments about the X- and Y-axes. Let us return to the problem in Example 3(a) by considering the next example.

Example 4. Find the polar moment of inertia of a circle about an axis through the center (Fig. 9.40).

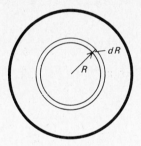

FIG. 9.40

Solution. Write $x^2 + y^2 = r^2$ and consider the area element in the form of a ring of width dR, the area of which is $2\pi R\, dR$. All of this area is at distance R from the axis; hence,

$$I_0 = \int_0^r R^2 (2\pi R)\, dR$$

$$= \frac{\pi R^4}{2} \bigg|_0^r = \frac{\pi r^4}{2} \text{ (area)}$$

$$= \frac{\rho \pi r^4}{2} = \frac{M r^2}{2} \text{ (plate)}$$

Clearly,

$$I_x = I_y = \tfrac{1}{2} I_0$$

$$= \frac{\pi r^4}{2} \text{ (area)}$$

$$= \frac{Mr^2}{4} \text{ (plate)}$$

This is the same result obtained in Example 3 (a), but is arrived at here with extreme ease.

It is also possible to get the moment of inertia about a tangent (Example 3 (b)) with ease.

Case III. Three-Dimensional Mass. The general case of a solid is discussed later in Chapter 11. For the present, if the mass considered is that of a solid of revolution about the Y-axis, we proceed to take disc elements perpendicular to the Y-axis. The mass of such an element will be $dm = \rho \pi x^2 \, dy$. The moment of inertia of this mass about the Y-axis is the polar moment of inertia of a circular plate about an axis through its center. By Example 4 above, the polar moment of inertia is equal to $\dfrac{Mr^2}{2}$, or $dI_y = \tfrac{1}{2} \rho \pi x^4 \, dy$. The total moment of inertia of the solid is the sum of the moments of inertia of all of the elementary discs, or

$$I_y = \tfrac{1}{2} \rho \pi \int x^4 \, dy \tag{7}$$

Example 5. The area of the ellipse $\dfrac{x^2}{a^2} + \dfrac{y^2}{b^2} = 1$ is revolved about the Y-axis. Find I_y for the solid thus generated.
Solution.

$$I_y = \frac{1}{2} \rho \pi \int x^4 \, dy$$

$$= \rho \pi \frac{a^4}{b^4} \int_0^b (b^2 - y^2)^2 \, dy$$

$$= \rho \pi \frac{a^4}{b^4} \left[b^4 y - \frac{2}{3} b^2 y^3 + \frac{1}{5} y^5 \right]_0^b$$

$$= \frac{8}{15} \rho \pi a^4 b$$

$$= \frac{8}{15} \pi a^4 b \quad (\rho = 1 \text{ for volume})$$

The volume generated is $\frac{4}{3} \pi a^2 b$; hence, $M = \rho V$ and $I_y = \frac{2}{5} M a^2$ (for solid of mass M).

The method used in Example 5 is sometimes called the **method of disc elements.** Another useful method makes use of *cylindrical elements.*

Example 6. Solve the problem in Example 5 using cylindrical elements.
Solution. The mass element is $\rho 2\pi xy \, dx$ and all of this mass is at the distance x from the Y-axis; therefore,

$$I_y = 4\pi\rho \int_0^a x^3 y \, dx$$

$$= 4\pi\rho \frac{b}{a} \int_0^a x^3 \sqrt{a^2 - x^2} \, dx$$

$$= 4\pi\rho \frac{b}{a} \left[-\frac{x^2}{3} (a^2 - x^2)^{3/2} \bigg|_0^a + \frac{2}{3} \int_0^a x (a^2 - x^2)^{3/2} \, dx \right]$$

(This integration was performed by parts with $u = x^2$ and $dv = x\sqrt{a^2 - x^2}$.)

$$= 4\pi\rho \frac{b}{a} \left[-\frac{x^2}{3} (a^2 - x^2)^{3/2} - \frac{2}{15} (a^2 - x^2)^{5/2} \right]_0^a$$

$$= \frac{8}{15} \rho \pi a^4 b$$

$$= \frac{8}{15} \pi a^4 b \quad \text{(for volume)}$$

$$= \frac{2}{5} M a^2 \quad \text{(for mass)}$$

Note that the integration by the method of discs was simpler than the integration in the second method where cylindrical elements were used. This is not always the case and the student should use the simpler method whichever it turns out to be.

Another important concept in mechanics is that of **radius of gyration** k defined as follows:

$$k = \sqrt{\frac{I}{M}} \tag{8}$$

This says that $I = Mk^2$, and so k can be interpreted as the fixed distance at which all of the mass M would have to be concentrated in order to yield the moment of inertia I. To calculate the radius of gyration, first calculate I (in terms of M) and substitute in (8).

PROBLEMS WITH SOLUTIONS

1. Find the moment of inertia of a uniformly thin rod of mass M and length L (a) about its end and (b) about its center.
   Solution.

   (a) $I_x = \int_0^L y^2 \frac{M}{L} dy = \frac{M}{L} \frac{y^3}{3} \Big|_0^L = \frac{ML^2}{3}$

   (b) $I_x = 2 \int_0^{L/2} y^2 \frac{M}{L} dy = \frac{2}{3} \frac{M}{L} y^3 \Big|_0^{L/2} = \frac{ML^2}{12}$

2. A rod 1 unit long extends along the X-axis from the origin to $(1, 0)$. If the density is $\rho = kx$, find I_y.

   Solution. $I_y = k \int_0^1 x^3 \, dx$

   $$= \frac{1}{4} k \ \text{(length)}$$

   $$dM = kx \, ds, \ ds = dx$$

   $$M = k \int_0^1 x \, dx = \frac{1}{2} k$$

   $$I_y = \frac{1}{2} M \ \text{(rod)}$$

3. A triangular plate of density $\rho = 1$ has vertices at $(0, 0)$, $(1, 0)$, and $(1, 1)$ (Fig. 9.41). Find I_x and I_y.

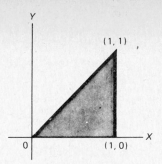

FIG. 9.41

Solution. $I_x = \displaystyle\int_0^1 y^2 \, dA = \int_0^1 y^2 (1 - y) \, dy$

$\quad\quad = \dfrac{1}{12} \text{ (area)}$

$\quad\quad = \dfrac{1}{6} M \text{ (plate)}$

$I_y = \displaystyle\int_0^1 x^2 y \, dx$

$\quad\quad = \displaystyle\int_0^1 x^3 \, dx$

$\quad\quad = \dfrac{1}{4} \text{ (area)}$

$\quad\quad = \dfrac{1}{2} M \text{ (plate)}$

4. A triangular plate of density $\rho = 1$ and with vertices at $(0,0)$, $(1, 0)$, and $(1, 1)$ is revolved about the Y-axis (Fig. 9.41). Find I_x and I_y for the resulting solid of revolution.

Solution. $I_x = \dfrac{1}{2} \pi \displaystyle\int x^4 \, dx$

$\quad\quad = \dfrac{\pi}{10} \text{ (area)}$

$$M = \frac{1}{3} \pi$$

$$I_x = \frac{3}{10} \text{ (solid)}$$

$$I_y = \frac{1}{2} \pi - \frac{1}{2} \pi \int y^4 \, dy$$

$$= \frac{2}{5} \pi \text{ (area)}$$

$$M = \frac{2}{3} \pi$$

$$I_y = \frac{3M}{5} \text{ (solid)}$$

5. The angular momentum of a solid about the X-axis is defined as $I_x \omega$, where ω is the angular velocity. From this definition, it is easy to see that I_x plays a role analogous to mass. Find the moment of inertia of a sphere of radius r about any diameter and find the angular momentum when the angular velocity is θ rad./sec.

Solution. $I_y = \dfrac{\pi}{2} \; \dfrac{M}{\frac{4}{3} \pi r^3} \displaystyle\int_{-r}^{r} (r^2 - y^2)^2 \, dy$

$$= \frac{3M}{8r^3} \left[r^4 y - \frac{2r^3 y^3}{3} + \frac{y^5}{5} \right]_{-r}^{r}$$

$$= \frac{3M}{8r^3} \cdot \frac{16r^5}{15} = \frac{2}{5} Mr^2$$

Angular momentum $= \dfrac{2\theta}{5} Mr^2$

6. Find the radius of gyration about the diameter of a homogeneous, solid sphere.
   Solution.

$$k = \sqrt{\frac{I}{M}}$$

$$= \sqrt{\frac{2Mr^2}{5M}} = \frac{r}{5} \sqrt{10}$$

PROBLEMS WITH ANSWERS

7. Find I_y for the line segment joining $(0,0)$ and $(L,0)$.

$$Ans. \quad \tfrac{1}{3}L^3$$

8. A plate of density ρ is in the form of the ellipse $\dfrac{x^2}{a^2} + \dfrac{y^2}{b^2} = 1$. Find I_x and I_y for this plate.

$$Ans. \quad I_x = \tfrac{1}{4}Mb^2, I_y = \tfrac{1}{4}Ma^2$$

9. A plate is in the form of a triangle with vertices at $(0,0)$, $(a,0)$, and $(0,b)$. Find I_x, I_y, and I_0.

$$Ans. \quad I_x = \tfrac{1}{6}Mb^2, I_y = \tfrac{1}{6}Ma^2, I_0 = \tfrac{1}{6}M(a^2+b^2)$$

10. Find the moment of inertia of a solid sphere of radius r about a diameter.

$$Ans. \quad \tfrac{2}{5}Mr^2 \text{ (See Example 5.)}$$

11. Find the moment of inertia of a solid cylinder of radius r and height h about (a) the axis and (b) a generator.

$$Ans. \quad \text{(a) } \tfrac{1}{2}Mr^2$$

$$\text{(b) } \tfrac{3}{2}Mr^2$$

12. Find I_y for the solid formed by revolving the area common to the two parabolas $y^2 = x$ and $y = x^2$ about the Y-axis.

$$Ans. \quad \tfrac{10}{27}M$$

9.13 Approximate Integration

It is impossible in some cases to express the indefinite integral in terms of the elementary functions. But it often happens, as in engineering problems, that only an approximation to the definite integral is needed. The only thing that may be known about a function may be just a table of values obtained through experimentation. In either case, it is possible to obtain an approximate value for $\int_a^b f(x)\,dx$.

Suppose this definite integral is interpreted as the area under $y = f(x)$ from $x = a$ to $x = b$ and suppose the interval $b - a$ is divided into n equal parts each of width $\Delta x = \dfrac{b-a}{n}$ by points of division x_i. Erect

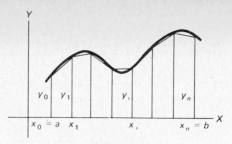

FIG. 9.42

ordinates y_i at these points and join the extremities by straight line segments (Fig. 9.42). Then the sum of the areas of the trapezoids thus formed will be an approximation to the area under the curve. This sum is

$$\frac{1}{2}(y_0 + y_1)\Delta x + \frac{1}{2}(y_1 + y_2)\Delta x + \cdots + \frac{1}{2}(y_{n-1} + y_n)\Delta x$$

$$= \frac{\Delta x}{2}(y_0 + 2y_1 + 2y_2 + \cdots + 2y_{n-1} + y_n)$$

This gives us the **trapezoidal rule**:

$$\int_{x_0}^{x_n} f(x)\,dx \cong \frac{\Delta x}{2}(y_0 + 2y_1 + 2y_2 + \cdots + 2y_{n-1} + y_n) \quad (1)$$

If $f(x)$ is given as a table, formula (1) may be applied if and only if the spacings between abscissa points are equal; if the spacings are unequal, the areas of the trapezoids can still be added together giving

$$\tfrac{1}{2}(y_0 + y_1)\Delta x_1 + \tfrac{1}{2}(y_1 + y_2)\Delta x_2 + \cdots + \tfrac{1}{2}(y_{n-1} + y_n)\Delta x_n$$

where $\Delta x_i = (x_i - x_{i-1})$. This will give the approximation sought.

Example 1. Evaluate $\int_0^2 x^2\,dx$ by the trapezoidal rule, using $n = 4$.
Solution. We note that this can be integrated, and so there would be no need to use a method of approximation. But as an exercise, we take $n = 4$ and $\Delta x = 0.5$ and write

$$\int_0^2 x^2\,dx \cong \frac{0.5}{2}\left(0 + \frac{1}{2} + 2 + \frac{9}{2} + 4\right) = \frac{11}{4}$$

The exact value is $\frac{8}{3}$ sq. units; the error is 0.08.

Example 2. Evaluate $\displaystyle\int_0^{\pi/2} \frac{\sin x}{x}\, dx$ by the trapezoidal rule, using $n = 3$.

Solution. The $\displaystyle\int \frac{\sin x}{x}\, dx$ cannot be expressed in terms of the elementary functions.

$$\int_0^{\pi/2} \frac{\sin x}{x}\, dx \cong \frac{\pi}{12}\left(1 + \frac{6}{\pi} + \frac{3\sqrt{3}}{\pi} + \frac{2}{\pi}\right) = 1.361$$

Example 3. In order to get an approximation to the area between a straight railroad track and a winding stream, measurements were made as in Fig. 9.43. Find approximately the area between the railroad and the stream.

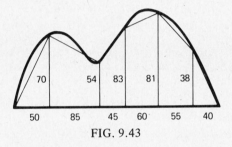

FIG. 9.43

Solution. $A \cong \frac{1}{2}(50)\,70 + \frac{1}{2}(70 + 54)(85)$

$$+ \frac{1}{2}(54 + 83)45 + \frac{1}{2}(83 + 81)60$$

$$+ \frac{1}{2}(81 + 38)55 + \frac{1}{2}(38)40 = 19{,}055 \text{ sq. ft.}$$

A definite integral may also be evaluated approximately by **Simpson's rule**:

$$\int_a^b f(x)\, dx \cong \frac{\Delta x}{3}\left[y_0 + 4(y_1 + y_3 + y_5 + \cdots)\right.$$

$$\left. + 2(y_2 + y_4 + y_6 + \cdots) + y_n\right] \qquad (2)$$

In applying Simpson's rule, it is necessary that the number of intervals n be even. The way to remember this rule is to note that when the first ordinate is called y_0 and the last ordinate y_n, n even, then the operations are

1. Add the first and last ordinates.

2. Add 4 times the sum of the ordinates with odd subscripts.

3. Add 2 times the sum of the ordinates with even subscripts (excluding, of course, y_0 and y_n).

4. Multiply the total sum by 1/3 of the common distance between ordinates.

This rule is developed by using parabolic arcs (instead of straight lines as in the trapezoidal rule) to approximate the curve. The parabolas used are those with axes parallel to the Y-axis, each one of them being passed through a consecutive set of three points on the curve of $y = f(x)$. Since each set of three points accounts for two intervals, the total number of intervals will be a multiple of 2, or an even number. In general, Simpson's rule gives a better approximation than the trapezoidal rule, where the number of intervals n is about the same.

Example 4. Apply Simpson's rule to the problem in Example 1.
Solution.

$$\int_0^2 x^2 \, dx = \frac{\Delta x}{3} \left[y_0 + 4(y_1 + y_3) + 2y_2 + y_4 \right]$$

$$= \frac{0.5}{3} \left[0 + 4 \left(\frac{1}{4} + \frac{9}{4} \right) + 2 + 4 \right] = \frac{8}{3} \text{ sq. units}$$

Note that Simpson's rule gives the exact answer in this case. (Simpson's rule gives the exact answer in case $y = ax^3 + bx^2 + cx + d$ with $n = 2$.)

Example 5. Evaluate $\int_0^{1.2} e^{-x^2} \, dx$ by Simpson's rule, using $n = 6$.
Solution.

$$\Delta x = 0.2, y_0 = 1, y_1 = 0.96079$$

$$y_2 = 0.85214, y_3 = 0.69768$$

$$y_4 = 0.52729, y_5 = 0.36788$$

$$y_6 = 0.23693 \text{ (From numerical tables of } e^{-t})$$

$$\int_0^{1.2} e^{-x^2} \, dx = 0.80675$$

PROBLEMS WITH SOLUTIONS

1. Estimate the value of $\int_0^2 \frac{dx}{\sqrt{1 + x^3}}$:

    a. By the trapezoidal rule, using $n = 2$

    b. By Simpson's rule, using $n = 2$

Solution.

a. $\displaystyle\int_0^2 \frac{dx}{\sqrt{1+x^3}} \cong \frac{1}{2}\left(1 + \frac{1}{2}\sqrt{2} + \frac{1}{3}\right) = 1.3738$

b. $\displaystyle\int_0^2 \frac{dx}{1+x^3} \cong \frac{1}{3}\left(1 + 2\sqrt{2} + \frac{1}{3}\right) = 1.3872$

2. Estimate the value of $\int_0^2 \sqrt{4+x^3}\,dx$:
   a. By the trapezoidal rule, using $n = 4$
   b. By Simpson's rule, using $n = 4$
   Solution.

   a. $\displaystyle\int_0^2 \sqrt{4+x^3}\,dx \cong \frac{5}{10}(1.000 + 2.031 + 2.236 + 2.716 + 1.732)$
   $$= 4.858$$

   b. $\displaystyle\int_0^2 \sqrt{4+x^3}\,dx \cong \frac{5}{30}(2.000 + 8.124 + 4.472 + 10.864 + 3.464)$
   $$= 4.821$$

3. Estimate $\int_0^{\pi/2} \sqrt{1+\cos^2 x}\,dx$ by Simpson's rule, using $n = 2$.

   Solution. $\displaystyle\int_0^{\pi/2} \sqrt{1+\cos^2 x}\,dx \cong \frac{\pi}{12}\left(\sqrt{2} + 4\sqrt{\frac{3}{2}} + 1\right)$
   $$= 1.9146$$

4. Use Simpson's rule to approximate the area of a circle of radius 1.
   Solution.

   $$A = 4\int_0^1 \sqrt{1-x^2}\,dx \cong 4\left\{\frac{0.1}{3}\,[1 + 4(0.995 + 0.954 + 0.866\right.$$
   $$+ 0.714 + 0.436) + 2(0.980 + 0.917$$
   $$+ 0.800 + 0.600) + 0] = 3.130$$

   To five places, the value is $\pi \cong 3.14159$.

PROBLEMS WITH ANSWERS

5. Find the area under the curve $y = \dfrac{1}{2+x}$ from $x = 0$ to $x = 2$ by
   (a) trapezoidal rule, using $n = 4$, (b) Simpson's rule, using $n = 4$,
   and (c) integration.

   Ans. (a) 0.69702
   (b) 0.69325
   (c) $\log 2 = 0.69315$

6. Evaluate $\int_0^4 e^{-x^2/2} \, dx$ by (a) trapezoidal rule with $\triangle x = \frac{1}{10}$ and
(b) Simpson's rule with $\triangle x = \frac{1}{10}$.

Ans.   (a) 0.3998
         (b) 0.3996

7. (a) Show that $4 \int_0^1 \dfrac{dx}{1 + x^2} = \pi$; (b) by applying Simpson's rule with
$n = 10$ to this integral show that $\pi \cong 3.1416$

8. A smooth curve is passed through the data of the following table.

x	0	1	2	3	4	5	6
y	0	1	1.2	1.6	2.3	2.4	0

Find the area under the curve by (a) the trapezoidal rule and
(b) Simpson's rule.

Ans.   (a) 8.5 sq. units
         (b) 9 sq. units

CHAPTER 10
PARTIAL DIFFERENTIATION

10.1 Partial Derivatives

In $z = f(x, y)$ let one of the independent variables, say y, be held fixed. When an increment is given to the other independent variable, the function z itself will experience a change. We write

$$z = f(x, y) \tag{1}$$

$$z + \Delta z = f(x + \Delta x, y) \tag{2}$$

$$\Delta z = f(x + \Delta x, y) - f(x, y) \tag{3}$$

$$\frac{\Delta z}{\Delta x} = \frac{f(x + \Delta x, y) - f(x, y)}{\Delta x} \tag{4}$$

This is the usual process of differentiation as applied to a function of a single variable; since y is held constant, z varies only with x. The **partial derivative** of z with respect to x is thus defined as

$$\frac{\partial z}{\partial x} = \lim_{\Delta x \to 0} \frac{f(x + \Delta x, y) - f(x, y)}{\Delta x} \tag{5}$$

Similarly,

$$\frac{\partial z}{\partial y} = \lim_{\Delta y \to 0} \frac{f(x, y + \Delta y) - f(x, y)}{\Delta y} \tag{6}$$

The curled ∂ is used so that this will clearly be distinguished from ordinary differentiation. The symbols f_x and f_y are also in general use for $\frac{\partial z}{\partial x}$ and $\frac{\partial z}{\partial y}$, respectively.

In the case of a function of a single independent variable $y = f(x)$ the differential dy was defined to be $dy = \frac{df}{dx} dx$. Where $z = f(x, y)$, the **total differential** dz is defined as

$$dz = \frac{\partial f}{\partial x} dx + \frac{\partial f}{\partial y} dy \tag{7}$$

255

If $z = f(x,y)$ and $x = x(t), y = y(t)$, then the **total derivative** of z with respect to t is

$$\frac{dz}{dt} = \frac{\partial f}{\partial x}\frac{dx}{dt} + \frac{\partial f}{\partial y}\frac{dy}{dt} \tag{8}$$

Straight d's are used here for $\dfrac{dz}{dt}, \dfrac{dx}{dt}$, and $\dfrac{dy}{dt}$, since t is the only independent variable; the partial symbols $\dfrac{\partial f}{\partial x}$, and $\dfrac{\partial f}{\partial y}$ are used to emphasize that, first of all, f is a function of two variables. In the event $z = f(x,y)$ and x and y are functions of two variables $x = x(u, v)$, and $y = y(u, v)$, then we write

$$\frac{\partial z}{\partial u} = \frac{\partial f}{\partial x}\frac{\partial x}{\partial u} + \frac{\partial f}{\partial y}\frac{\partial y}{\partial u} \tag{9}$$

$$\frac{\partial z}{\partial v} = \frac{\partial f}{\partial x}\frac{\partial x}{\partial v} + \frac{\partial f}{\partial y}\frac{\partial y}{\partial v} \tag{10}$$

For a function u of n variables, $u = f(x_1, x_2, \cdots, x_n)$, there are n first partial derivatives $\dfrac{\partial u}{\partial x_1}, \dfrac{\partial u}{\partial x_2}, \cdots, \dfrac{\partial u}{\partial x_n}$. The differential du is

$$du = \frac{\partial f}{\partial x_1}dx_1 + \frac{\partial f}{\partial x_2}dx_2 + \cdots + \frac{\partial f}{\partial x_n}dx_n \tag{11}$$

and relations analogous to those in (8), (9), and (10) may be written down.

Geometrically, $\dfrac{\partial z}{\partial x}$ represents the slope of the curve cut from the surface $z = f(x,y)$ by the plane $y = $ constant (Fig. 10.1). Evidently $\dfrac{\partial z}{\partial y}$ represents the slope of the plane curve $z = f(x, y)$, where $x = $ const. As in Fig. 10.2, the total differential dz represents the increment of the z-coordinate of the tangent plane to the surface $z = f(x,y)$ when x and y undergo independent changes; note the single variable case (Section 8.10).

Higher partial derivatives are obtained when the first partial derivatives are differentiated. Thus, we have $\dfrac{\partial^2 f}{\partial x^2}, \dfrac{\partial^2 f}{\partial x\, \partial y}, \dfrac{\partial^2 f}{\partial y^2}, \dfrac{\partial^3 f}{\partial x^3}$, etc. The symbol $\dfrac{\partial^2 f}{\partial x\, \partial y}$ is read *the second partial derivative of f, first with re-*

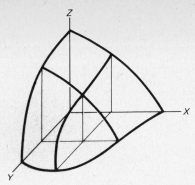

FIG. 10.1

spect to x, then with respect to y. Some authors write this $\dfrac{\partial^2 f}{\partial y\, \partial x}$, al-though this usually stands for *the second partial derivative of f, first with respect to y, then with respect to x.* But since, for most functions that are met with in the first course in the calculus the order of differentiation is immaterial, i.e., $\dfrac{\partial^2 f}{\partial x\, \partial y} = \dfrac{\partial^2 f}{\partial y\, \partial x}$, we shall pay little attention to the order in which the differentiation takes place. The symbols $f_{xx}, f_{xy}, f_{yy}, f_{xxy}$, etc., are also frequently used.

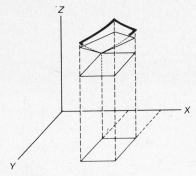

FIG. 10.2

Example 1. Given $z = e^x \sin y$, find $\dfrac{\partial z}{\partial x}, \dfrac{\partial z}{\partial y}, \dfrac{\partial^2 z}{\partial x^2}, \dfrac{\partial^2 z}{\partial x\, \partial y} = \dfrac{\partial^2 z}{\partial y\, \partial x}$, and $\dfrac{\partial^2 z}{\partial y^2}$.

Solution.
$$\frac{\partial z}{\partial x} = e^x \sin y \text{ and } \frac{\partial z}{\partial y} = e^x \cos y$$

$$\frac{\partial^2 z}{\partial x^2} = e^x \sin y$$

$$\frac{\partial^2 z}{\partial x \partial y} = e^x \cos y = \frac{\partial^2 z}{\partial y \partial x}$$

$$\frac{\partial^2 z}{\partial y^2} = -e^x \sin y$$

Example 2. Given $z = x^2 y + e^{xy}$, $x = t^2$, and $y = \log t$, find $\dfrac{dz}{dt}$.

Solution.
$$\frac{dz}{dt} = \frac{\partial z}{\partial x} \frac{dx}{dt} + \frac{\partial z}{\partial y} \frac{dy}{dt}$$

$$= (2xy + ye^{xy})2t + (x^2 + xe^{xy})\frac{1}{t}$$

Example 3. Given $z = x^2 + y^2$, where $x = \rho \cos \theta, y = \rho \sin \theta$, find $\dfrac{\partial z}{\partial \rho}$ and $\dfrac{\partial z}{\partial \theta}$.

Solution.
$$\frac{\partial z}{\partial \rho} = \frac{\partial z}{\partial x} \frac{\partial x}{\partial \rho} + \frac{\partial z}{\partial y} \frac{\partial y}{\partial \rho}$$

$$= 2x \cos \theta + 2y \sin \theta$$

$$= 2\rho \cos^2 \theta + 2\rho \sin^2 \theta = 2\rho$$

$$\frac{\partial z}{\partial \theta} = \frac{\partial z}{\partial x} \frac{\partial x}{\partial \theta} + \frac{\partial z}{\partial y} \frac{\partial y}{\partial \theta}$$

$$= -2x\rho \sin \theta + 2y \rho \cos \theta$$

$$= -2\rho^2 \sin \theta \cos \theta + 2\rho^2 \sin \theta \cos \theta = 0$$

PROBLEMS WITH SOLUTIONS

1.   Given $z = \log(x^2 + y^2)$, find

$$\frac{\partial z}{\partial x}, \frac{\partial z}{\partial y}, \frac{\partial^2 z}{\partial y \partial x} = \frac{\partial^2 z}{\partial x \partial y}, \frac{\partial^2 z}{\partial x^2}, \text{ and } \frac{\partial^2 z}{\partial y^2}$$

Solution.

$$\frac{\partial z}{\partial x} = \frac{2x}{x^2 + y^2}, \frac{\partial z}{\partial y} = \frac{2y}{x^2 + y^2}$$

$$\frac{\partial^2 z}{\partial x\, \partial y} = \frac{4xy}{(x^2 + y^2)^2} = \frac{\partial^2 z}{\partial y\, \partial x}, \frac{\partial^2 z}{\partial x^2} = \frac{-2x^2 + 2y^2}{(x^2 + y^2)^2}$$

$$\frac{\partial^2 z}{\partial y^2} = \frac{2x^2 - 2y^2}{(x^2 + y^2)^2}$$

2. Given $z = x^2 + e^y \cos x + \text{Tan}^{-1} xy$, $x = \sin t$, and $y = \log \sin t$, find $\frac{dz}{dt}$.

Solution. Since $\dfrac{dz}{dt} = \dfrac{\partial z}{\partial x} \dfrac{dx}{dt} + \dfrac{\partial z}{\partial y} \dfrac{dy}{dt}$, we get

$$\frac{dz}{dt} = \left(2x - e^y \sin x + \frac{y}{1 + x^2 y^2}\right) \cos t + \left(e^y \cos x + \frac{x}{1 + x^2 y^2}\right) \cot t$$

3. Given $z = x^2 - 2y^2$, $x = \rho \sin \theta$, and $y = \rho e^{-\theta}$, find $\dfrac{\partial z}{\partial \rho}$ and $\dfrac{\partial z}{\partial \theta}$.

Solution. $\dfrac{\partial z}{\partial \rho} = \dfrac{\partial z}{\partial x} \dfrac{\partial x}{\partial \rho} + \dfrac{\partial z}{\partial y} \dfrac{\partial y}{\partial \rho}$

$$= 2x \sin \theta - 4ye^{-\theta}$$

$$= 2\rho \sin^2 \theta - 4\rho e^{-2\theta}$$

$$\frac{\partial z}{\partial \theta} = \frac{\partial z}{\partial x} \frac{\partial x}{\partial \theta} + \frac{\partial z}{\partial y} \frac{\partial y}{\partial \theta}$$

$$= 2x\rho \cos \theta + 4y\rho e^{-\theta}$$

$$= \rho^2 \sin \theta \cos \theta + 4\rho^2 e^{-2\theta}$$

PROBLEMS WITH ANSWERS

4. Given $z = x^2 y^3 + \text{Tan}^{-1} \dfrac{y}{x}$, find $\dfrac{\partial z}{\partial x}$ and $\dfrac{\partial z}{\partial y}$.

Ans. $2xy^3 - \dfrac{y}{x^2 + y^2}$, $3x^2 y^2 + \dfrac{x}{x^2 + y^2}$

5. Given $z = x \log y + x \sin y$, find dz.

Ans. $(\log y + \sin y)\, dx + \left(\dfrac{x}{y} + x \cos y\right) dy$

6. Given $u = xyz$, find $\dfrac{\partial u}{\partial x}, \dfrac{\partial^2 u}{\partial x\, \partial y}$, and $\dfrac{\partial^2 u}{\partial z^2}$.　　Ans. $yz, z, 0$

7. Given $u = xyz$, $x = e^t$, $y = t$, and $z = \dfrac{1}{t}$, find $\dfrac{du}{dt}$.   *Ans.* e^t

8. Given $z = x^2 - y^2$, $x = \rho \cos \theta$, and $y = \rho \sin \theta$, find $\dfrac{\partial z}{\partial \rho}$ and $\dfrac{\partial z}{\partial \theta}$.

Ans. $2\rho \cos 2\theta$, $-2\rho^2 \sin 2\theta$

10.2 Implicit Differentiation

Let $z = f(x, y) = 0$ so that in reality y is given implicitly as a function of x; then

$$dz = \frac{\partial f}{\partial x} dx + \frac{\partial f}{\partial y} dy = 0 \tag{1}$$

Hence,

$$\frac{dy}{dx} = - \frac{\dfrac{\partial f}{\partial x}}{\dfrac{\partial f}{\partial y}} \tag{2}$$

Relation (2) affords a simple way of computing y', where $f(x, y) = 0$. This is essentially what was done in Section 5.6, although this notation was not used at the time.

Now consider a functional relation $F(x, y, z) = 0$ which defines z implicitly as a function of x and y. Since, in computing $\dfrac{\partial z}{\partial x}$, y is held constant so that z is considered a function of x only, we may apply (2) in finding $\dfrac{\partial z}{\partial x}$; thus, modified for $F(x, y, z) = 0$, (2) becomes

$$\frac{\partial z}{\partial x} = - \frac{\dfrac{\partial F}{\partial x}}{\dfrac{\partial F}{\partial z}} \tag{3}$$

Similarly,

$$\frac{\partial z}{\partial y} = - \frac{\dfrac{\partial F}{\partial y}}{\dfrac{\partial F}{\partial z}} \tag{4}$$

Example 1. Given $x^2 + y^2 \tan x + e^y = 0$, find $\dfrac{dy}{dx}$.

Solution.

$$\frac{dy}{dx} = -\frac{f_x}{f_y}$$

$$= -\frac{2x + y^2 \sec^2 x}{2y \tan x + e^y}$$

Example 2. Given $x^2 + y^2 \tan z - xe^z = 0$, find $\dfrac{\partial z}{\partial x}$ and $\dfrac{\partial z}{\partial y}$.

Solution.

$$\frac{\partial z}{\partial x} = -\frac{\dfrac{\partial F}{\partial x}}{\dfrac{\partial F}{\partial z}}$$

$$= \frac{2x - e^z}{y^2 \sec^2 z - xe^z}$$

$$\frac{\partial z}{\partial y} = -\frac{\dfrac{\partial F}{\partial y}}{\dfrac{\partial F}{\partial z}}$$

$$= \frac{2y \tan z}{y^2 \sec^2 z - xe^z}$$

PROBLEMS WITH SOLUTIONS

1. Given $x \log y + y \cos x + xe^y = 0$, find $\dfrac{dy}{dx}$.

    Solution.

$$\frac{dy}{dx} = -\frac{f_x}{f_y} = -\frac{\log y - y \sin x + e^y}{\dfrac{x}{y} + \cos x + xe^y}$$

2. Given $x^2 \sin z + y^2 z^2 - ze^x = 0$, find $\dfrac{\partial z}{\partial x}$ and $\dfrac{\partial z}{\partial y}$.

    Solution.

$$\frac{\partial z}{\partial x} = -\frac{\dfrac{\partial F}{\partial x}}{\dfrac{\partial F}{\partial z}} = -\frac{2x \sin z - ze^x}{x^2 \cos z + 2y^2 z - e^x}$$

$$\frac{\partial z}{\partial y} = -\frac{\dfrac{\partial F}{\partial y}}{\dfrac{\partial F}{\partial z}} = -\frac{2yz^2}{x^2 \cos z + 2y^2 z - e^x}$$

PROBLEMS WITH ANSWERS

3. Given $\dfrac{x^2}{a^2} + \dfrac{y^2}{b^2} + \dfrac{z^2}{c^2} = 1$, find $\dfrac{\partial z}{\partial x}$ and $\dfrac{\partial z}{\partial y}$.   *Ans.* $-\dfrac{c^2 x}{a^2 z}, -\dfrac{c^2 y}{b^2 z}$

4. Given $x \log y + y \log z + z \log x = 0$, find $\dfrac{\partial z}{\partial x}$ and $\dfrac{\partial z}{\partial y}$.

$$Ans. \quad -\frac{z^2 + xz \log y}{xy + xz \log x}, \; -\frac{xz + yz \log z}{y^2 + yz \log x}$$

10.3 Small Errors

The theory parallels that of the single variable case (Section 8.11). For small increments, the differential dz is approximately equal to the increment $\Delta z;$ that is,

$$\Delta z \cong dz = \frac{\partial z}{\partial x} dx + \frac{\partial z}{\partial y} dy \tag{1}$$

Example 1. The sides of a rectangular plate 6 ft. $\times$ 9 ft. are measured to be 6.01 ft. and 8.98 ft., respectively. Approximately what is the error made in the computed area?

Solution. $A = xy$

$$dA = y \, dx + x \, dy$$
$$= 6(-0.02) + 9(0.01)$$
$$= -0.03 \text{ sq. ft.}$$

Example 2. In determining the acceleration due to gravity, the formula $g = \dfrac{2s}{t^2}$ is sometimes used. Find the approximate maximum percentage error in the computed value of g, if s and t are measured only to within 1%.

Solution. $\log g = \log 2 + \log s - 2 \log t$

$$\frac{dg}{g} = \frac{ds}{s} - \frac{2}{t} dt$$
$$= 0.01 - 2(0.01)$$

Therefore, the maximum percentage error in g is $(0.01 + 0.02) = 3\%$, since the signs are to be taken so that $\dfrac{dg}{g}$ is largest.

PROBLEMS WITH SOLUTIONS

1. The generalized gas law is $PV = nRT$, where P = pressure, V = volume, T = absolute temperature, and n and R are constants of the system. What is the approximate maximum percentage error made in computing the volume, if the measurements of P and T are measured only to 1%?

   Solution.
   $$\log V = \log nR + \log T - \log P$$
   $$100\,\frac{dV}{V} = 100\,\frac{dT}{T} - 100\,\frac{dP}{P}$$
   $$\max\left(100\,\frac{dV}{V}\right) = \left|100\,\frac{dT}{T}\right| + \left|-100\,\frac{dP}{P}\right|$$
   $$= 1 + 1 = 2$$

   Hence, the maximum percentage error is 2%.

2. The stretching Δl of a wire of radius a and initial length l under a force F is given by $\Delta l = \dfrac{Fl}{aM}$, where M (Young's modulus) is a constant characteristic of the material of which the wire is made. Suppose there is at most a 1% error in F, a $\frac{1}{2}\%$ error in l, a 1% error in a, and M is exact. What is the maximum relative error in Δl?

   Solution.
   $$\log \Delta l = \log F + \log l - \log a - \log M$$
   $$\frac{d(\Delta l)}{\Delta l} = \frac{dF}{F} + \frac{dl}{l} - \frac{da}{a}$$
   $$\max\left[\frac{d(\Delta l)}{\Delta l}\right] = \left|\frac{dF}{F}\right| + \left|\frac{dl}{l}\right| + \left|-\frac{da}{a}\right|$$
   $$= 0.01 + 0.005 + 0.01 = 0.025\%$$

3. The moment of inertia about a diameter of a solid sphere of radius r and mass M is $I = \frac{2}{5}Mr^2$. What is the relative error made by using $M = 100 \pm 0.02$ and $r = 2 \pm 0.03$?

   Solution.
   $$\log I = \log \frac{2}{5} + \log M + 2\log r$$
   $$\frac{dI}{I} = \frac{dM}{M} + \frac{2\,dr}{r}$$

$$= \frac{0.02}{100} + \frac{0.06}{2}$$

$$= 0.0002 + 0.03 \cong 0.03$$

4. A surveyor determines that two legs of an obtuse triangle of terrain are 120.0 ± 0.1 ft. and 100.0 ± 0.2 ft., while the angle included between them is $120° \pm 1°$. If he uses the law of cosines to find the third side of the triangle, what is his approximate maximum error? *Solution.* The law of cosines states: $z^2 = x^2 + y^2 - 2xy \cos \theta$; therefore,

$$dz = \frac{1}{\sqrt{x^2 + y^2 - 2xy \cos \theta}} [(x - y \cos \theta) \, dx + (y - x \cos \theta) \, dy$$

$$+ xy \sin \theta \, d\theta]$$

$$\cong \frac{1}{190.79} \left\{ (170)(0.1) + (160)(0.2) + \frac{10,392\pi}{180} \right\}$$

$$\cong \frac{1}{190.79} \left\{ 17 + 32 + 181.8 \right\} \cong 1.2$$

PROBLEMS WITH ANSWERS

5. In order to determine the surface area of a conical filter, the slant height s and the radius r are measured as $s = 3 \pm 0.02$ in. and $r = 2 \pm 0.01$ in. What is the approximate maximum error made in computing the surface area S from $S = \pi rs$?

Ans. 0.07π sq. in.

6. The period of a pendulum is given by $t = \pi \sqrt{\dfrac{l}{g}}$ (see Problem 10, Section 8.11). What is the maximum percentage error made in computing g from this formula, if the percentage error in t and l are $\frac{1}{2}\%$ and 1%, respectively?

Ans. 2%

7. The volume of a rectangular box is given by $V = xyz$. What is the error made in computing the volume where the measurements are $x = 10 \pm 0.03$, $y = 6 \pm 0.01$, and $z = 8 \pm 0.02$?

Ans. error $\leqslant 3.44$ cu. units

8. The acceleration of a particle down an inclined plane is given by $a = g \sin \alpha$, where α is the angle of inclination of the plane. Suppose $g = 32 \pm 0.01$ and $\alpha = 30° \pm 1°$. What is the approximate maximum error in the computed value of a?

Ans. 0.49

10.4   Tangent Plane and Normal Line to a Surface

Let the equation of the plane tangent to the surface $F(x,y,z) = 0$ at the point $P(x_0, y_0, z_0)$ be

$$z - z_0 = A(x - x_0) + B(y - y_0) \tag{1}$$

This plane will be determined by the two tangent lines $z - z_0 = A(x - x_0), y = y_0$ and $z - z_0 = B(y - y_0), x = x_0$ gotten by cutting the plane (1) with the planes $y = y_0$ and $x = x_0$, respectively. But the slopes A and B of these two lines at P are $\dfrac{\partial z}{\partial x}$ and $\dfrac{\partial z}{\partial y}$ evaluated at (x_0, y_0, z_0).

Further, $\dfrac{\partial z}{\partial x} = -\dfrac{F_x}{F_z}$ and $\dfrac{\partial z}{\partial y} = -\dfrac{F_y}{F_z}$ (Section 10.2). Therefore, the **equation of the tangent plane** at P is

$$z - z_0 = -\frac{(F_x)_0}{(F_z)_0}(x - x_0) - \frac{(F_y)_0}{(F_z)_0}(y - y_0) \tag{2}$$

or, more symmetrically,

$$\left(\frac{\partial F}{\partial x}\right)_0 (x - x_0) + \left(\frac{\partial F}{\partial y}\right)_0 (y - y_0) + \left(\frac{\partial F}{\partial z}\right)_0 (z - z_0) = 0 \tag{3}$$

$$\text{(tangent plane)}$$

The symbol $\left(\dfrac{\partial F}{\partial x}\right)_0$ means that $\dfrac{\partial F}{\partial x}$ is first computed and then evaluated at $P(x_0, y_0, z_0)$.

If $F(x, y, z) = 0$ is solved for z, i.e., $z = f(x, y)$, then we have a special case of (3) which becomes

$$z - z_0 = \left(\frac{\partial f}{\partial x}\right)_0 (x - x_0) + \left(\frac{\partial f}{\partial y}\right)_0 (y - y_0) \tag{4}$$

Since $\left(\dfrac{\partial F}{\partial x}\right)_0, \left(\dfrac{\partial F}{\partial y}\right)_0$, and $\left(\dfrac{\partial F}{\partial z}\right)_0$ are the **direction numbers** (see Chapter 12) of a line perpendicular to the plane (3), the **equations of the line normal** to (3), and hence normal to the surface $F(x, y, z) = 0$, are (Fig. 10.3)

$$\frac{x - x_0}{\left(\dfrac{\partial F}{\partial x}\right)_0} = \frac{y - y_0}{\left(\dfrac{\partial F}{\partial y}\right)_0} = \frac{z - z_0}{\left(\dfrac{\partial F}{\partial z}\right)_0} \quad \text{(normal line)} \tag{5}$$

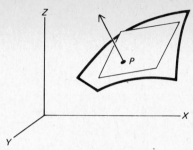

FIG. 10.3

For the surface $z = f(x, y)$, the equations of the normal line (5) reduce as a special case to

$$\frac{x - x_0}{\left(\frac{\partial f}{\partial x}\right)_0} = \frac{y - y_0}{\left(\frac{\partial f}{\partial y}\right)_0} = \frac{z - z_0}{-1} \tag{6}$$

Example 1. Find the equations of the tangent plane and normal line to $x^2 + 2y^2 - 4z^2 + 1 = 0$ at the point $(1, -1, -1)$.

Solution. $\dfrac{\partial F}{\partial x} = 2x, \dfrac{\partial F}{\partial y} = 4y$, and $\dfrac{\partial F}{\partial z} = -8z$

The equation of the tangent plane is

$$(x - 1) - 2(y + 1) + 4(z + 1) = 0$$

The equations of the normal line are

$$\frac{x - 1}{1} = \frac{y + 1}{-2} = \frac{z + 1}{4}$$

Example 2. Find the equations of the tangent plane and normal line to $z = xy - x^3 + y^2$ at the point $(0, -1, 1)$.

Solution. $\dfrac{\partial z}{\partial x} = y - 3x^2, \dfrac{\partial z}{\partial y} = x + 2y$

The equation of the tangent plane is

$$(z - 1) = -x - 2(y + 1)$$

The equations of the normal line are

$$\frac{x}{1} = \frac{y + 1}{2} = \frac{z - 1}{1}$$

PROBLEMS WITH SOLUTIONS

1. Find the equations of the tangent plane and normal line to the sphere $x^2 + y^2 + z^2 = 1$ at the point $(1/2, 1/2, \sqrt{2}/2)$, and show that the normal line passes through the center of the sphere.

    Solution. $\qquad f(x, y, z) = x^2 + y^2 + z^2 - 1 = 0$

    $$\frac{\partial F}{\partial x} = 2x, \quad \frac{\partial F}{\partial y} = 2y, \text{ and } \frac{\partial F}{\partial z} = 2z$$

    Hence, the equation of the tangent plane is

    $$\left(x - \frac{1}{2}\right) + \left(y - \frac{1}{2}\right) + \sqrt{2}\left(z - \frac{\sqrt{2}}{2}\right) = 0$$

    The normal line has the equations

    $$x - \frac{1}{2} = y - \frac{1}{2} = \frac{\sqrt{2}}{2}\left(z - \frac{\sqrt{2}}{2}\right)$$

    Setting $x = y = z = 0$ yields $-\frac{1}{2} = -\frac{1}{2} = -\frac{1}{2}$, which shows that the normal line passes through the center of the sphere.

2. Find the equations of the tangent plane and normal line to the surface $x + y + z = 1$ at the point $(1/3, 1/3, 1/3)$.

    Solution. $\qquad \dfrac{\partial F}{\partial x} = \dfrac{\partial F}{\partial y} = \dfrac{\partial F}{\partial z} = 1$

    Hence, the tangent plane is $\left(x - \frac{1}{3}\right) + \left(y - \frac{1}{3}\right) + \left(z - \frac{1}{3}\right) = 0$, or $x + y + z = 1$. The normal line is given by

    $$x - \tfrac{1}{3} = y - \tfrac{1}{3} = z - \tfrac{1}{3}$$

3. Find the equations of the tangent plane and normal line to the surface $x^2 - \dfrac{y^2}{4} = z - 1$ at the point $(0, 0, 1)$.

    Solution.

    $$\frac{\partial F}{\partial x} = 2x, \quad \frac{\partial F}{\partial y} = -\frac{y}{2}, \text{ and } \frac{\partial F}{\partial z} = -1$$

    Hence, the tangent plane is $(z - 1) = 0$. The equations of the normal line are $x = 0, y = 0$.

4. Find the equations of the tangent plane and normal line to the elliptic paraboloid $\dfrac{x^2}{a^2} + \dfrac{y^2}{b^2} - cz = 0$ at the point $\left(a, b, \dfrac{2}{c}\right)$.

Solution.

$$\frac{\partial F}{\partial x} = \frac{2x}{a^2}, \frac{\partial F}{\partial y} = \frac{2y}{b^2}, \frac{\partial F}{\partial z} = -c$$

The equation of the tangent plane is

$$\frac{2}{a}(x - a) + \frac{2}{b}(x - b) - (cz - 2) = 0$$

Equations of the normal line are

$$\frac{a(x - a)}{2} = \frac{b(y - b)}{2} = \frac{z - 2/c}{-c}$$

PROBLEMS WITH ANSWERS

5. Find the equations of the tangent plane and normal line to $xyz - 1 = 0$ at $\left(2, 3, \frac{1}{6}\right)$.

$$\text{Ans.} \quad 3x + 2y + 36z - 18 = 0, \frac{x - 2}{3} = \frac{y - 3}{2} = \frac{z - \frac{1}{6}}{36}$$

6. Find the equations of the tangent plane and normal line to $z = x^{1/2} + y^{1/2}$ at the point $(4, 9, 5)$.

$$\text{Ans.} \quad 3x + 2y - 12z + 30 = 0, \frac{x - 4}{3} = \frac{y - 9}{2} = \frac{z - 5}{-12}$$

10.5 Tangent Line and Normal Plane to a Skew Curve

A space curve not lying in a plane is called a *skew curve*. The equations of such a curve C will be in parametric form (see also Chapter 12)

$$x = f(t), \ y = g(t), \ z = h(t) \tag{1}$$

Let the two points $P(x, y, z)$ and $Q(x + \Delta x, y + \Delta y, z + \Delta z)$ lie on the curve (1). The **direction cosines** of PQ are proportional to $\Delta x, \Delta y$, and Δz. In the limit, as $Q \longrightarrow P$, the secant line PQ becomes the tangent line at P with direction numbers dx, dy, and dz, or $\frac{dx}{dt}, \frac{dy}{dt}$, and $\frac{dz}{dt}$. Therefore, the **equations of the line tangent** to the skew curve at (x_0, y_0, z_0) are

$$\frac{x - x_0}{\left(\dfrac{dx}{dt}\right)_0} = \frac{y - y_0}{\left(\dfrac{dy}{dt}\right)_0} = \frac{z - z_0}{\left(\dfrac{dz}{dt}\right)_0} \quad \text{(tangent line)} \tag{2}$$

The **equation of the plane normal** to the curve C at (x_0, y_0, z_0) is

$$\left(\frac{dx}{dt}\right)_0 (x - x_0) + \left(\frac{dy}{dt}\right)_0 (y - y_0) + \left(\frac{dz}{dt}\right)_0 (z - z_0) = 0 \qquad (3)$$

If the skew curve C is given as the intersection of two surfaces $F(x, y, z) = 0$ and $G(x, y, z) = 0$, then the tangent line to C lies in the two tangent planes to $F = 0$ and $G = 0$. At $P(x_0, y_0, z_0)$, the equations of the tangent planes to $F = 0$ and $G = 0$ are, respectively (Fig. 10.4),

$$\left.\begin{array}{l}\left(\dfrac{\partial F}{\partial x}\right)_0 (x - x_0) + \left(\dfrac{\partial F}{\partial y}\right)_0 (y - y_0) + \left(\dfrac{\partial F}{\partial z}\right)_0 (z - z_0) = 0 \\[4mm] \left(\dfrac{\partial G}{\partial x}\right)_0 (x - x_0) + \left(\dfrac{\partial G}{\partial y}\right)_0 (y - y_0) + \left(\dfrac{\partial G}{\partial z}\right)_0 (z - z_0) = 0\end{array}\right\}\begin{array}{l}(4)\\[6mm]\text{tangent}\\\text{line}\\(5)\end{array}$$

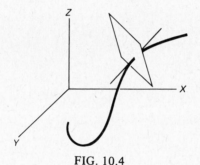

FIG. 10.4

Together, (4) and (5) are the equations of the tangent line. They may be written in symmetric form by computing the direction numbers

$$l : m : n = \begin{vmatrix}\left(\dfrac{\partial F}{\partial y}\right)_0 & \left(\dfrac{\partial F}{\partial z}\right)_0 \\[4mm] \left(\dfrac{\partial G}{\partial y}\right)_0 & \left(\dfrac{\partial G}{\partial z}\right)_0\end{vmatrix} : - \begin{vmatrix}\left(\dfrac{\partial F}{\partial x}\right)_0 & \left(\dfrac{\partial F}{\partial z}\right)_0 \\[4mm] \left(\dfrac{\partial G}{\partial x}\right)_0 & \left(\dfrac{\partial G}{\partial z}\right)_0\end{vmatrix} : \begin{vmatrix}\left(\dfrac{\partial F}{\partial x}\right)_0 & \left(\dfrac{\partial F}{\partial y}\right)_0 \\[4mm] \left(\dfrac{\partial G}{\partial x}\right)_0 & \left(\dfrac{\partial G}{\partial y}\right)_0\end{vmatrix}$$

and writing

$$\frac{x - x_0}{l} = \frac{y - y_0}{m} = \frac{z - z_0}{n} \qquad (6)$$

The equation of the normal plane would be

$$l(x - x_0) + m(y - y_0) + n(z - z_0) = 0 \quad \text{(normal plane)} \qquad (7)$$

Example 1. Find the equations of the tangent line and normal plane to the space curve $x = t, y = 2t^2, z = t^3 - t$ at the point for which $t = 1$.

Solution. $\quad\quad \dfrac{dx}{dt} = 1, \dfrac{dy}{dt} = 4t$, and $\dfrac{dz}{dt} = 3t^2 - 1$

The equations of the tangent line are

$$\frac{x - 1}{1} = \frac{y - 2}{4} = \frac{z}{2}$$

The equation of the normal plane is

$$(x - 1) + 4(y - 2) + 2z = 0$$

Example 2. Find the equations of the tangent line and normal plane to the curve of intersection of $z = x^2 + y^2$ and $x^2 + y^2 = 2$ at the point $(1, 1, 2)$.

Solution.

$$F \equiv x^2 + y^2 - z = 0$$

$$G \equiv x^2 + y^2 - 2 = 0$$

$$F_x = 2x, F_y = 2y, F_z = -1$$

$$G_x = 2x, G_y = 2y, G_z = 0$$

The equations of the tangent line are

$$2(x - 1) + 2(y - 1) - (z - 2) = 0$$

$$(x - 1) + (y - 1) = 0$$

Here $l : m : n = 1 : -1 : 0$ and the line is parallel to the xy-plane; hence, the symmetric form would not be used. The equation of the normal plane is $(x - 1) - (y - 1) = 0$.

PROBLEMS WITH SOLUTIONS

In Problems 1 through 4, find the equations of the tangent line and normal plane to the following skew curves at the indicated points.

1. $y = x^2, x^2 + z^2 = 1$, at $(1/2, 1/4, \frac{1}{2}\sqrt{3}\,)$.

Solution.

$$F(x, y, z), \ y - x^2 = 0 \text{ and } G(x, y, z) = x^2 + z^2 - 1 = 0$$

$$\frac{\partial F}{\partial x} = -2x, \frac{\partial F}{\partial y} = 1, \text{ and } \frac{\partial F}{\partial z} = 0$$

$$\frac{\partial G}{\partial x} = 2x, \frac{\partial G}{\partial y} = 0, \text{ and } \frac{\partial G}{\partial z} = 2z$$

Hence,

$$l : m : n = \begin{vmatrix} 1 & 0 \\ 0 & \sqrt{3} \end{vmatrix} : - \begin{vmatrix} -1 & 0 \\ 1 & \sqrt{3} \end{vmatrix} : \begin{vmatrix} -1 & 1 \\ 1 & 0 \end{vmatrix}$$

$$= \sqrt{3} : \sqrt{3} : -1$$

Tangent line: $\dfrac{x - 1/2}{\sqrt{3}} = \dfrac{y - 1/4}{\sqrt{3}} = \dfrac{\sqrt{3}}{2} - z$

Normal plane: $\sqrt{3}\left(x - \dfrac{1}{2}\right) + \sqrt{3}\left(y - \dfrac{1}{4}\right) + \left(\dfrac{\sqrt{3}}{2} - z\right) = 0$

2. $z = x^2 + y^2$, $z^2 + x^2 = 5$, at $(-1, 1, 2)$.
    Solution.

(a) $G(x, y, z) = z - x^2 - y^2 = 0$ and $F(x, y, z) = z^2 + x^2 - 5 = 0$

$$\frac{\partial G}{\partial x} = -2x, \quad \frac{\partial G}{\partial y} = -2y, \text{ and } \frac{\partial G}{\partial z} = 1$$

$$\frac{\partial F}{\partial x} = 2x, \quad \frac{\partial F}{\partial y} = 0, \text{ and } \frac{\partial F}{\partial z} = 2z$$

At $(-1, 1, 2)$,

$$\frac{\partial G}{\partial x} = 2, \quad \frac{\partial G}{\partial y} = -2, \text{ and } \frac{\partial G}{\partial z} = 1$$

$$\frac{\partial F}{\partial x} = -2, \quad \frac{\partial F}{\partial y} = 0, \text{ and } \frac{\partial F}{\partial z} = 4$$

Hence, we have

$$l : m : n = \begin{vmatrix} -2 & 1 \\ 0 & 4 \end{vmatrix} : - \begin{vmatrix} 2 & 1 \\ -2 & 4 \end{vmatrix} : \begin{vmatrix} 2 & -2 \\ -2 & 0 \end{vmatrix}$$

$$= -8 : -10 : -4$$

Tangent line: $\dfrac{x + 1}{4} = \dfrac{y - 1}{5} = \dfrac{z - 2}{2}$

Normal plane: $4(x + 1) + 5(y - 1) + 2(z - 2) = 0$

3. $x(t) = t$, $y(t) = t^2$, $z(t) = t^3$, at $(1, 1, 1)$.
    Solution.

$$\frac{dx}{dt} = 1, \quad \frac{dy}{dt} = 2t, \text{ and } \frac{dz}{dt} = 3t^2$$

Tangent line: $\dfrac{x-1}{1} = \dfrac{y-1}{2} = \dfrac{z-1}{3}$

Normal plane: $(x-1) + 2(y-1) + 3(z-1) = 0$
Hence, $x + 2y + 3z = 6$

4. $x(t) = t$, $y(t) = 2t$, $z(t) = 3t$, at the point where $t = -1$.
Solution.

$\dfrac{dx}{dt} = 1$, $\dfrac{dy}{dt} = 2$, and $\dfrac{dz}{dt} = 3$

Tangent line: $\dfrac{x+1}{1} = \dfrac{y+2}{2} = \dfrac{z+3}{3}$

Normal plane: $x + 1 + 2(y+2) + 3(z+3) = 0$
Hence, $x + 2y + 3z = -6$

PROBLEMS WITH ANSWERS

Find the equations of (a) the tangent line and (b) the normal plane to the following skew curves at the indicated points.

5. $x = t^2 + t$, $y = 2t - 8$, $z = t^4 - 6t^2$, at the point for which $t = 2$.

Ans. (a) $\dfrac{x-6}{5} = \dfrac{y+4}{2} = \dfrac{z+8}{8}$

(b) $5x + 2y + 8z + 42 = 0$

6. $x = 2e^t$, $y = e^{-t}$, $z = 3e^t + e^{-t}$, at $(2, 1, 4)$.

Ans. (a) $\dfrac{x-2}{2} = \dfrac{y-1}{-1} = \dfrac{z-4}{2}$

(b) $2x - y + 2z - 11 = 0$

7. $z = x^2 + y^2$, $z^2 + x^2 = 5$, at $(-1, 1, 2)$.

Ans. (a) $\dfrac{x+1}{4} = \dfrac{y-1}{5} = \dfrac{z-2}{2}$

(b) $4(x+1) + 5(y-1) + 2(z-2) = 0$

10.6 Related Rates

If, in a function of several variables, say $z = f(x,y)$, the variables x and y are themselves functions of another variable t, then z is a function of t and may be differentiated with respect to t (Section 10.1).

$$\frac{dz}{dt} = \frac{\partial f}{\partial x}\frac{dx}{dt} + \frac{\partial f}{\partial y}\frac{dy}{dt} \tag{1}$$

This relation is useful in solving problems in related rates. (See Section 8.8 for the one independent variable case.)

Example. The radius r of a cone is decreasing at the rate of 2 ft./sec., and the height h is increasing at the rate of 3 ft./sec. How fast is the volume changing when $r = 6$ ft. and $h = 10$ ft.?

Solution.
$$V = \tfrac{1}{3} \pi r^2 h$$

$$\frac{dV}{dt} = \frac{1}{3} \pi \left(2rh \frac{dr}{dt} + r^2 \frac{dh}{dt} \right)$$

$$= \tfrac{1}{3} \pi \left[2 \times 6 \times 10(-2) + 36 \times 3 \right]$$

$$= -44 \pi \text{ cu. ft./sec.}$$

PROBLEMS WITH SOLUTIONS

1. Two cars leave the same point at the same time. One travels due east at 60 mi./hr.; the other travels due south at 45 mi./hr. After 4 hr., how fast is the distance between the cars increasing?

   Solution. Since the distance z separating the cars is the hypotenuse of a right triangle, $z = \sqrt{x^2 + y^2}$. Hence,

$$\frac{dz}{dt} = \frac{1}{\sqrt{x^2 + y^2}} \left(x \frac{dx}{dt} + y \frac{dy}{dt} \right)$$

$$x(t) = 60t, \frac{dx}{dt} = 60, \text{ and } x(4) = 240$$

$$y(t) = 45t, \frac{dy}{dt} = 45, \text{ and } y(4) = 180$$

   Therefore, after 4 hr.,

$$\frac{dz}{dt} = \frac{1}{\sqrt{240^2 + 180^2}} (240 \times 60 + 180 \times 45)$$

$$= \frac{1}{300} (22,500) = 75 \text{ mi./hr.}$$

2. A plane flying 500 mi./hr. due east on a level course at an altitude of 1 mile flies directly over a ship speeding due north at 50 mi./hr. How fast are they separating 1 hr. later?

   Solution.

$$z = \sqrt{x^2 + y^2 + 1}$$

$$\frac{dz}{dt} = \frac{1}{\sqrt{x^2 + y^2 + 1}} \left(x \frac{dx}{dt} + y \frac{dy}{dt} \right)$$

$$= \frac{1}{502.5} (50 \times 50 + 500 \times 500) \cong 502.5 \text{ mi./hr.}$$

3. The length and width of a rectangular prism are decreasing at the rates of 2 units/sec. and 3 units/sec., respectively. How fast must the height be increasing if the volume remains constant? (Let $x, y,$ and z represent the dimensions of the box at time t.)

Solution.
$$0 = \frac{dV}{dt} = yz \frac{dx}{dt} + xz \frac{dy}{dt} + xy \frac{dz}{dt}$$

$$= -yz(2) - xz(3) + xy \frac{dz}{dt}$$

$$\frac{dz}{dt} = \frac{z(2y + 3x)}{xy}$$

4. The volume of a gas under pressure is directly proportional to the absolute temperature and inversely proportional to the pressure. Write the expression for the rate of change of volume.

Solution. $V = \dfrac{kT}{P}$

Thus, $\dfrac{\partial V}{\partial T} = \dfrac{k}{P}$ and $\dfrac{\partial V}{\partial P} = -\dfrac{kT}{P^2}$

Therefore, $\dfrac{dV}{dt} = \dfrac{k}{P} \dfrac{dT}{dt} - \dfrac{kT}{P^2} \dfrac{dP}{dt}$

5. How fast is the area of a triangle changing when sides originally a and b are changing at the same rate, and the angle θ between these two sides is changing at 2 rad./sec.?

Solution. $A = \dfrac{1}{2} ab \sin \theta$

$$\frac{dA}{dt} = \frac{1}{2} \left(b \sin \theta \frac{da}{dt} + a \sin \theta \frac{db}{dt} + ab \cos \theta \frac{d\theta}{dt} \right)$$

$$= \frac{1}{2} \left[\frac{da}{dt} (a + b) \sin \theta + 2ab \cos \theta \right]$$

PROBLEMS WITH ANSWERS

6. How fast is the area of a triangle increasing if the altitude increases at the rate of 3 in./sec. while the base increases at the rate of 4 in./sec.?

Ans. $\dfrac{3}{2} b + 2h$ sq. in./sec.

7. At any time t, the dimensions of a rectangular box are x, y, and z. If the volume remains constant, and if y and z each increase at the rate of 2 ft./sec., how must x change?

Ans. x must decrease at the rate of $\dfrac{2x(y+z)}{yz}$ ft./sec.

10.7 Directional Derivative

We have seen that $\dfrac{\partial z}{\partial x}$ and $\dfrac{\partial z}{\partial y}$ give the rate of change of z in the directions of the X- and Y-axes. The rate of change of z in any direction can be found as follows: Let P be a point on the surface $z = f(x,y)$, and pass a plane through P perpendicular to the xy-plane (Fig. 10.5). This

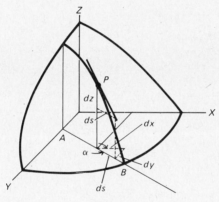

FIG. 10.5

plane cuts out a curve on $z = f(x,y)$, and the slope of this curve is the quantity sought. The trace AB of this plane in the xy-plane will specify the direction. Let the **direction cosines** of AB be $\cos \alpha$ and $\cos \beta = \sin \alpha$. Now,

$$\frac{dz}{ds} = \frac{\partial f}{\partial x}\frac{dx}{ds} + \frac{\partial f}{\partial y}\frac{dy}{ds} \tag{1}$$

where s is measured in the direction AB, but $\dfrac{dx}{ds} = \cos \alpha$ and $\dfrac{dy}{ds} = \sin \alpha$. Therefore, the rate of change of z in the direction AB is

$$\frac{dz}{ds} = \frac{\partial f}{\partial x}\cos \alpha + \frac{\partial f}{\partial y}\sin \alpha \tag{2}$$

This is called the **directional derivative** of z and represents the slope of the tangent line to the surface at any point P in the direction determined by α. When the point P is particularized, the partial derivatives $\dfrac{\partial f}{\partial x}$ and $\dfrac{\partial f}{\partial y}$ can be evaluated. (See also Chapter 12.)

Example. Find the rate of change of $z = x^2 - y^2$ in the direction $\alpha = 30°$ at the point $(2, 1, 3)$.

Solution.

$$\frac{\partial z}{\partial x} = 2x \text{ and } \frac{\partial z}{\partial y} = -2y$$

$$\frac{dz}{ds} = 2x \cos 30° - 2y \sin 30°$$

$$= 2\sqrt{3} - 1$$

PROBLEMS WITH SOLUTIONS

1. Find the directional derivative of z in the direction α at the given point $P(x, y, z)$.

   a. $z = ax^2 + by^2$, $\alpha = 60°$, and $P(1, 1, a + b)$

   b. $z = \sin xy$, $\alpha = 30°$, and $P(\pi, \frac{1}{2}, 1)$

   Solution.

   a. $\dfrac{\partial z}{\partial x} = 2ax$ and $\dfrac{\partial z}{\partial y} = 2by$

   $$\frac{dz}{ds} = 2a\left(\frac{1}{2}\right) + 2b\left(\frac{\sqrt{3}}{2}\right) = a + b\sqrt{3}$$

   b. $\dfrac{\partial z}{\partial x} = y \cos xy$ and $\dfrac{\partial z}{\partial y} = x \cos xy$

   $$\frac{dz}{ds} = 0(\cos 30°) + 0(\sin 30°) = 0$$

2. Show that the definition of the directional derivative is consistent with the definition of the partial derivative by finding dz/ds for $\alpha = 0°$ and $90°$.

   Solution. $\dfrac{dz}{ds} = \dfrac{\partial z}{\partial x} \cos \alpha + \dfrac{\partial z}{\partial y} \sin \alpha = \dfrac{\partial z}{\partial x}$, for $\alpha = 0°$

   (direction of the X-axis)

$$\frac{dz}{ds} = \frac{\partial z}{\partial y}, \text{ for } \alpha = 90^0 \text{ (direction of the } Y\text{-axis)}$$

PROBLEMS WITH ANSWERS

3. Find the slope of the tangent line whose projection line in the xy-plane makes $45°$ with the X-axis, the surface being $z = x^2 + 2y^2$ and the point, $(1, -1, 3)$.

Ans. $-\sqrt{2}$

4. Show that the angle α which produces a maximum value for $\dfrac{dz}{ds}$ is

$$\alpha = \text{Tan}^{-1} \frac{\dfrac{\partial z}{\partial y}}{\dfrac{\partial z}{\partial x}}$$

Hence, show that

$$\left| \frac{dz}{ds} \right|_{max} = \sqrt{\left(\frac{\partial z}{\partial x}\right)^2 + \left(\frac{\partial z}{\partial y}\right)^2}$$

5. Assume that $\left| \dfrac{dz}{ds} \right|_{max} = \sqrt{\left(\dfrac{\partial z}{\partial x}\right)^2 + \left(\dfrac{\partial z}{\partial y}\right)^2}$

This function is the **gradient** of z and is often represented as ∇z (read "del z"). Find ∇z for (a) $z = xy$, (b) $z = e^{x+y}$, (c) $z = ax + by$, and (d) $z = \sin xy$.

Ans. (a) $\dfrac{\partial z}{\partial x} = y, \dfrac{\partial z}{\partial y} = x, \nabla z = \sqrt{x^2 + y^2}$

(b) $\dfrac{\partial z}{\partial x} = \dfrac{\partial z}{\partial y} = e^{x+y}, \nabla z = \sqrt{2e^{2x+y}} = e^{x+y}\sqrt{2}$

(c) $\dfrac{\partial z}{\partial x} = a, \dfrac{\partial z}{\partial y} = b, \nabla z = \sqrt{a^2 + b^2}$

(d) $\dfrac{\partial z}{\partial x} = y\cos xy, \dfrac{\partial z}{\partial y} = x\cos xy, \nabla z = \cos xy\sqrt{x^2 + y^2}$

10.8 Maxima and Minima

A function of two independent variables $z = f(x,y)$ is a *maximum* at a point (a,b) if $f(x,y) < f(a,b)$ for all values of x and y in the neighborhood of (a, b). Similarly, the function is a *minimum* if $f(x, y) > f(a, b)$ for all values of x and y in the neighborhood of (a, b).

For $z = f(x, y)$ to have an *extreme* (maximum or minimum) at a point (a,b), it is necessary that $\dfrac{\partial f}{\partial x} = 0$ and $\dfrac{\partial f}{\partial y} = 0$ at (a, b). But this condition is not sufficient. In order to examine $z = f(x, y)$ for maxima and minima the procedure is as follows:

1. Compute $\dfrac{\partial f}{\partial x}, \dfrac{\partial f}{\partial y}, \dfrac{\partial^2 f}{\partial x^2}, \dfrac{\partial^2 f}{\partial x\, \partial y}$, and $\dfrac{\partial^2 f}{\partial y^2}$.

2. Solve simultaneously $\dfrac{\partial f}{\partial x} = 0$ and $\dfrac{\partial f}{\partial y} = 0$, and let a pair of (critical) values satisfying these equations be (x_0, y_0).

3. Evaluate $\Delta = \dfrac{\partial^2 f}{\partial x^2} \dfrac{\partial^2 f}{\partial y^2} - \left(\dfrac{\partial^2 f}{\partial x\, \partial y} \right)^2$ at (x_0, y_0).

4. Then $z = f(x_0, y_0)$ will be:

$$\text{Maximum if } \Delta > 0 \text{ and } \frac{\partial^2 f}{\partial x^2} \left(\text{or } \frac{\partial^2 f}{\partial y^2} \right) < 0$$

$$\text{Minimum if } \Delta > 0 \text{ and } \frac{\partial^2 f}{\partial x^2} \left(\text{or } \frac{\partial^2 f}{\partial y^2} \right) > 0$$

$$\text{Neither if } \Delta < 0$$

$$\text{Test fails if } \Delta = 0.$$

(Review discussion of maxima and minima for $y = f(x)$ in Sections 8.4 and 8.5.)

Example 1. Examine $z = x^2 + xy + y^2 - y$ for maxima and minima.

Solution.

$$\frac{\partial z}{\partial x} = 2x + y \text{ and } \frac{\partial z}{\partial y} = x + 2y - 1$$

$$\frac{\partial^2 z}{\partial x^2} = 2, \frac{\partial^2 z}{\partial x\, \partial y} = 1, \text{ and } \frac{\partial^2 z}{\partial y^2} = 2$$

$$\left. \begin{array}{r} 2x + y = 0 \\ x + 2y = 1 \end{array} \right\} \quad x = -\frac{1}{3} \text{ and } y = \frac{2}{3}$$

This is the only critical point. At this point, $\Delta > 0$ and $\dfrac{\partial^2 z}{\partial x^2} > 0$, so z is

a minimum; the minimum value of $z = -\dfrac{1}{3}$.

Example 2. Examine $z = x^2 - y^2$ for maxima and minima.

Solution.

$$\frac{\partial z}{\partial x} = 2x \text{ and } \frac{\partial z}{\partial y} = -2y$$

$$\frac{\partial^2 z}{\partial x^2} = 2, \quad \frac{\partial^2 z}{\partial x\,\partial y} = 0, \text{ and } \frac{\partial^2 z}{\partial y^2} = -2$$

The critical point is the origin, but $\Delta = -4 < 0$; therefore, the point is neither a maximum nor a minimum. Such a point is called a **saddle point** and corresponds to a *point of inflection* on a curve $y = f(x)$.

PROBLEMS WITH SOLUTIONS

1. Examine the following functions for critical points:
   a. $z = x^2 + xy + y^2 - y$,
   b. $z = \sin xy + \cos xy$,
   c. $z = \exp \frac{1}{2}(x^2 + y^2)$
   d. $z = x^2 + y^2$
   e. $z = (x - 2)^2 + (y + 3)^2$
   f. $z = Ax + By$

   Solution.

   a. $\dfrac{\partial z}{\partial x} = 2x + y$ and $\dfrac{\partial z}{\partial y} = x + 2y - 1$

   $\dfrac{\partial^2 z}{\partial x^2} = 2, \quad \dfrac{\partial^2 z}{\partial y^2} = 2, \text{ and } \dfrac{\partial^2 z}{\partial x\,\partial y} = 1$

   Thus,

   $$\left.\begin{array}{r} 2x + y = 0 \\ x + 2y = 1 \end{array}\right\} \quad x = -\frac{1}{3} \text{ and } y = \frac{2}{3}$$

   $$\Delta = 2 \times 2 - 1 > 0 \text{ and } \frac{\partial^2 f}{\partial x^2} > 0$$

   The only critical point is $(-1/3, 2/3)$, and z is a minimum there; thus, the minimum value of z is $-1/3$.

   b. $\dfrac{\partial z}{\partial x} = y \cos xy - y \sin xy$ and $\dfrac{\partial z}{\partial y} = x \cos xy - x \sin xy$

   $\dfrac{\partial^2 z}{\partial x^2} = -y^2 \sin xy - y^2 \cos xy, \quad \dfrac{\partial^2 z}{\partial y^2} = -x^2 \sin xy - x^2 \cos xy,$

and $\dfrac{\partial^2 z}{\partial x\,\partial y} = -yx\sin xy + \cos xy - (yx\cos xy + \sin xy)$

$$= -xy(\sin xy + \cos xy) + \cos xy - \sin xy$$

$$\left.\begin{array}{l} y(\cos xy - \sin xy) = 0 \\[2mm] x(\cos xy - \sin xy) = 0 \end{array}\right\} \begin{array}{l} x = 0 \text{ and } y = 0 \text{ or any point} \\[2mm] \text{for which } xy = \dfrac{\pi}{4} \,. \end{array}$$

Case 1. $x = 0$, $y = 0$, and $\Delta = 0 - 1 < 0$; therefore, $(0, 0, 1)$ yields neither a maximum nor a minimum. This is a saddle point and corresponds to a point of inflection on a plane curve $y = f(x)$.

Case 2. $xy = \dfrac{\pi}{4}$.

$$\Delta = 2x^2 y^2 - \left(\frac{\pi}{4}\sqrt{2}\right)^2 = \frac{\pi^2}{8} - \frac{\pi^2}{8} = 0$$

Hence, all points where $xy = \dfrac{\pi}{4}$ are saddle points.

c. $\dfrac{\partial z}{\partial x} = -x\exp-\dfrac{1}{2}(x^2 + y^2)$ $\dfrac{\partial z}{\partial y} = -y\exp-\dfrac{1}{2}(x^2 + y^2)$

$\dfrac{\partial^2 z}{\partial x^2} = -x\left(-x\exp-\dfrac{1}{2}(x^2 + y^2)\right) - \exp-\dfrac{1}{2}(x^2 + y^2)$

$$= (x^2 - 1)\exp-\dfrac{1}{2}(x^2 + y^2)$$

By symmetry, $\dfrac{\partial^2 z}{\partial y^2} = (y^2 - 1)\exp-\dfrac{1}{2}(x^2 + y^2)$

$$\dfrac{\partial^2 z}{\partial x\,\partial y} = xy\exp-\dfrac{1}{2}(x^2 + y^2)$$

$$\left.\begin{array}{l} x\exp-\dfrac{1}{2}\ (x^2 + y^2) = 0 \\[3mm] y\exp-\dfrac{1}{2}\ (x^2 + y^2) = 0 \end{array}\right\} x = 0 \text{ and } y = 0$$

Thus, $\Delta = 1 - 0 = 1$; hence, $(\Delta > 0)$. Since $\dfrac{\partial^2 z}{\partial x^2} < 0$, $(0, 0, 1)$ is a point at which z is a maximum.

d. $\dfrac{\partial z}{\partial x} = 2x$ and $\dfrac{\partial z}{\partial y} = 2y$

$\dfrac{\partial^2 z}{\partial x^2} = 2$, $\dfrac{\partial^2 z}{\partial y^2} = 2$, and $\dfrac{\partial^2 z}{\partial x \partial y} = 0$

$\left. \begin{array}{c} 2x = 0 \\ 2y = 0 \end{array} \right\}$ $x = 0$ and $y = 0$

$\Delta = 2(2) - 0 = 4$; hence, $(0, 0, 0)$ is a minimum point.

e. $\dfrac{\partial z}{\partial x} = 2(x - 2)$ and $\dfrac{\partial z}{\partial y} = 2(y + 3)$

$\dfrac{\partial^2 z}{\partial x^2} = 2$, $\dfrac{\partial^2 z}{\partial y^2} = 2$, and $\dfrac{\partial^2 z}{\partial x \partial y} = 0$

$\left. \begin{array}{c} 2(x - 2) = 0 \\ 2(y + 3) = 0 \end{array} \right\}$ $x = 2$ and $y = -3$

$$\Delta = 4 - 0 = 4 \text{ and } \dfrac{\partial^2 z}{\partial x^2} > 0$$

Hence, $(2, -3, 0)$ is a minimum point.

f. $z = Ax + By$

$\dfrac{\partial z}{\partial x} = A$, $\dfrac{\partial z}{\partial y} = B$, and $\dfrac{\partial^2 z}{\partial x^2} = \dfrac{\partial^2 z}{\partial y^2} = \dfrac{\partial^2 z}{\partial x \partial y} = 0$

Since $\dfrac{\partial z}{\partial x}$ and $\dfrac{\partial z}{\partial y}$ are nonzero constants, there are no critical points.

2. A rectangular box without a top is to have a given volume. What are the dimensions of the box if it is to have least surface area?
Solution.

$$V = xyz$$

$$S = xy + 2xz + 2yz$$

$$= xy + \frac{2V}{y} + \frac{2V}{x}$$

$$\frac{\partial S}{\partial x} = y - \frac{2V}{x^2} \text{ and } \frac{\partial S}{\partial y} = x - \frac{2V}{y^2}$$

Hence, $\left. \begin{array}{c} x^2 y = 2V \\ xy^2 = 2V \end{array} \right\}$ $x = y = \sqrt[3]{2V}$

and

$$z = \frac{V}{xy} = \frac{Vx}{x^2 y} = \frac{Vx}{2V} = \frac{x}{2}.$$

$$\frac{\partial^2 S}{\partial x^2} = \frac{4V}{x^3}, \quad \frac{\partial^2 S}{\partial y^2} = \frac{4V}{y^3}, \quad \text{and} \quad \frac{\partial^2 S}{\partial x \partial y} = 1$$

$$\Delta = \frac{16V^2}{x^3 y^3} - 1 = 3 \quad \text{and} \quad \frac{\partial^2 S}{\partial x^2} > 0$$

Hence, $(\sqrt[3]{2V}, \sqrt[3]{2V}, \frac{1}{2}\sqrt[3]{2V})$ is a minimum point. The coordinates give the dimensions of the box with least surface area.

3. A plane with positive intercepts passes through (x_0, y_0, z_0). As-sume that the point (x_0, y_0, z_0) lies in the first octant. If this plane cuts off a volume in the first octant, determine the plane so that the volume is minimal.

   Solution. $V = \dfrac{abc}{6}$, where $\dfrac{x}{a} + \dfrac{y}{b} + \dfrac{z}{c} = 1$ is the equation of the plane. Thus,

$$V = \frac{a^2 b^2 z_0}{6[ab - bx_0 - ay_0]}$$

$$\frac{\partial V}{\partial a} = \frac{1}{6}\left[\frac{(ab - bx_0 - ay_0)\, 2ab^2 z_0 - a^2 b^2 z_0 (b - y_0)}{(ab - bx_0 - ay_0)^2}\right] = 0$$

$$\frac{\partial V}{\partial b} = \frac{1}{6}\left[\frac{(ab - bx_0 - ay_0)\, 2a^2 b z_0 - a^2 b^2 z_0 (a - x_0)}{(ab - bx_0 - ay_0)^2}\right] = 0$$

Assume that $(ab - bx_0 - ay_0) \neq 0$, multiply $\dfrac{\partial V}{\partial a}$ by a and $\dfrac{\partial V}{\partial b}$ by b, and subtract. Then,

$$\frac{-a^3 b^2 z_0 (b - y_0) + a^2 b^3 z_0 (a - x_0)}{(ab - bx_0 - ay_0)^2}$$

Hence, $a^2 b^3 (a - x_0) = a^3 b^2 (b - y_0)$

Thus, $b = \dfrac{ay_0}{x_0}$

Substituting this value of b in $\dfrac{\partial V}{\partial b} = 0$ leads to $a = 3x_0$, $b = 3y_0$, $c = 3z_0$. Testing these values for a, b, and c, we find they yield a minimum volume.

PROBLEMS WITH ANSWERS

4. Examine $z + xy + \dfrac{1}{x} + \dfrac{1}{y} = 0$ for maxima and minima.

   Ans. maximum $(1, 1, -3)$

5. Examine $4(x - 1)^2 + 4(y - 1)^2 + (z - 2)^2 - 4 = 0$ for maxima and minima.

$Ans.$ maximum $(1, 1, 4)$, minimum $(1, 1, 0)$

6. Find the volume of the largest rectangular parallelepiped that can be inscribed in a sphere of radius r.

$Ans.$ $\dfrac{8r^2}{3\sqrt{3}}$

CHAPTER 11
MULTIPLE INTEGRATION

11.1 Repeated Integration

Since the integral of a function of x is itself a function of x, say $F(x) + C_1$, it may also be integrated, and we write

$$\int f(x) \, dx = F(x) + C_1$$

$$\int [F(x) + C_1] \, dx = G(x) + C_1 x + C_2 \qquad (1)$$

Hence,

$$\int \left[\int f(x) \, dx \right] dx = \int \int f(x) \, dx \, dx$$
$$= \int \int f(x) \, (dx)^2 \qquad (2)$$

The process of integration may be repeated any number of times.

$$\int \left\{ \int \left[\int f(x) \, dx \right] dx \right\} dx = \int \int \int f(x) \, dx \, dx \, dx$$
$$= \int \int \int f(x) \, (dx)^3 \qquad (3)$$

The integrals (2) and (3) are called *iterated*, or repeated, integrals; more often they are called *double* and *triple* integrals, respectively. The n-fold iterated integral is written

$$\int \int \cdots \int f(x) \, (dx)^n \qquad (4)$$

Example 1. Find $\int \int \sin 2x \, dx \, dx$.

Solution. $\int \int \sin 2x \, dx \, dx = \int (-\frac{1}{2} \cos 2x + C_1) \, dx$
$$= -\frac{1}{4} \sin 2x + C_1 x + C_2$$

Note that there are two constants of integration C_1 and C_2, and that when the first constant of integration is integrated, we get the $C_1 x$ term.

Example 2. Find $\int \int \int (x^2 + e^{-3x}) \, (dx)^3$.

Solution. $\int \int \int (x^2 + e^{-3x}) (dx)^3 = \int \int \left(\frac{x^3}{3} - \frac{1}{3} e^{-3x} + C_1 \right) (dx)^2$
$$= \int \left(\frac{x^4}{12} + \frac{1}{9} e^{-3x} + C_1 x + C_2 \right) dx$$

$$= \frac{x^5}{60} - \frac{1}{27}e^{-3x} + \frac{C_1 x^2}{2} + C_2 x + C_3$$

For iterated definite integrals the procedure is the same. The integration is performed from the inside out; that is to say, $\int_c^d \int_a^b f(x)\,(dx)^2$ means that $f(x)$ is to be integrated first between the limits a and b, and then this constant is to be integrated between the limits c and d.

Example 3. Evaluate $\int_{-1}^{2} \int_0^2 (x^2 - \sin x)\,dx\,dx$.

Solution. $\displaystyle \int_{-1}^{2} \int_0^2 (x^2 - \sin x)\,dx\,dx = \int_{-1}^{2} \left[\frac{x^3}{3} + \cos x \right]_0^2 dx$

$$= \int_{-1}^{2} \left(\tfrac{8}{3} + \cos 2 - 1 \right) dx$$

$$= \left(\tfrac{5}{3} + \cos 2 \right) x \Big|_{-1}^{2}$$

$$= 5 + 3 \cos 2$$

The problem in Example 3 would have had the same answer even though we had written it in the form

$$\int_{-1}^{2} \left[\int_0^2 (t^2 - \sin t)\,dt \right] dw$$

since t and w are only *dummy variables* in this process of integrating and evaluating between definite limits. If x and y are independent variables, we can generalize the notation used in (2) to compute the iterated integral of a function of x and y; thus,

$$\iint f(x, y)\,dx\,dy \tag{5}$$

or

$$\int_c^d \int_a^b f(x, y)\,dx\,dy \tag{6}$$

In (5) and (6) the integration is first performed with respect to x; that is, the inside integration symbol goes with the first differential symbol, in this case dx. Some authors couple the first integration symbol with the first differential, and the student, when referring to other works on the calculus, should determine which system the particular writer is using. Occasionally, the notation $\int dx \int f(x, y)\, dy$ is used where integration takes place first with respect to y. Since, in (6), after the first integration with respect to x, there is to be a second integration with respect to y, a and b could be thought of as functions of y. Hence, we are led to the consideration of such integrals as

$$\int_c^d \int_{a(y)}^{b(y)} f(x, y)\, dx\, dy \tag{7}$$

The triple integral corresponding to (7) can be written

$$\int_c^d \int_{\alpha(z)}^{\beta(z)} \int_{a(y,z)}^{b(y,z)} f(x, y, z)\, dx\, dy\, dz \tag{8}$$

Example 4. Evaluate $\displaystyle\int_0^1 \int_0^y (x + y^2)\, dx\, dy$.

Solution.
$$\int_0^1 \int_0^y (x + y^2)\, dx\, dy = \int_0^1 \left[\frac{x^2}{2} + xy^2 \right]_0^y dy$$

$$= \int_0^1 \left(\frac{y^2}{2} + y^3 \right) dy$$

$$= \frac{y^3}{6} + \frac{y^4}{4} \Big|_0^1 = \frac{5}{12}$$

Example 5. Evaluate $A = \displaystyle\int_0^\pi \int_{-z}^z \int_0^{yz} (x + \sin y - z^3)\, dx\, dy\, dz$.

Solution. $A = \displaystyle\int_0^\pi \int_{-z}^z \left[\frac{x^2}{2} + x \sin y - xz^3 \right]_0^{yz} dy\, dz$

$$= \int_0^\pi \int_{-z}^z \left(\frac{y^2 z^2}{2} + yz \sin y - yz^4 \right) dy\, dz$$

$$= \int_0^\pi \left[\frac{y^3 z^2}{6} + z(\sin y - y \cos y) - \frac{y^2 z^4}{2} \right]_{-z}^z dz$$

$$= \int_0^\pi \left[\frac{z^5}{3} + 2z(\sin z - z \cos z) \right] dz$$

$$= \frac{z^6}{18} + 2(\sin z - z \cos z) - 2 \left\{ 2z \cos z + (z^2 - 2) \sin z \right\} \Big|_0^\pi$$

$$= \frac{\pi^6}{18} + 6\pi$$

PROBLEMS WITH SOLUTIONS

Evaluate the integrals in Problems 1 through 4.

1. $\displaystyle \int_0^\infty \int_0^{\pi/2} \rho \exp(-\rho^2/2)\, d\theta\, d\rho = \int_0^\infty \left[\rho \exp(-\rho^2/2)\theta \right]_0^{\pi/2} d\rho$

$$= \frac{\pi}{2} \int_0^\infty \rho \exp(-\rho^2/2)\, d\rho$$

$$= -\frac{\pi}{2} \exp(-\rho^2/2) \Big|_0^\infty = \frac{\pi}{2}$$

2. $\displaystyle \int_0^\pi \int_{-z}^z \left(\frac{y^2 z^2}{2} + yz \sin y - yz^4 \right) dy\, dz$

$$= \int_0^\pi \left[\frac{y^3 z^2}{6} + z(\sin y - y \cos y) - \frac{y^2 z^4}{2} \right]_{-z}^z dz$$

$$= \int_0^\pi \left[\frac{z^5}{3} + 2z(\sin z - z \cos z) \right] dz$$

$$= \frac{z^6}{18} + 2(\sin z - z \cos z) - 2[2z \cos z + (z^2 - 2) \sin z] \Big|_0^\pi$$

$$= \frac{\pi^6}{18} + 6\pi$$

3. $\displaystyle \int_0^{2\pi} \int_0^{\pi/2} \int_0^{a \cos \theta} \rho^2 \sin \theta\, d\rho\, d\theta\, d\phi$

$$= \frac{1}{3} a^3 \int_0^{2\pi} \int_0^{\pi/2} \cos^3 \theta \sin \theta \, d\theta \, d\phi$$

$$= \frac{a^3}{12} \int_0^{2\pi} d\phi = \frac{\pi a^3}{6}$$

4. $\displaystyle\int_0^1 \int_y^{y+1} \int_1^{x^2} dz \, dx \, dy = \int_0^1 \int_y^{y+1} (x^2 - 1) \, dx \, dy$

$$= \int_0^1 \left(\frac{x^3}{3} - x \right) \Bigg|_y^{y+1} dy$$

$$= \int_0^1 \left(y^2 + y - \frac{2}{3} \right) dy$$

$$= \frac{y^3}{3} + \frac{y^2}{2} - \frac{2y}{3} \Bigg|_0^1 = \frac{1}{6}$$

PROBLEMS WITH ANSWERS

Evaluate the integrals in Problems 5 through 7.

5. $\displaystyle 4 \int_0^a \int_0^{\sqrt{a^2 - y^2}} \frac{a}{\sqrt{a^2 - y^2}} \, dz \, dy.$ *Ans.* $4a^2$

6. $\displaystyle \int_0^2 \int_0^{\sqrt{4 - x^2}} dy \, dx.$ *Ans.* π

7. $\displaystyle \int_{-1}^1 \int_0^x \int_1^{x+y} dz \, dy \, dx.$ *Ans.* 1

11.2 Plane Area By Double Integration

Consider the area A bounded by a closed curve, and suppose that the interval (a, b) (Fig. 11.1) is subdivided by the $n + 1$ points $x_1 = a, x_2,$ $\cdots, x_i, x_{i+1}, \cdots x_{n+1} = b$, and the interval (c, d) is subdivided by the $m + 1$ points $y_1 = c, y_2, \cdots, y_j, y_{j+1}, \cdots, y_{m+1} = d$. Erect ordinates at the points x_i and abscissas at points y_i thus covering the area A with a

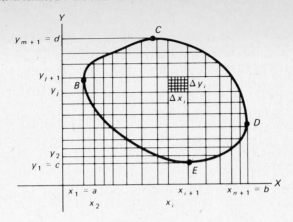

FIG. 11.1

net of lines forming small rectangular areas. Set $\Delta x_i = \left| x_{i+1} - x_i \right|$ and $\Delta y_j = \left| y_{j+1} - y_j \right|$. Then the area of one of these rectangles is, $\Delta A_{ij} = \Delta x_i \, \Delta y_j$. The sum of all of these areas as i and j independently take on values from 1 to n and 1 to m, respectively, and will be an approximation to the area A; that is,

$$A \cong \sum_{i=1}^{n} \sum_{j=1}^{m} \Delta x_i \, \Delta y_j \tag{1}$$

A theorem of integral calculus states that

$$A = \lim_{\substack{n,m \to \infty \\ \Delta x_i, \, \Delta y_j \to 0}} \sum_{i=1}^{n} \sum_{j=1}^{m} \Delta x_i \, \Delta y_j \tag{2}$$

$$= \int_{c}^{d} \int_{\phi_1(y)}^{\phi_2(y)} dx \, dy \tag{3}$$

$$= \int_{a}^{b} \int_{f_1(x)}^{f_2(x)} dy \, dx \tag{4}$$

where $x = \phi_1(y)$ is the equation of the curve CBE; $x = \phi_2(y)$ is the equation of CDE; $y = f_1(x)$ is the equation of BED; and $y = f_2(x)$ is the equation of BCD.

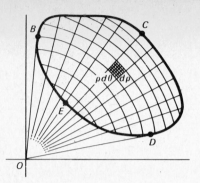

FIG. 11.2

The corresponding formulas in polar coordinates are (Fig. 11.2)

$$A = \int_{\alpha}^{\beta} \int_{\phi_1(\theta)}^{\phi_2(\theta)} \rho \, d\rho \, d\theta \tag{5}$$

$$= \int_{\rho_1}^{\rho_2} \int_{f_1(\rho)}^{f_2(\rho)} \rho \, d\theta \, d\rho \tag{6}$$

where $\rho = \phi_1(\theta)$ is the equation of the curve DEB; $\rho = \phi_2(\theta)$ is the equation of BCD; $\theta = f_1(\rho)$ is the equation of EDC; and $\theta = f_2(\rho)$ is the equation of EBC.

In (3), where the integration is performed first with respect to x, a row of elementary rectangles is obtained. The second integration sums up all such rows (Fig. 11.3). In (4), integrating first with respect to y, we obtain a column of elementary rectangular areas. The second integration sums up all such columns (Fig. 11.4).

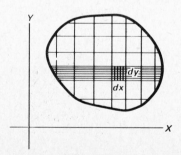

FIG. 11.3

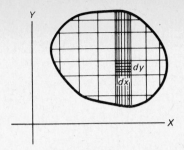

FIG. 11.4

Similarly, in polar coordinates (5) sums, first, elements lying in one radial strip, then, second, all such strips (Fig. 11.5), and (6) sums, first, elements lying in one circular ring, then, second, all such rings (Fig. 11.6).

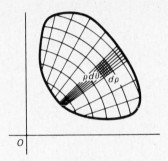

FIG. 11.5

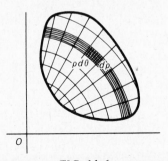

FIG. 11.6

It is not possible to say offhand which order of integration will be the simpler in a given problem of finding the area of a closed region, since so much depends upon the way in which the functions determining the boundary curves behave. In all cases, the figures should be drawn in order that the geometry be made clear. The algebraic computation should agree with the geometry of the configuration. After the geometry is understood, the simpler order of integration can be determined. The student will find it good practice to solve a few problems both ways; this will give him more than just a check on his answers.

Example 1. Find by double integration the total area enclosed by the two curves $y = 2x$ and $y^3 = 2x$ (Fig. 11.7).

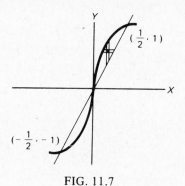

FIG. 11.7

Solution. The points of intersection are $(-\frac{1}{2}, -1)$, $(0, 0)$, and $(\frac{1}{2}, 1)$. By symmetry, the common area (not regarding sign) is twice that lying in the first quadrant. It should be clear that it will make no difference which order of integration we choose. Formulas (3) and (4) become, respectively,

$$A = 2 \int_0^1 \int_{y^3/2}^{y/2} dx \, dy$$

$$= 2 \int_0^{1/2} \int_{2x}^{(2x)^{1/3}} dy \, dx$$

Evaluating the first, we get

$$A = 2 \int_0^1 (\tfrac{1}{2}y - \tfrac{1}{2}y^3)\,dy$$

$$= \tfrac{1}{4} \text{ sq. unit (total area)}$$

Evaluating the second, we get

$$A = 2 \int_0^{1/2} [(2x)^{1/3} - 2x]\,dx = \frac{1}{4}\text{ sq. unit}$$

The enclosed area lying in the first quadrant is $\tfrac{1}{8}$ sq. unit.

Example 2. Find by double integration the area enclosed by $y = x^2$ and $x + y - 2 = 0$ (Fig. 11.8).

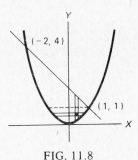

FIG. 11.8

Solution. The points of intersection are $(-2, 4)$ and $(1, 1)$. Integrating first with respect to y and then with respect to x, we get

$$A = \int_{-2}^1 \int_{x^2}^{2-x} dy\,dx$$

$$= \int_{-2}^1 [2 - x - x^2]\,dx = \frac{9}{2}\text{ sq. units}$$

If we take $dx\,dy$ as the element of area where the integration is to be performed first with respect to x, we must write the area as the sum of two double integrals; thus,

$$A = \int_0^1 \int_{-\sqrt{y}}^{\sqrt{y}} dx\,dy + \int_1^4 \int_{-\sqrt{y}}^{2-y} dx\,dy$$

The reason is clear since a row element stretches first of all ($y = 0$ to $y = 1$) between the two branches of the same curve $y = x^2$, and, only later, ($y = 1$ to $y = 4$) stretches from $y = x^2$ to $x + y - 2 = 0$. But, of course, the answer is the same only the method is harder.

Example 3. Find by double integration the area inside the circle $\rho = a \cos \theta$ and outside the cardioid $\rho = a(1 - \cos \theta)$. (See Fig. 11.9 and Problem 7, Section 9.4.)

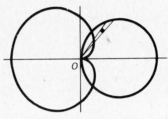

FIG. 11.9

Solution. It is better to integrate with respect to ρ first, since ρ runs directly from the one curve to the other.

$$A = 2 \int_0^{\pi/3} \int_{a(1-\cos\theta)}^{a\cos\theta} \rho \, d\rho \, d\theta$$

$$= \int_0^{\pi/3} [\rho^2]_{a(1-\cos\theta)}^{a\cos\theta} \, d\theta$$

$$= \int_0^{\pi/3} [a^2 \cos^2 \theta - a^2(1 - \cos \theta)^2] \, d\theta$$

$$= a^2 \int_0^{\pi/3} (2 \cos \theta - 1) \, d\theta$$

$$= a^2 [2 \sin \theta - \theta]_0^{\pi/3}$$

$$= \frac{a^2}{3} (3\sqrt{3} - \pi) \text{ sq. units}$$

PROBLEMS WITH SOLUTIONS

In Problems 1 through 7, find the area indicated by double integration.

1. The area enclosed by $y = x^3$ and $y = x^2$.

Solution. The points of intersection are $(0,0)$ and $(1,1)$.

$$A = \int_0^1 \int_{x^3}^{x^2} dy \, dx$$

$$= \int_0^1 (x^2 - x^3) \, dx$$

$$= \frac{x^3}{3} - \frac{x^4}{4} \Big|_0^1 = \frac{1}{12}$$

2. The area enclosed by the hyperbola $\dfrac{x^2}{a^2} - \dfrac{y^2}{b^2} = 1$ and the line $x = x_0 \, (> a)$.
Solution.

$$A = 2 \int_a^{x_0} \int_0^{b\sqrt{x^2 - a^2}/a} dy \, dx$$

$$= \frac{2b}{a} \int_a^{x_0} \sqrt{x^2 - a^2} \, dx$$

$$= \frac{2b}{a} \left[\frac{x}{2} \sqrt{x^2 - a^2} - \frac{a^2}{2} \log \; x + \sqrt{x^2 - a^2} \right]_a^{x_0}$$

$$= \frac{2b}{a} \left[\frac{x_0}{2} \sqrt{x_0^2 - a^2} - \frac{a^2}{2} \log \left(x_0 + \sqrt{x_0^2 - a^2} \right) + \frac{a^2}{2} \log x_0 \right]$$

3. The area under the "Witch of Agnesi" given by $y = \dfrac{8}{4 + x^2}$ from -2 to $+2$.
Solution.

$$A = 2 \int_0^2 \int_0^{8/(4+y^2)} dy \, dx$$

$$= 2 \int_0^2 \frac{8}{4 + x^2} \, dx$$

$$= 8 \; \text{Tan}^{-1} \frac{x}{2} \Big|_0^2 = 4\pi$$

4. The area under the catenary $y = \frac{1}{2}(e^x + e^{-x})$ from $x = -r$ to $x = +r$.
    Solution.

$$A = 2 \int_0^r \int_0^{(e^x + e^{-x})/2} dy \, dx$$

$$= 2 \int_0^r \frac{1}{2}(e^x + e^{-x}) \, dx$$

$$= \int_0^r (e^x + e^{-x}) \, dx = (e^x - e^{-x}) \bigg|_0^r = e^r - e^{-r}$$

$$= 2 \sinh r$$

5. The area bounded by the coordinate axes and $x^{1/2} + y^{1/2} = a^{1/2}$.
    Solution.

$$A = \int_0^a \int_0^{(a^{1/2} - x^{1/2})^2} dy \, dx$$

$$= \int_0^a (a^{1/2} - x^{1/2})^2 \, dx$$

$$= ax - 2a^{1/2} \frac{x^{3/2}}{3/2} + \frac{x^2}{2} \bigg|_0^a$$

$$= a^2 - \frac{4}{3}a^2 + \frac{a^2}{2} = \frac{1}{6}a^2$$

6. The area swept out by the radius vector of the logarithmic spiral,
    $\rho = 2e^{\theta/2}$ in one half revolution.
    Solution.

$$A = \int_0^\pi \int_0^{2e^{\theta/2}} \rho \, d\rho \, d\theta$$

$$= 2 \int_0^\pi e^\theta \, d\theta = 2(e^\pi - 1)$$

7. The area of the two loops of the lemniscate $\rho^2 = a^2 \cos 2\theta, a > 0$.
    Make use of symmetry of the curve, and take smallest range of
    integration.

Solution.

$$A = 4 \int_0^{\pi/4} \int_0^{a\sqrt{\cos 2\theta}} \rho \, d\rho \, d\theta$$

$$= 2a^2 \int_0^{\pi/4} \cos 2\theta \, d\theta$$

$$= a^2 \sin 2\theta \Big|_0^{\pi/4} = a^2$$

PROBLEMS WITH ANSWERS

Find the indicated area by double integration.

8. The area between $y^2 = 4x$ and $2x - y - 4 = 0$.

$$Ans. \quad A = \int_{-2}^{4} \int_{y^2/4}^{(y/2)+2} dx \, dy = 9$$

9. The area between $x^{1/2} + y^{1/2} = a^{1/2}$ and $x + y = a$.

$$Ans. \quad A = \int_0^a \int_{(a^{1/2}-x^{1/2})^2}^{a-x} dy \, dx = \frac{1}{3} a^2$$

10. The area between $y = 6x - x^3$ and $y = x^3 - 2x$.

$$Ans. \quad A = 2 \int_0^2 \int_{x^3-2x}^{6x-x^3} dy \, dx = 16$$

11. The area between $y = xe^{-x}$, $x = 1$, and $y = 0$.

$$Ans. \quad A = \int_0^1 \int_0^{xe^{-x}} dy \, dx = 1 - \frac{2}{e}$$

12. The area between $y = \frac{4}{x}$ and $y = 5 - x$.

$$Ans. \quad A = \int_1^4 \int_{4/x}^{5-x} dy \, dx = \frac{15}{2} - 4 \log 4$$

13. The area between $y = x$, $y = 2x$, and $y = 6 - x$.

$$Ans. \quad A = \int_0^2 \int_x^{2x} dy \, dx + \int_2^3 \int_x^{6-x} dy \, dx = 3$$

14.   The area enclosed by the cardioid $\rho = 2a\,(1 - \cos\theta)$.

$$Ans. \quad A = 2 \int_0^\pi \int_0^{2a(1-\cos\theta)} \rho\,d\rho\,d\theta = 6\pi a^2$$

15.   The area of one leaf of the four-leaved rose $\rho = a \sin 2\theta$.

$$Ans. \quad A = \int_0^{\pi/2} \int_0^{a\sin 2\theta} \rho\,d\rho\,d\theta = \frac{\pi a^2}{8}$$

11.3   Surface Area by Double Integration

In Section 9.8 we treated areas of surfaces of revolution. The following formula holds for the area of a curved surface $z = f(x, y)$.

$$S = \iint \sqrt{\left(\frac{\partial z}{\partial x}\right)^2 + \left(\frac{\partial z}{\partial y}\right)^2 + 1}\; dy\,dz \qquad (1)$$

This integral is to be evaluated over the projected area in the xy-plane. If the surface area in question cannot be projected onto the xy-plane (as would be the case if the area were that of a cylinder $f(x, y) = 0$, which is perpendicular to the xy-plane), the area should be projected onto another coordinate plane and formula (1) modified accordingly. Polar coordinates should be used if the integration in (1) is thus simplified.

Example 1.   Find the surface area cut from the plane $x + 2y + z = 1$ by the coordinate planes.

Solution. $z = 1 - x - 2y$

$$\frac{\partial z}{\partial x} = -1 \ \text{ and } \ \frac{\partial z}{\partial y} = -2$$

$$S = \int_0^1 \int_0^{(1-x)/2} \sqrt{6}\; dy\,dx$$

$$= \frac{\sqrt{6}}{2} \int_0^1 (1-x)\,dx = \frac{\sqrt{6}}{4} \ \text{ sq. unit}$$

Example 2.   Find the surface area cut from the surface $2z = x^2 + y^2$ by the two planes $z = 0$ and $z = \frac{1}{2}$.

Solution. $\dfrac{\partial z}{\partial x} = x \ \text{ and } \ \dfrac{\partial z}{\partial y} = y$

$$S = 4 \int_0^1 \int_0^{\sqrt{1-x^2}} \sqrt{x^2 + y^2 + 1} \; dy \, dx$$

The integration will be simplified by using polar coordinates:

$$S = 4 \int_0^{\pi/2} \int_0^1 \sqrt{\rho^2 + 1} \; \rho \, d\rho \, d\theta$$

$$= \frac{4}{3} \int_0^{\pi/2} (2^{3/2} - 1) \, d\theta$$

$$= \frac{2\pi}{3} (2^{3/2} - 1) \text{ sq. units}$$

PROBLEMS WITH SOLUTIONS

1. Find the area of the portion of the surface of the cylinder $z = \frac{1}{2} x^2$ which lies in the first octant and inside the cylinder $y = 1 - x^2$ (Fig. 11.10).

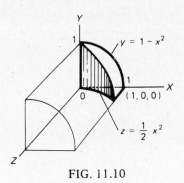

FIG. 11.10

Solution. For $z = \frac{1}{2} x^2$, $\dfrac{\partial z}{\partial x} = x$ and $\dfrac{\partial z}{\partial y} = 0$

$$S = \int_0^1 \int_0^{1-x^2} \sqrt{1 + x^2} \; dy \, dx$$

$$= \int_0^1 y \sqrt{1 + x^2} \, \Big|_0^{1-x^2} \, dx$$

$$= \int_0^1 (1 + x^2 - x^2 \sqrt{1 + x^2}) \, dx$$

$$= \frac{1}{8} \left[\sqrt{2} + 5 \log (1 + \sqrt{2}) \right]$$

2. Find the area of the surface of the cylinder $x^2 + y^2 = \rho x$ intercepted by the sphere $x^2 + y^2 + z^2 = \rho^2$ (Fig. 11.11).

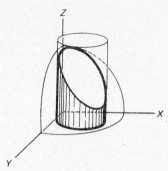

FIG. 11.11

Solution. For the cylinder $x^2 + y^2 = \rho x$, $\dfrac{\partial y}{\partial x} = \dfrac{\rho - 2x}{2y}$ and $\dfrac{\partial y}{\partial z} = 0$.

$$S = 4 \int_0^\rho \int_0^{\sqrt{\rho^2 - \rho x}} \left[1 + \left(\frac{\rho - 2x}{2y} \right)^2 \right]^{\frac{1}{2}} dz \, dx$$

$$= 2\rho \int_0^\rho \int_0^{\sqrt{\rho^2 - \rho x}} \frac{dz \, dx}{\sqrt{\rho x - x^2}}$$

$$= 2\rho \int_0^\rho \frac{\sqrt{\rho^2 - \rho x}}{\sqrt{\rho x - x^2}} \, dx$$

$$= 2\rho \int_0^\rho \sqrt{\frac{\rho}{x}} \, dx = 4\rho^2$$

3. Find the surface area cut from the sphere $x^2 + y^2 + z^2 = 9$ by the cylinder $\rho = 3 \cos 2\theta$ (Fig. 11.12).
   Solution. $z = \sqrt{9 - x^2 - y^2}$

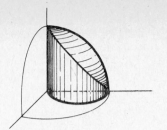

FIG. 11.12

$$\frac{\partial z}{\partial x} = \frac{-x}{\sqrt{9 - x^2 - y^2}}$$

$$\frac{\partial z}{\partial y} = \frac{-y}{\sqrt{9 - x^2 - y^2}}$$

$$\sqrt{1 + \left(\frac{\partial z}{\partial x}\right)^2 + \left(\frac{\partial z}{\partial y}\right)^2} = \sqrt{1 + \frac{x^2 + y^2}{9 - x^2 - y^2}}$$

$$= \frac{3}{\sqrt{9 - x^2 - y^2}}$$

$$= \frac{3}{\sqrt{9 - \rho^2}}$$

$$S = 2\int_0^{\pi/4} \int_0^{3\cos 2\theta} \frac{3}{\sqrt{9 - \rho^2}} \rho \, d\rho \, d\theta = 9\left(\frac{\pi}{2} - 1\right)$$

4. Find the surface area of a cone of height h and radius r.

   Solution. $x^2 + y^2 - \dfrac{r^2 z^2}{h^2} = 0$

   $$z^2 = \frac{h^2}{r^2}(x^2 + y^2)$$

   $$z = \frac{h}{r}\sqrt{x^2 + y^2}$$

   $$\frac{\partial z}{\partial x} = \frac{hx}{r\sqrt{x^2 + y^2}} \text{ and } \frac{\partial z}{\partial y} = \frac{hy}{r\sqrt{x^2 + y^2}}$$

   $$S = 4\int_0^r \int_0^{\sqrt{r^2 - y^2}} \sqrt{\frac{h^2(x^2 + y^2)}{r^2(x^2 + y^2)} + 1} \; dx \, dy$$

$$S = \frac{4}{r} \int_0^r \int_0^{\sqrt{r^2-y^2}} \sqrt{h^2 + r^2} \; dx \, dy$$

$$= \pi r \sqrt{r^2 + h^2} = \pi r s, \text{ where } s = \text{the slant height}$$

PROBLEMS WITH ANSWERS

5. Find the surface area of a sphere of radius r.   *Ans.* $4\pi r^2$

6. Find the surface area of a cylinder of radius r and height h (excluding the ends).   *Ans.* $2\pi rh$

7. Find the surface area of the volume common to the two cylinders $x^2 + y^2 = a^2$ and $x^2 + z^2 = a^2$.   *Ans.* $16a^2$

8. Find the area of the surface of the sphere $x^2 + y^2 + z^2 = r^2$ intercepted by cylinder $x^2 + y^2 = rx$.   *Ans.* $2(\pi - 2)r^2$

11.4 Volumes by Double and Triple Integration

If the element of area $dy \; dx$ in the xy-plane is projected vertically upward to the surface $z = F(x, y)$, an elementary column (Fig. 11.13)

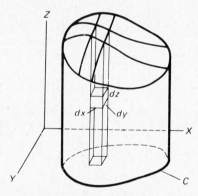

FIG. 11.13

is formed whose height is z and whose volume is $F(x, y) \; dy \; dx$. The sum of all such volume elements taken over the region R bounded by the closed curve C in the xy-plane gives the volume bounded above by the surface $z = F(x, y)$ and below by the xy-plane; thus, this volume is contained in the vertical cylinder whose xy-trace is the curve C.

This sum is

$$V = \int_a^b \int_{f_1(x)}^{f_2(x)} F(x, y)\, dy\, dx \qquad (1)$$

We could begin with an elementary "cube" $dz\, dy\, dx$ and sum all such elements within the region considered in order to obtain the volume. This would give

$$V = \int_a^b \int_{f_1(x)}^{f_2(x)} \int_0^{F(x,y)} dz\, dy\, dx \text{ (rectangular coordinates)} \qquad (2)$$

In the first integration, z travels from the xy-plane ($z = 0$) up to the surface $z = F(x, y)$. This reduces (2) to (1) and the procedure thereafter is the same in both cases.

It will be necessary to make appropriate modifications in (1) and (2) when different types of volumes are considered. If the volume enclosed by the two surfaces $z = F_1(x, y)$ and $z = F_2(x, y)$ is desired, then (2) is modified to read (Fig. 11.14).

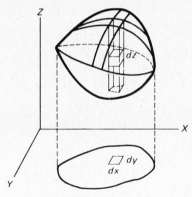

FIG. 11.14

$$V = \int_a^b \int_{f_1(x)}^{f_2(x)} \int_{F_1(x,y)}^{F_2(x,y)} dz\, dy\, dx \text{ (rectangular coordinates)} \qquad (3)$$

The curve C whose equations by pieces are $y = f_1(x)$ and $y = f_2(x)$ is, in this case, the projection of the curve of intersection of the two surfaces.

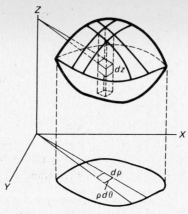

FIG. 11.15

In cylindrical coordinates, the volume element is $\rho \, dz \, d\rho \, d\theta$, and formula (3) becomes (Fig. 11.15).

$$V = \int_{\alpha}^{\beta} \int_{\phi_1(\theta)}^{\phi_2(\theta)} \int_{\Phi_1(\rho,\theta)}^{\Phi_2(\rho,\theta)} \rho \, dz \, d\rho \, d\theta \quad \text{(cylindrical coordinates)} \quad (4)$$

The equations of the bounding surfaces are $z = \Phi_1(\rho, \theta)$ and $z = \Phi_2(\rho, \theta)$; $\rho = \phi_1(\theta)$ and $\rho = \phi_2(\theta)$ are the equations of the projection of the curve of intersection of the surfaces onto the $\rho\theta$-plane.

The volume element in spherical coordinates (Fig. 11.16) is $r^2 \sin \phi$

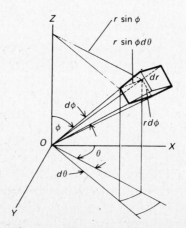

FIG. 11.16

$dr\, d\theta\, d\phi$, and the formula for the volume common to the two surfaces $r = \Psi_1(\theta, \phi)$ and $r = \Psi_2(\theta, \phi)$ is

$$V = \int_{\gamma}^{\delta} \int_{g_1(\phi)}^{g_2(\phi)} \int_{\Psi_1(\theta,\phi)}^{\Psi_2(\theta,\phi)} r^2 \sin\phi\, dr\, d\theta\, d\phi \quad \text{(spherical coordinates)} \quad (5)$$

Here r travels from surface to surface while θ and ϕ sweep over the solid angle of the cone with vertex at the origin and passing through the the curve of intersection of the two surfaces. Some illustrations will help to show how these limits of integration are found in a given problem.

Example 1. Find by triple integration the volume of a sphere of radius a using (a) rectangular coordinates, (b) cylindrical coordinates, and (c) spherical coordinates (Fig. 11.17).

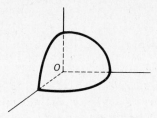

FIG. 11.17

Solution.
(a) The equation of the sphere is $x^2 + y^2 + z^2 = a^2$. For reasons of symmetry, we work only with the part lying in the first octant.

$$V = 8 \int_0^a \int_0^{\sqrt{a^2-x^2}} \int_0^{\sqrt{a^2-x^2-y^2}} dz\, dy\, dx$$

In the first integration, z sweeps from surface ($z = 0$) to surface $z = \sqrt{a^2 - x^2 - y^2}$. In the second integration, y travels from curve ($y = 0$) to curve $y = \sqrt{a^2 - x^2}$. This last form is gotten by setting $z = 0$ in the equation of the surface since this gives the curve of intersection of the surface with the xy-plane. In the final integration, x goes from 0 to a, the extent of the figure in the x-direction.

$$V = 8 \int_0^a \int_0^{\sqrt{a^2-x^2}} \sqrt{a^2 - x^2 - y^2} \, dy \, dx$$

$$= 8 \int_0^a \left[\frac{y}{2} \sqrt{a^2 - x^2 - y^2} \right.$$

$$\left. + \frac{a^2 - x^2}{2} \sin^{-1} \frac{y}{\sqrt{a^2 - x^2}} \right]_0^{\sqrt{a^2-x^2}} dx$$

$$= 2\pi \int_0^a (a^2 - x^2) \, dx$$

$$= 2\pi \left[a^2 x - \frac{x^3}{3} \right]_0^a = \frac{4}{3} \pi a^3 \text{ cu. units}$$

(b) The equation of the sphere is $\rho^2 + z^2 = a^2$.

$$V = 8 \int_0^{\pi/2} \int_0^a \int_0^{\sqrt{a^2-\rho^2}} \rho \, dz \, d\rho \, d\theta$$

$$= 8 \int_0^{\pi/2} \int_0^a \rho \sqrt{a^2 - \rho^2} \, d\rho \, d\theta$$

$$= 8 \int_0^{\pi/2} \left[-\frac{1}{3} (a^2 - \rho^2)^{3/2} \right]_0^a d\theta$$

$$= \frac{8}{3} \int_0^{\pi/2} a^3 \, d\theta = \frac{4}{3} \pi a^3 \text{ cu. units}$$

(c) The equation of the sphere is $r = a$.

$$V = 8 \int_0^{\pi/2} \int_0^{\pi/2} \int_0^a r^2 \sin \phi \, dr \, d\theta \, d\phi$$

$$= \frac{8}{3} a^3 \int_0^{\pi/2} \int_0^{\pi/2} \sin \phi \, d\theta \, d\phi$$

$$= \frac{4}{3} \pi a^3 \int_0^{\pi/2} \sin \phi \, d\phi = \frac{4}{3} \pi a^3 \text{ cu. units}$$

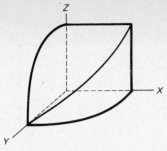

FIG. 11.18

Example 2. Find the volume common to the two cylinders $x^2 + y^2 = a^2$ and $y^2 + z^2 = a^2$ (Fig. 11.18).

Solution. Again we work with the part of the volume lying in the first octant. Since the curve of intersection lies on the cylinders, it will project into $x^2 + y^2 = a^2$ in the xy-plane.

$$V = 8 \int_0^a \int_0^{\sqrt{a^2-x^2}} \int_0^{\sqrt{a^2-y^2}} dz \, dy \, dx$$

$$= 8 \int_0^a \int_0^{\sqrt{a^2-x^2}} \sqrt{a^2 - y^2} \, dy \, dx$$

The integration can be performed in this order, but it is simpler to interchange the order of integration here and write

$$V = 8 \int_0^a \int_0^{\sqrt{a^2-y^2}} \sqrt{a^2 - y^2} \, dx \, dy$$

$$= 8 \int_0^a \left[x \sqrt{a^2 - y^2} \right]_0^{\sqrt{a^2-y^2}} dy$$

$$= 8 \int_0^a (a^2 - y^2) \, dy = \frac{16}{3} a^3 \text{ cu. units}$$

Example 3. Find the volume cut from the elliptic paraboloid $z = x^2 + 4y^2$ by the plane $z = 1$ (Fig. 11.19).

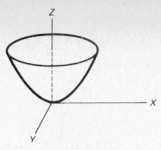

FIG. 11.19

Solution.

$$V = 4 \int_0^1 \int_0^{\sqrt{1-x^2}/2} \int_{x^2+4y^2}^1 dz\, dy\, dx$$

$$= 4 \int_0^1 \int_0^{\sqrt{1-x^2}/2} (1 - x^2 - 4y^2)\, dy\, dx$$

$$= 4 \int_0^1 \left[y(1-x^2) - \frac{4}{3}y^3 \right]_0^{\sqrt{1-x^2}/2} dx$$

$$= 4 \int_0^1 \frac{1}{3}(1 - x^2)^{3/2}\, dx$$

$$= \frac{1}{3} \left[x(1-x^2)^{3/2} + \frac{3}{2}x(1-x^2)^{1/2} + \frac{3}{2}\sin^{-1} x \right]_0^1$$

(Formula 48, Appendix B)

$$= \frac{\pi}{4} \text{ cu. unit}$$

Example 4. Find the volume bounded above by the cone $z = k\rho$ and below by the xy-plane, and lying in the cylinder on one loop of $\rho = \cos 2\theta$ (Fig. 11.20).

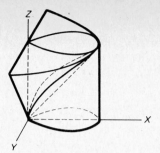

FIG. 11.20

Solution.

$$V = 2\int_0^{\pi/4} \int_0^{\cos 2\theta} \int_0^{k\rho} \rho\, dz\, d\rho\, d\theta$$

$$= 2\int_0^{\pi/4} \int_0^{\cos 2\theta} k\rho^2\, d\rho\, d\theta$$

$$= \frac{2k}{3}\int_0^{\pi/4} \cos^3 2\theta\, d\theta$$

$$= \frac{k}{9} \sin 2\theta\, (\cos^2 2\theta + 2)\, \Big|_0^{\pi/4} = \frac{2}{9} k \text{ cu. units}$$

Example 5. Find the volume inside the cone $\phi = \alpha$ and the sphere $r = a$ (Fig 11.21).

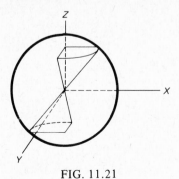

FIG. 11.21

Solution.

$$V = 2 \int_0^\alpha \int_0^{2\pi} \int_0^a r^2 \sin\phi \, dr \, d\theta \, d\phi$$

$$= \frac{2}{3} \int_0^\alpha \int_0^{2\pi} a^3 \sin\phi \, d\theta \, d\phi$$

$$= \frac{4\pi a^3}{3} \int_0^\alpha \sin\phi \, d\phi$$

$$= -\frac{4\pi a^3}{3} \left[\cos\phi\right]_0^\alpha$$

$$= \frac{4}{3} \pi a^3 (1 - \cos\alpha) \text{ cu. units}$$

PROBLEMS WITH SOLUTIONS

In Problems 1 through 6, find the indicated volume by *double* integration.

1. $x^2 + y^2 = r^2$, $z = 0$ and $z = h$.

Solution. $V = 4 \int_0^r \int_0^{\sqrt{r^2 - x^2}} h \, dy \, dx$

$$= 4 \int_0^r h \sqrt{r^2 - x^2} \, dx$$

$$= 4h \left[\frac{x}{r} \sqrt{r^2 - x^2} + \frac{r^2}{2} \sin^{-1}\frac{x}{r}\right]_0^r = \pi r^2 h$$

2. $\dfrac{x^2}{a^2} + \dfrac{y^2}{b^2} = 1$, $z = 0$ and $z = h$.

Solution. $V = 4 \int_0^a \int_0^{b\sqrt{a^2 - x^2}/a} h \, dy \, dx$

$$= \frac{4bh}{a} \int_0^a \sqrt{a^2 - x^2} \, dx = \pi abh$$

3. $z = x^2 + y^2$ and $z = h^2$.

Solution. $V = 4 \displaystyle\int_0^h \int_0^{\sqrt{h^2-x^2}} (h^2 - x^2 - y^2)\, dy\, dx$

$$= 4 \int_0^h \left[(h^2 - x^2)\, y - \frac{y^3}{3} \right]_0^{\sqrt{h^2-x^2}} dx$$

$$= \frac{8}{3} \int_0^h (h^2 - x^2)^{3/2}\, dx$$

$$= \frac{8}{3} \left[\frac{x}{4} (h^2 - x^2)^{3/2} + \frac{3}{8} h^2 x (h^2 - x^2)^{1/2} \right.$$

$$\left. + \frac{3}{8} h^4 \operatorname{Sin}^{-1} \frac{x}{h} \right]_0^h = \frac{\pi}{2} h^4$$

4. One of the wedges cut from $x^2 + y^2 = r^2$ by $z = 0$ and $z = x$ (Fig. 11.22).

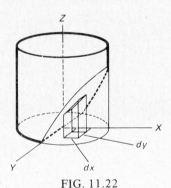

FIG. 11.22

Solution. $V = 2 \displaystyle\int_0^r \int_0^{\sqrt{r^2-x^2}} x\, dy\, dx$

$$= \int_0^r (r^2 - x^2)\, dx = \frac{2}{3} r^3$$

5. A tetrahedron bounded by $\dfrac{x}{a} + \dfrac{y}{b} + \dfrac{z}{c} = 1$, $x = 0$, $y = 0$, and $z = 0$.

Solution. $V = \displaystyle\int_0^a \int_0^{b(1-x/a)} c\left(-\dfrac{x}{a} - \dfrac{y}{b}\right) dy\, dx$

$\qquad = c\displaystyle\int_0^a \left(1 - \dfrac{x}{a}\right)y - \dfrac{y^2}{2b}\Bigg|_0^{b(1-x/a)} dx$

$\qquad = \dfrac{bc}{2}\displaystyle\int_0^a \left(1 - \dfrac{x}{a}\right)^2 dx = \dfrac{abc}{6}$

6. $y^2 + z = 1$, $x + y = 1$, $x = 0$, and $z = 0$ (Fig. 11.23).

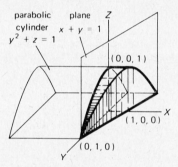

FIG. 11.23

Solution. $V = \displaystyle\int_0^2 \int_{-1}^{1-x} (1 - y^2)\, dy\, dx$

$\qquad = \displaystyle\int_{-1}^1 \int_0^{1-y} (1 - y^2)\, dx\, dy$

$\qquad = \displaystyle\int_{-1}^1 (1 - y^2)(1 - y)\, dy$

$\qquad = \left[y + \dfrac{y^2}{2} - \dfrac{y^3}{3} + \dfrac{y^4}{4}\right]_{-1}^1 = \dfrac{4}{3}$

In Problems 7 through 10, find the indicated volume by *triple* integration.

7. The inside of $x^2 + y^2 = 4$ bounded above by $y + z = 3$ and below by $z = 0$ (Fig. 11.24).

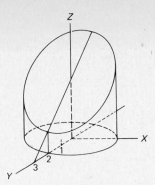

FIG. 11.24

Solution. $V = 4 \int_0^2 \int_0^{\sqrt{4-x^2}} \int_0^{3-y} dz\, dy\, dx$

$= 4 \int_0^2 \int_0^{\sqrt{4-x^2}} (3 - y)\, dy\, dx$

$= 4 \int_0^2 \left[3y - \frac{y^2}{2} \right]_0^{\sqrt{4-x^2}} dx$

$= 24 \int_0^2 \sqrt{4 - x^2}\, dx = 12\pi$

8. The volume bounded by the two elliptic paraboloids $z = 4 - x^2 - \frac{1}{4} y^2$ and $z = 3x^2 + \frac{1}{4} y^2$ (Fig. 11.25). The cylinder passing through the curve of intersection (eliminating z) is $1 = x^2 + \frac{y^2}{8}$.

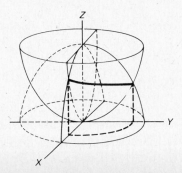

FIG. 11.25

Solution. $V = 4 \int_0^1 \int_0^{2\sqrt{2}\ \sqrt{1-x^2}} \int_{3x^2+(y^2/4)}^{4-x^2-(y^2/4)} dz\ dy\ dx$

$= 4 \int_0^1 \int_0^{2\sqrt{2}\ \sqrt{1-x^2}} \left(4 - 4x^2 - \frac{1}{2}y^2 \right) dy\ dx$

$= 4 \int_0^1 \left[4(1 - x^2)\ y - \frac{1}{6}y^3 \right]_0^{2\sqrt{2}\ \sqrt{1-x^2}} dx$

$= 4 \int_0^1 \left(8\sqrt{2} - \frac{8}{3}\ \sqrt{2} \right) (1 - x^2)^{3/2}\ dx$

$= 4 \left(8\sqrt{2}\ - \frac{8}{3}\sqrt{2} \right) \left[\frac{x}{4}\ (1 - x^2)^{3/2} \right.$

$\left. + \frac{3}{8}x\ (1 - x^2)^{1/2} + \frac{3}{8}\ \mathrm{Sin}^{-1}\ x \right]_0^1 = 4\sqrt{2}\pi$

9. The spherical cap cut from the sphere $x^2 + y^2 + z^2 = 1$ by the plane $x + y + z = 1$.

Solution. $V = \frac{\pi}{6} - \int_0^1 \int_0^{1-x} \int_0^{1-x-y} dz\ dy\ dx$

$= \frac{\pi}{6} - \int_0^1 \int_0^{1-x} (1 - x - y)\ dy\ dx$

$= \frac{\pi}{6} - \int_0^1 \left. y - xy - \frac{y^2}{2} \right|_0^{1-x} dx$

$= \frac{\pi}{6} - \int_0^1 \frac{(1 - x)^2}{2}\ dx = \frac{\pi - 1}{6}$

10. The hyperboloid of one sheet $\frac{x^2}{a^2} + \frac{y^2}{b^2} - \frac{z^2}{c^2} = 1$ from $z = 0$ to H.

Solution. $V = 4\displaystyle\int_0^H \int_0^{b\sqrt{z^2+c^2}/c} \int_0^{a\sqrt{1+(z^2/c^2)-(y^2/b^2)}} dx\, dy\, dz$

$$= \frac{4a}{b}\int_0^H \int_0^{b\sqrt{z^2+c^2}/c} \sqrt{\frac{b^2}{c^2}(c^2+z^2)-y^2}\; dy\, dz$$

$$= \frac{4a}{b}\int_0^H \left[\frac{y}{2}\sqrt{\frac{b^2}{c^2}(c^2+z^2)-y^2}\right.$$

$$\left. + \frac{b^2}{2c^2}(c^2+z^2)\,\text{Sin}^{-1}\frac{cy}{b\sqrt{c^2+z^2}}\right]_0^{b\sqrt{z^2+c^2}/c} dz$$

$$= \frac{\pi ab}{c^2}\int_0^H (c^2+z^2)\,dz = \pi abH\left[1+\frac{1}{3}\left(\frac{H}{c}\right)^2\right]$$

PROBLEMS WITH ANSWERS

In Problems 11 through 15, find the volumes described by triple integration.

11. The volume bounded by $x = 0$, $y = 0$, $z = 0$, and $\dfrac{x}{a}+\dfrac{y}{b}+\dfrac{z}{c} = 1$.

Ans. $\frac{1}{6}abc$

12. The volume inside the cylinder $\rho = a\cos\theta$ and the sphere $\rho^2 + z^2 = a^2$.

Ans. $\dfrac{4}{3}a^3\left(\dfrac{\pi}{2}-\dfrac{2}{3}\right)$

13. The ellipsoid $\dfrac{x^2}{a^2}+\dfrac{y^2}{b^2}+\dfrac{z^2}{c^2} = 1$.

Ans. $\frac{4}{3}\pi abc$

14. The spherical wedge made by cutting a slice from a sphere of radius a by two planes passing through a diameter and making an angle α with each other.

Ans. $\frac{2}{3}\alpha a^3$

15. The volume cut off from the paraboloid $z = x^2 + y^2$ by the plane $z - y = 0$.

Ans. $\dfrac{\pi}{32}$

11.5   Centroids and Moments

The deviation and application of the following formulas should be wholly within the grasp of the student of the calculus at this stage in his study. They make use only of the definitions of center of mass and moment of inertia and of the ideas of multiple integration. Again, the student is warned not to try to memorize these formulas, but rather to spend the equivalent time on the fundamental principles involved in them. The notations used are self-explanatory.

For Plane Areas (density = σ)

Centroid: Rectangular and Polar Coordinates

$$\overline{x} = \frac{\int\int \sigma x \, dy \, dx}{\int\int \sigma \, dy \, dx} \tag{1}$$

$$\overline{y} = \frac{\int\int \sigma y \, dy \, dx}{\int\int \sigma \, dy \, dx} \tag{2}$$

$$\overline{x} = \frac{\int\int \sigma \rho^2 \, \cos \theta \, d\rho \, d\theta}{\int\int \sigma \rho \, d\rho \, d\theta} \tag{3}$$

$$\overline{y} = \frac{\int\int \sigma \rho^2 \, \sin \theta \, d\rho \, d\theta}{\int\int \sigma \rho \, d\rho \, d\theta} \tag{4}$$

From (3) and (4), $\overline{\rho}$ and $\overline{\theta}$ can be determined: $\overline{\rho} = \sqrt{\overline{x}^2 + \overline{y}^2}$ and $\overline{\theta} = \text{Tan}^{-1} \dfrac{\overline{y}}{\overline{x}}$.

Moment of Inertia: Rectangular and Polar Coordinates

$$I_x = \int\int \sigma y^2 \, dy \, dx \tag{5}$$

$$I_y = \int\int \sigma x^2 \, dy \, dx \tag{6}$$

$$I_0 = \int\int \sigma (x^2 + y^2) \, dy \, dx \tag{7}$$

$$= I_x + I_y$$

$$I_x = \int\int \sigma \rho^3 \, \sin^2 \theta \, d\rho \, d\theta \tag{8}$$

$$I_y = \int\int \sigma \rho^3 \, \cos^2 \theta \, d\rho \, d\theta \tag{9}$$

$$I_0 = \int\int \sigma\rho^3 \, d\rho \, d\theta \tag{10}$$

For Volumes (density = σ)

Centroid: Rectangular and Cylindrical Coordinates

$$\bar{x} = \frac{\int\int\int \sigma x \, dz \, dy \, dx}{\int\int\int \sigma \, dz \, dy \, dx} \tag{11}$$

$$\bar{y} = \frac{\int\int\int \sigma y \, dz \, dy \, dx}{\int\int\int \sigma \, dz \, dy \, dx} \tag{12}$$

$$\bar{z} = \frac{\int\int\int \sigma z \, dz \, dy \, dx}{\int\int\int \sigma \, dz \, dy \, dx} \tag{13}$$

$$\bar{x} = \frac{\int\int\int \sigma\rho^2 \cos\theta \, dz \, d\rho \, d\theta}{\int\int\int \sigma\rho \, dz \, d\rho \, d\theta} \tag{14}$$

$$\bar{y} = \frac{\int\int\int \sigma\rho^2 \sin\theta \, dz \, d\rho \, d\theta}{\int\int\int \sigma\rho \, dz \, d\rho \, d\theta} \tag{15}$$

$$\bar{z} = \frac{\int\int\int \sigma\rho z \, dz \, d\rho \, d\theta}{\int\int\int \sigma\rho \, dz \, d\rho \, d\theta} \tag{16}$$

Moment of Inertia with Respect to Coordinate Axes

$$I_x = \int\int\int \sigma(y^2 + z^2) \, dz \, dy \, dx \tag{17}$$

$$I_y = \int\int\int \sigma(x^2 + z^2) \, dz \, dy \, dx \tag{18}$$

$$I_z = \int\int\int \sigma(x^2 + y^2) \, dz \, dy \, dx \tag{19}$$

For geometric areas and volumes, consider $\sigma = 1$. Where density is a constant (homogeneous masses), σ may be brought out from under the sign of integration.

Example 1. Find by triple integration the moment of inertia of a solid sphere of radius a about a diameter. (See Problem 10, Section 9.12.)

Solution. We use spherical coordinates. Although we did not include such a formula in the above set, it should be clear that

$$I_d = 8 \int_0^{\pi/2} \int_0^{\pi/2} \int_0^a \sigma r^2 \sin^2 \phi \cdot r^2 \sin \phi \, dr \, d\theta \, d\phi$$

$$= 8\sigma \frac{a^5}{5} \int_0^{\pi/2} \int_0^{\pi/2} \sin^3 \phi \, d\theta \, d\phi$$

$$= 8\sigma \pi \frac{a^5}{10} \int_0^{\pi/2} \sin^3 \phi \, d\phi$$

$$= \tfrac{8}{15} \pi \sigma a^2 \text{ (volume, } \sigma = 1)$$

$$= \tfrac{2}{5} M a^2 \text{ (mass)}$$

Example 2. Find the c.g. of the tetrahedron formed by the coordinate planes and the plane $\dfrac{x}{a} + \dfrac{y}{b} + \dfrac{z}{c} = 1$.

Solution. The volume is $\frac{1}{6} abc$.

$$\bar{x} = \frac{6}{abc} \int_0^a \int_0^{b(1-x/a)} \int_0^{c(1-x/a-y/b)} x \, dz \, dy \, dx$$

$$= \frac{6}{abc} \int_0^a \int_0^{b(1-x/a)} cx \left(1 - \frac{x}{a} - \frac{y}{b}\right) dy \, dx$$

$$= \frac{6}{abc} \int_0^a \frac{bcx}{2} \left(1 - \frac{x}{a}\right)^2 dx$$

$$= \frac{3}{a} \int_0^a \left(x - \frac{2}{a}x^2 + \frac{x^3}{a^2}\right) dx = \frac{1}{4} a$$

By symmetry, $\bar{y} = \frac{1}{4} b$ and $\bar{z} = \frac{1}{4} c$.

The c.g. of a composite body can be found from the formula

$$\bar{x} = \frac{m_1 \bar{x}_1 + m_2 \bar{x}_2 + \cdots + m_n \bar{x}_n}{m_1 + m_2 + \cdots + m_n}$$

where $\bar{x}_i$ is the c.g. of the ith part m_i of the total mass $m_1 + m_2 + \cdots + m_n$ considered.

Example 3. Find the c.g. of the plate shown in Fig. 11.26.

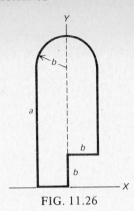

FIG. 11.26

Solution. For the rectangle,

$$\overline{x} = 0 \quad \text{and} \quad \overline{y} = \frac{a}{2}$$

For the semicircular top,

$$\overline{x} = 0 \quad \text{and} \quad \overline{y} = a + \frac{4b}{3\pi}$$

(See Example 3, Section 9.11.) For the square cut out:

$$\overline{x} = \tfrac{1}{2}b \quad \text{and} \quad \overline{y} = \tfrac{1}{2}b$$

The c.g. $(\overline{x}, \overline{y})$ of the composite body will be at

$$\overline{x} = \frac{(2ab)(0) + \frac{1}{2}\pi b^2(0) - b^2(\frac{1}{2}b)}{2ab + \frac{1}{2}\pi b^2 - b^2}$$

$$= \frac{-b^2}{4a + b(\pi - 2)}$$

$$\overline{y} = \frac{(2ab)\left(\frac{1}{2}a\right) + \frac{1}{2}\pi b^2\left(a + \frac{4b}{3\pi}\right) - b^2\left(\frac{1}{2}b\right)}{2ab + \frac{1}{2}\pi b^2 - b^2}$$

$$= \frac{2a^2 + b\left(\pi a + \frac{4b}{3}\right) - 1}{4a + b(\pi - 2)}$$

For determining the moment of inertia of a composite body we make use of the following *transfer* or

PARALLEL AXIS THEOREM. *The moment of inertia I_L of a mass M with respect to a line L equals the moment of inertia about the line parallel to L and passing through the c.g., plus the mass M times the square of the distance d between the two lines.* That is,

$$I_L = I_g + Md^2 \qquad\qquad (20)$$

Example 4. Find the moment of inertia of a circular area about a tangent. (See Example 3, Section 9.12.)

Solution.
$$I_L = I_g + Md^2$$

$$= \frac{\pi r^4}{4} + \pi r^2 \cdot r^2$$

$$= \tfrac{5}{4}\pi r^4 \ (\text{area})$$

$$= \frac{Mr^2}{4} + Mr^2$$

$$= \tfrac{5}{4}Mr^2 \ (\text{plate of mass } M)$$

Example 5. Find the moment of inertia of a solid cylinder of radius r and height h about a generator.

Solution.
$$I_g = \tfrac{1}{2}Mr^2$$
$$I_L = \tfrac{1}{2}Mr^2 + Mr^2$$
$$= \tfrac{3}{2}Mr^2$$

Example 6. Find the moment of inertia of a solid cylinder of radius r and height h about a diameter of the base (Fig. 11.27).

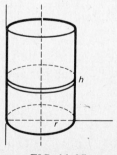

FIG. 11.27

Solution. Take disc elements of mass $\rho\pi r^2\, dy$. The moment of inertia of this disc about a diameter is

$$\frac{Mr^2}{4}, \quad \text{or} \quad \frac{\rho\pi r^2\, dy}{4}\, r^2$$

Now we use the transfer theorem to get the moment about the diameter of the base of the cylinder. Since the disc element is y units above the base, this is

$$I_{\text{disc}} = \frac{\rho\pi r^2\, dy}{4}\, r^2 + \rho\pi r^2\, dy \cdot y^2$$

$$= \tfrac{1}{4}\rho\pi r^2 (r^2 + 4y^2)\, dy$$

Hence,

$$I_x = \frac{1}{4}\pi r^2 \int_0^b \rho (r^2 + 4y^2)\, dy$$

$$= \frac{1}{4}\pi r^2 \rho \left(r^2 h + \frac{4h^3}{3} \right)$$

$$= \rho\pi r^2 h \left(\frac{r^2}{4} + \frac{h^3}{3} \right) \text{(volume)}$$

$$= M \left(\frac{r^2}{4} + \frac{h^2}{3} \right) \text{(mass)}$$

It should be emphasized that this parallel axis theorem applies only to parallel axes one of which passes through the center of gravity.

There are two theorems, known as the theorems of Pappus, which are of great use in the calculus.

PAPPUS' THEOREM I. *When a plane area is revolved about a coplanar axis not cutting the area, the volume generated is equal to the product of the area and the length of the path described by the center of gravity of the area.*

PAPPUS' THEOREM II. *When a plane curve is revolved about a coplanar axis not cutting the curve, the surface area generated is equal to the product of the length of the curve and the length of the path of the center of gravity of the curve.*

The reason for the condition that the line about which the revolution takes place shall not cut the area or curve is apparent, since, otherwise, dual volumes and surface areas would be generated. Algebraically, the theorems would still be true and could be applied even in this exceptional case if appropriate use was made of signs.

Example 7. A torus (doughnut) is generated by revolving the circle $(x - a)^2 + y^2 = b^2$, $(b < a)$, about the Y-axis. Find (a) the volume and (b) the surface area of this solid.

Solution. The c.g. of the area and also of the curve is at the center of the circle; hence,

$$V = \pi b^2 (2\pi a) \qquad\qquad\qquad\text{(a)}$$

$$= 2\pi^2 ab^2 \text{ cu. units}$$

$$S = 2\pi b (2\pi a) \qquad\qquad\qquad\text{(b)}$$

$$= 4\pi^2 ba \text{ sq. units}$$

Example 8. Use Pappus' theorems to find (a) the volume and (b) the lateral area of a right circular cone of height h and radius r.

Solution.
(a) Consider the triangle with vertices at $(0, 0)$, $(r, 0)$, and $(0, h)$. The c.g. of this triangle considered as an area (plate) is

$$\overline{x} = \tfrac{1}{3}r \quad \text{and} \quad \overline{y} = \tfrac{1}{3}h$$

By revolving this triangle about the Y-axis, the cone is generated. (See Problems 6 and 7, Section 9.11.)

$$V = \tfrac{1}{2}rh\left(\tfrac{1}{3}\pi r\right)$$

$$= \tfrac{1}{3}\pi r^2 h \text{ cu. units}$$

(b) In order to get the surface area of the cone, we merely revolve the hypotenuse of the triangle about the Y-axis. The c.g. of this curve is at $\overline{x} = \tfrac{1}{2}r$ and $\overline{y} = \tfrac{1}{2}h$; hence,

$$S = \sqrt{r^2 + h^2}\left(\frac{2\pi r}{2}\right)$$

$$= \pi r \sqrt{r^2 + h^2} \text{ sq. units}$$

PROBLEMS WITH SOLUTIONS

1. Prove Pappus' Theorem I (Fig. 11.28).

Solution. $V = 2\pi \displaystyle\int_a^b Ly \, dy$

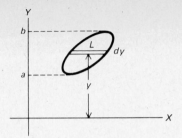

FIG. 11.28

But $A\overline{y} = \displaystyle\int_a^b L y \, dy$

Therefore, $V = 2\pi\overline{y} \times A$

2. Prove Pappus' Theorem II (Fig. 11.29).

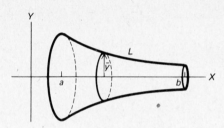

FIG. 11.29

Solution. $S = 2\pi \displaystyle\int_a^L y \, ds$

$\overline{y} = \dfrac{\displaystyle\int_a^b y \, ds}{L}$

Therefore, $S = 2\pi\overline{y} \times L$

3. The circle $(x - b)^2 + y^2 = r^2$, $b > r$, is revolved about the Y-axis (Fig. 11.30). Show by Pappus' theorems that:
   a. $V = 2b\pi^2 r^2$
   b. $S = 4b\pi^2 r$

Solution.
   a. $V = 2\pi b \cdot \pi r^2$
   b. $S = 2\pi b \cdot 2\pi r$

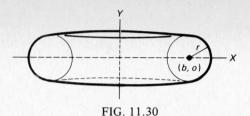

FIG. 11.30

4. Show by Pappus' theorems that (a) the centroid of a semicircular area of radius r is $\overline{x} = \dfrac{4r}{3\pi}$ and $\overline{y} = 0$, where the equation of the semicircle is $x = \sqrt{r^2 - y^2}$, and (b) the centroid of a semicircular arc is $\overline{x} = \dfrac{2r}{\pi}$ and $\overline{y} = 0$.

Solution.

(a) $\overline{x} = \dfrac{V}{A \cdot 2\pi} = \dfrac{\frac{4}{3}\pi r^3}{\dfrac{\pi r^2}{2} \cdot 2\pi}$

(b) $\overline{x} = \dfrac{S}{L \cdot 2\pi} = \dfrac{4\pi r^2}{\pi r \cdot 2\pi}$

5. Find the volume and surface area generated by revolving the area (perimeter) of an equilateral triangle about a side of length a.
   Solution. $V = \frac{1}{4}\pi a^3$ and $S = 3\pi a^2$

6. Find the centroid of a thin plate in the form of one quadrant of the ellipse $\dfrac{x^2}{a^2} + \dfrac{y^2}{b^2} = 1$ in which the density varies as the product xy.
   Solution.

$$\overline{x} = \frac{\int_0^a \int_0^{b\sqrt{a^2 - x^2}/a} kx^2\, y\; dy\; dx}{\int_0^a \int_0^{b\sqrt{a^2 - x^2}/a} kxy\; dy\; dx} = \frac{8}{15}a$$

Similarly, $\overline{y} = \frac{8}{15}b$.

7. Find the centroid of the wedge-shaped solid cut from the cylinder $x^2 + y^2 = a^2$ by the planes $z = 0$ and $\dfrac{z}{b} + \dfrac{x}{a} = 1$ (Fig. 11.31).

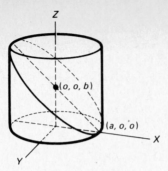

FIG. 11.31

Solution. The moments with respect to the respective coordinate axes are as follows:

$$M_{xy} = \int_{-a}^{a} \int_{-\sqrt{a^2-x^2}}^{\sqrt{a^2-x^2}} \int_{0}^{b(1-x/a)} z \, dz \, dy \, dx$$

$$= b^2 \int_{-a}^{a} \left(1 - \frac{x}{a}\right)^2 \sqrt{a^2 - x^2} \, dx = \frac{5\pi a^2 b^2}{8}$$

$$M_{zy} = \int_{-a}^{a} \int_{-\sqrt{a^2-x^2}}^{+\sqrt{a^2-x^2}} \int_{0}^{b(1-x/a)} x \, dz \, dy \, dx$$

$$= 2b \int_{-a}^{a} x\left(1 - \frac{x}{a}\right) \sqrt{a^2 - x^2} \, dx = -\frac{\pi a^3 b}{4}$$

$$M_{xz} = 0$$

Hence, $\overline{x} = \dfrac{M_{zy}}{V} = -\dfrac{\pi a^3 b/4}{\pi a^2 b} = -\dfrac{1}{4}a$

$$\overline{y} = 0$$

$$\overline{z} = \frac{5\pi a^2 b^2}{8\pi a^2 b} = \frac{5}{8}b$$

PROBLEMS WITH ANSWERS

8. Find the c.g. of the solid bounded by the hyperboloid $z^2 = 1 + \rho^2$ and the upper half of the cone $z^2 = 2\rho^2$.

Ans. $\overline{x} = 0 = \overline{y}, \ \overline{z} = \frac{3}{8}(1 + \sqrt{2})$

9. Find the moment of inertia of a rectangular bar of length L, width a, and thickness b about an axis parallel to b and passing through the c.g.

$$Ans. \quad \frac{M}{12}(a^2 + L^2)$$

10. Find the moment of inertia of a solid circular cylinder of radius r and height h about an axis perpendicular to the axis of the cylinder and passing through the c.g. (See Example 6.)

$$Ans. \quad \frac{M}{12}(3r^2 + h^2)$$

11. Find the c.g. of the area of the ellipse $b^2 x^2 + a^2 y^2 = a^2 b^2$ lying in the first quadrant.

$$Ans. \quad \overline{x} = \frac{4a}{3\pi}, \quad \overline{y} = \frac{4b}{3\pi}$$

12. The triangular plate with sides 3, 4, and 5 is placed in the xy-plane so that its c.g. is at $(2, 3)$. Find the volume generated when this triangular plate is rotated about the X-axis.   *Ans.* 36π cu. units

13. Find the moment of inertia of a solid sphere about a tangent line.

$$Ans. \quad \tfrac{7}{5}Mr^2$$

14. A cylindrical pencil $\frac{1}{8}$ in. in radius and 7 in. in overall length has a sharp conical point 1 in. long. Find the c.g. measured from the point. (See Example 6, Section 9.11.)   *Ans.* $\frac{291}{76} = 3.83$ in.

15. Find the moment of inertia of the pencil in Problem 14 about the axis.

$$Ans. \quad \frac{3\rho\pi}{20{,}480}$$

11.6 Liquid Pressure and Force

The following theorem is most useful and has wide applications in mechanics and hydrodynamics. It connects area, c.g., and force.

LIQUID PRESSURE THEOREM. *When a plate of plane area A is submerged vertically, the total force on one side is equal to the product of the area of the plate, the depth of the c.g. of the plate, and w the weight of the liquid per unit volume.* To put this another way: *the total force equals the area times the pressure at the c.g.*

Example. The center of a circular floodgate of radius 2 ft. in a reservoir is at a depth of 6 ft. Find the total force on the floodgate. (See Example 2, Section 9.10.)

Solution.
$$F = 4\pi(6)w$$
$$= 24\pi w \text{ lbs.}$$

PROBLEMS WITH SOLUTIONS

1. Prove the liquid pressure theorem.

    Solution. $F = w \int_0^h yL(y)\,dy$

    $$\overline{y} = \frac{\int_0^h yL(y)\,dy}{\int_0^h L(y)\,dy} = \frac{\int_0^h yL(y)\,dy}{A}$$

    $$F = A \times w\overline{y}$$

    Note that the total force equals the product of the area and pressure at the c.g.

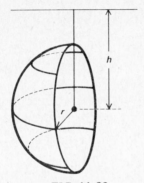

FIG. 11.32

2. A hollow hemisphere is suspended in water (Fig. 11.32). Find the total pressure on one side.
    Solution. $F = 4\pi r^2 \times wh$

PROBLEMS WITH ANSWERS

3. A semicircular plate of radius r ft. is submerged vertically until its c.g. is at a depth of h ft. Find the total pressure on one side.
    Ans. $\frac{1}{2}\pi wr^2 h$ lbs.

4. The total force on one side of a vertically submerged elliptical plate $\left(\dfrac{x^2}{9} + \dfrac{y^2}{4} = 1\right)$ is $120\pi w$ lbs. At what depth is the c.g.? *Ans.* 20 ft.

CHAPTER 12
VECTORS

12.1 Vector Algebra

The **vector** $A = \overrightarrow{PQ}$ is a **directed line segment** from an *initial* point P to a *terminal* point Q (Fig. 12.1). A vector, therefore, has a **magnitude** (length) and a **direction**. The magnitude of A is denoted by $|A|$ or by $|\overrightarrow{PQ}|$; the direction is usually given in terms of θ, the angle the vector makes with the *positive X*-axis. The vector whose length is 0 is denoted by the zero vector O, which has no direction. Two vectors $A = \overrightarrow{PQ}$ and $B = \overrightarrow{RS}$ are equal if and only if they have the *same* magnitude and direction. If $A = \overrightarrow{PQ}$ and $B = \overrightarrow{QP}$, we write $A = -B$.

The sum of two vectors $\overrightarrow{PQ}$ and $\overrightarrow{RS}$ is defined as follows: place $\overrightarrow{RS}$ so that R coincides with Q (Fig. 12.2), then

$$\overrightarrow{PQ} + \overrightarrow{RS} = \overrightarrow{PS} \tag{1}$$

Vector addition is *commutative* and *associative*; that is,

$$A + B = B + A \text{ (commutative law)}$$

$$A + (B + C) = (A + B) + C \text{ (associative law)}$$

We write $A + A = 2A$, and, in general, we consider the vector A multiplied by a **scalar** k (k a real number) and write kA. This represents a vector of magnitude $|k|\,|A|$ in the direction of A. The *distributive law* applies when vectors are multiplied by scalars; that is,

$$k(A + B) = kA + kB \quad \text{(Fig. 12.3)}$$

$$(k_1 + k_2)A = k_1 A + k_2 A \quad \text{(Fig. 12.4)}$$

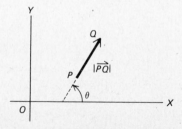

FIG. 12.1

328

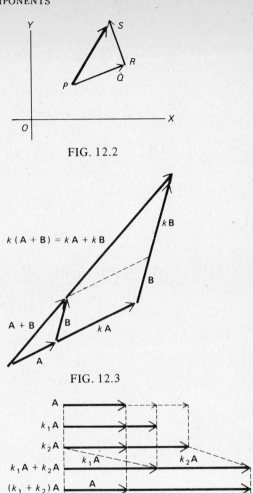

FIG. 12.2

FIG. 12.3

FIG. 12.4

12.2 Vector Components

A vector may be translated, or moved to a straight line (not rotated) parallel to itself and retain its magnitude and direction. The vector $\mathbf{A} = \overrightarrow{OP}$ with its initial point at the *origin* of a rectangular coordinate system is called the **position vector** of the point P (Fig. 12.5). Let $\mathbf{i}$ and $\mathbf{j}$ be *unit* vectors (each of unit length) with directions parallel to the X- and

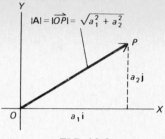

FIG. 12.5

Y-axes, respectively. Then the vector **A** can be written in component form:

$$\mathbf{A} = a_1 \mathbf{i} + a_2 \mathbf{j} \tag{1}$$

where **i** and **j** are the unit components in the x and y directions, respectively, and a_1 and a_2 are arbitrary scalars. The *magnitude* of **A** is defined as

$$|\mathbf{A}| = \sqrt{a_1^2 + a_2^2} \tag{2}$$

Then the sum of two vectors **A** and **B** (Fig. 12.6) is

$$\mathbf{A} + \mathbf{B} = (a_1 \mathbf{i} + a_2 \mathbf{j}) + (b_1 \mathbf{i} + b_2 \mathbf{j}) = (a_1 + b_1)\mathbf{i} + (a_2 + b_2)\mathbf{j}$$

where the component form of **A** is $a_1 \mathbf{i} + a_2 \mathbf{j}$ and the direction of **A** is given by $\theta = \tan^{-1} \dfrac{a_2}{a_1}$; similarly for **B**. The vector **A** can also be written in the form

$$\mathbf{A} = \mathbf{i} |\mathbf{A}| \cos \theta + \mathbf{j} |\mathbf{A}| \sin \theta \tag{3}$$

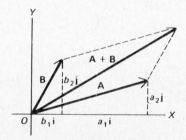

FIG. 12.6

The unit vector **e** is represented as

$$\mathbf{e} = \frac{\mathbf{A}}{|\mathbf{A}|} = \mathbf{i} \cos\theta + \mathbf{j} \sin\theta$$

$$= \frac{\mathbf{A}}{\sqrt{a_1^2 + a_2^2}} = \frac{a_1\mathbf{i} + a_2\mathbf{j}}{\sqrt{a_1^2 + a_2^2}} \tag{4}$$

The right-hand member of the last equation may also be written as

$$\frac{(a_1, a_2)}{\sqrt{a_1^2 + a_2^2}}$$

12.3 Derivative of a Vector

Let t be a scalar and C be a curve given in parametric form $x = f(t)$, $y = g(t)$ and let the position vector **R** of a point P on C be given in component form by

$$\mathbf{R}(t) = f(t)\mathbf{i} + g(t)\mathbf{j} \tag{1}$$

Then the derivative of **R** with respect to t is

$$\mathbf{R}'(t) = \frac{d\mathbf{R}}{dt} = \frac{df}{dt}\mathbf{i} + \frac{dg}{dt}\mathbf{j} \tag{2}$$

If $\mathbf{R}'(t) \neq 0$ at P, then $\mathbf{R}'(t)$ is a vector tangent to C at P pointing in the direction of motion of P along C as t increases (Fig. 12.7). The *length* of the vector $\mathbf{R}'(t)$ is

$$|\mathbf{R}'(t)| = \sqrt{\left(\frac{df}{dt}\right)^2 + \left(\frac{dg}{dt}\right)^2} \tag{3}$$

(evaluated at t associated with the point P)

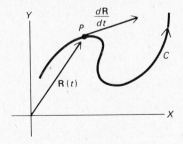

FIG. 12.7

Thus, $\mathbf{R}(t)$ is the *position* vector of P, $\mathbf{R}'(t)$ is the *velocity* vector $\mathbf{V}(t)$ of P along the curve, and $\mathbf{A}(t) = d\mathbf{V}/dt = \mathbf{R}''(t)$ is the *acceleration* vector.

12.4   The Scalar Product

The **scalar product** of two vectors (also called the *dot product*) is

$$\mathbf{A} \cdot \mathbf{B} = |\mathbf{A}|\,|\mathbf{B}|\cos\theta \tag{1}$$

where $|\mathbf{A}|$ and $|\mathbf{B}|$ are the magnitudes of $\mathbf{A}$ and $\mathbf{B}$ and θ is the angle between $\mathbf{A}$ and $\mathbf{B}$. Vectors $\mathbf{A}$ and $\mathbf{B}$ are perpendicular if and only if $\mathbf{A} \cdot \mathbf{B} = 0$, assuming that neither $\mathbf{A}$ nor $\mathbf{B}$ is the zero vector. This follows since $\cos\dfrac{\pi}{2} = 0$ (Fig. 12.8). Vectors $\mathbf{A}$ and $\mathbf{B}$ are *parallel* if and only if

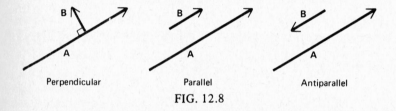

Perpendicular   Parallel   Antiparallel

FIG. 12.8

$\mathbf{A} \cdot \mathbf{B} = |\mathbf{A}|\,|\mathbf{B}|$ (here $\theta = 0$). Vectors $\mathbf{A}$ and $\mathbf{B}$ are *antiparallel* if and only if $\mathbf{A} \cdot \mathbf{B} = -|\mathbf{A}||\mathbf{B}|$, where $\theta = \pi$. In general, the angle θ is given by

$$\cos\theta = \frac{\mathbf{A} \cdot \mathbf{B}}{|\mathbf{A}|\,|\mathbf{B}|}$$

Note that

$$\mathbf{i} \cdot \mathbf{i} = \mathbf{j} \cdot \mathbf{j} = 1$$
$$\mathbf{j} \cdot \mathbf{i} = \mathbf{i} \cdot \mathbf{j} = 0$$

Where $\mathbf{A} = a_1\mathbf{i} + a_2\mathbf{j}$ and $\mathbf{B} = b_1\mathbf{i} + b_2\mathbf{j}$, the scalar product can be written in the form

$$\mathbf{A} \cdot \mathbf{B} = (a_1\mathbf{i} + a_2\mathbf{j}) \cdot (b_1\mathbf{i} + b_2\mathbf{j}) = a_1b_1 + a_2b_2$$

For any vectors $\mathbf{A}$, $\mathbf{B}$, and $\mathbf{C}$, the following laws hold.

$$k(\mathbf{A} \cdot \mathbf{B}) = (k\mathbf{A}) \cdot \mathbf{B} = \mathbf{A} \cdot (k\mathbf{B})$$
$$\mathbf{A} \cdot (\mathbf{B} + \mathbf{C}) = \mathbf{A} \cdot \mathbf{B} + \mathbf{A} \cdot \mathbf{C}$$

PROBLEMS WITH SOLUTIONS

1. Prove that $|A + B| \leqslant |A| + |B|$ and $|A - B| \geqslant ||A| - |B||$.
   Solution. These follow from the triangle inequality which states that *the length of one side of a triangle is less than or equal to the sum of the lengths of the other two sides.*

2. Given $A = 2i - 3j$, $B = i + 2j$, and $C = 4j$. Show that (a) $A + B = B + A$ and (b) $A + (B + C) = (A + B) + C$.
   Solution.
   (a) $A + B = (2 + 1)i + (-3 + 2)j = 3i - j$
       $B + A = (1 + 2)i + (2 - 3)j = 3i - j$
   (b) $A + (B + C) = 2i - 3j + (i + 6j) = 3i + 3j$
       $(A + B) + C = (3i - j) + 4j = 3i + 3j$

3. Show that $k(A + B) = kA + kB$.
   Solution. This follows from the properties of similar triangles (see Fig. 12.3).

4. Show that $(k_1 + k_2)A = k_1 A + k_2 A$.
   Solution. See Fig. 12.4.

5. Find the magnitudes of (a) $A = 3i + 4j$, (b) $B = 4i - 3j$, (c) $C = -2i + j$, (d) $D = 6i$, and (e) $E = -i - 7j$. Also find the direction angle θ (the angle between the vector and the positive X-axis).
   Solution.
   (a) $|A| = \sqrt{9 + 16} = 5$, $\theta = \mathrm{Tan}^{-1}\frac{4}{3}$ (quadrant I)
   (b) $|B| = \sqrt{16 + 9} = 5$, $\theta = \mathrm{Tan}^{-1}(-\frac{3}{4})$ (quadrant IV)
   (c) $|C| = \sqrt{4 + 1} = \sqrt{5}$, $\theta = \mathrm{Tan}^{-1}(-\frac{1}{2})$ (quadrant II)
   (d) $|D| = \sqrt{36} = 6$, $\theta = 0$
   (e) $|E| = \sqrt{1 + 49} = 5\sqrt{2}$, $\theta = \mathrm{Tan}^{-1}(\frac{1}{7})$ (quadrant III)

6. Given the two points $A(a_1, a_2)$ and $B(b_1, b_2)$ in the plane, find the components of the vector $\overline{AB}$ (Fig. 12.9).

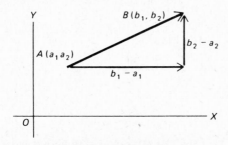

FIG. 12.9

Solution. The vector $\vec{AB}$ has components $(b_1 - a_1)$ and $(b_2 - a_2)$ so that $\vec{AB} = (b_1 - a_1)\mathbf{i} + (b_2 - a_2)\mathbf{j}$.

7. By vector methods, find the coordinates of the midpoint P of the line segment AB with coordinates $A(a_1, a_2)$ and $B(b_1, b_2)$ (Fig. 12.10).

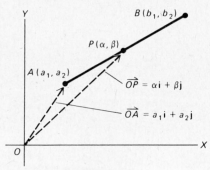

FIG. 12.10

Solution. If $\vec{OP} = \alpha\mathbf{i} + \beta\mathbf{j}$, we must find α and β such that $\vec{OP} = \vec{OA} + \frac{1}{2}\vec{AB}$ or

$$\alpha\mathbf{i} + \beta\mathbf{j} = (a_1\mathbf{i} + a_2\mathbf{j}) + \frac{1}{2}(b_1 - a_1)\mathbf{i} + (b_2 - a_2)\mathbf{j}$$

$$= \frac{a_1 + b_1}{2}\mathbf{i} + \frac{a_2 + b_2}{2}\mathbf{j}$$

Therefore, the coordinates of P are $\left(\dfrac{a_1 + b_1}{2}, \dfrac{a_2 + b_2}{2}\right)$.

8. Find the derivative of each of the following vector functions:
   (a) $\mathbf{R}(t) = e^t\mathbf{i} + (\sin t^2)\mathbf{j}$, (b) $\mathbf{S}(t) = \mathbf{i} \log \sin t + \mathbf{j} \tan t$, (c) $\mathbf{U}(t) = \mathbf{i} + t\mathbf{j}$, and (d) $\mathbf{W}(t) = \mathbf{i} t \cos t + \mathbf{j} \operatorname{Tan}^{-1} t$.
   Solution.
   (a) $\mathbf{R}'(t) = e^t\mathbf{i} + (2t \cos t^2)\mathbf{j}$
   (b) $\mathbf{S}'(t) = \mathbf{i} \cot t + \mathbf{j} \sec^2 t$
   (c) $\mathbf{U}'(t) = \mathbf{j}$
   (d) $\mathbf{W}'(t) = (\cos t - t \sin t)\mathbf{i} + \dfrac{1}{1 + t^2}\mathbf{j}$

9. For each of the following position vectors of a moving particle, find the velocity and acceleration vectors:

(a) $\mathbf{R}(t) = \mathbf{i}\, 3 \cos t - \mathbf{j}\, 5 \sin t$, (b) $\mathbf{S}(t) = t^2 \mathbf{i} - (1 - t)\mathbf{j}$, (c) $\mathbf{U}(t) = e^t \mathbf{i} - 3e^{-t} \mathbf{j}$, and (d) $\mathbf{W}(t) = \dfrac{1}{1 + t} \mathbf{i} + \dfrac{1}{1 - t} \mathbf{j}$.

Solution.
(a) $\mathbf{V}(t) = \mathbf{R}'(t) = -\mathbf{i}\, 3 \sin t - \mathbf{j}\, 5 \cos t$
$\quad \mathbf{A}(t) = \mathbf{R}''(t) = -\mathbf{i}\, 3 \cos t + \mathbf{j}\, 5 \sin t$

(b) $\mathbf{V}(t) = \mathbf{S}'(t) = 2t\mathbf{i} + \mathbf{j}$
$\quad \mathbf{A}(t) = \mathbf{S}''(t) = 2\mathbf{i}$

(c) $\mathbf{V}(t) = \mathbf{U}'(t) = e^t \mathbf{i} + 3e^{-t}\mathbf{j}$
$\quad \mathbf{A}(t) = \mathbf{U}''(t) = e^t \mathbf{i} - 3e^{-t}\mathbf{j}$

(d) $\mathbf{V}(t) = \mathbf{W}'(t) = \dfrac{-1}{(1 + t)^2}\mathbf{i} + \dfrac{1}{(1 - t)^2}\mathbf{j}$

$\quad \mathbf{A}(t) = \mathbf{W}''(t) = \dfrac{2}{(1 + t)^3}\mathbf{i} + \dfrac{2}{(1 - t)^3}\mathbf{j}$

10. Find the scalar products (a) $\mathbf{A} \cdot \mathbf{B}$, (b) $\mathbf{A} \cdot \mathbf{C}$, (c) $\mathbf{A} \cdot \mathbf{D}$, (d) $\mathbf{A} \cdot \mathbf{E}$, and (e) $\mathbf{E} \cdot \mathbf{F}$, where

$$\mathbf{A} = 2\mathbf{i} + 3\mathbf{j} \qquad\qquad \mathbf{B} = 3\mathbf{i} - 2\mathbf{j}$$
$$\mathbf{C} = 4\mathbf{i} + 6\mathbf{j} \qquad\qquad \mathbf{D} = -6\mathbf{i} - 9\mathbf{j}$$
$$\mathbf{E} = \mathbf{i} + \mathbf{j} \qquad\qquad \mathbf{F} = (2 - \sqrt{3})\mathbf{i} - \mathbf{j}$$

Also find the cosine of the angle between the vectors.

Solution. (a) $\mathbf{A} \cdot \mathbf{B} = (2)(3) + (3)(-2) = 0$

$$\cos \theta = \frac{0}{\sqrt{13}\sqrt{13}} = 0$$

(**A** and **B** are perpendicular.)

(b) $\mathbf{A} \cdot \mathbf{C} = (2)(4) + (3)(6) = 26$

$$\cos \theta = \frac{26}{\sqrt{13}\sqrt{52}} = 1$$

(**A** and **C** are parallel.)

(c) $\mathbf{A} \cdot \mathbf{D} = (2)(-6) + (3)(-9) = -39$

$$\cos \theta = \frac{-39}{\sqrt{13}\sqrt{117}} = -1$$

(**A** and **D** are antiparallel.)

(d) $\mathbf{A} \cdot \mathbf{E} = (2)(1) + (3)(1) = 5$

$$\cos \theta = \frac{5}{\sqrt{13}\sqrt{2}}$$

(e) $\mathbf{E} \cdot \mathbf{F} = (1)(2 - \sqrt{3}) + (1)(-1) = 1 - \sqrt{3}$

$$\cos \theta = \frac{1 - \sqrt{3}}{\sqrt{2}\sqrt{8 - 4\sqrt{3}}} = \frac{1}{2}$$

12.5 Vectors in Three Dimensions

Much of the theory of vectors in two dimensions carries over into three dimensions in an obvious manner. The vector,

$$\mathbf{R} = \overrightarrow{OP} = x\mathbf{i} + y\mathbf{j} + z\mathbf{k} \tag{1}$$

is the position vector of the point P (Fig. 12.11). Note the right-hand coordinate system in the figure. It is desirable to use such a system in

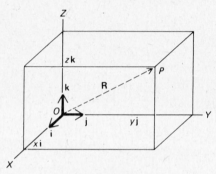

FIG. 12.11

dealing with vectors in three space. The vector sum of $\mathbf{A} = a_1\mathbf{i} + a_2\mathbf{j} + a_3\mathbf{k}$ and $\mathbf{B} = b_1\mathbf{i} + b_2\mathbf{j} + b_3\mathbf{k}$ is

$$\mathbf{A} + \mathbf{B} = (a_1 + b_1)\mathbf{i} + (a_2 + b_2)\mathbf{j} + (a_3 + b_3)\mathbf{k}$$

Also, we have

$$c\mathbf{A} = ca_1\mathbf{i} + ca_2\mathbf{j} + ca_3\mathbf{k} \ (c \ \text{a constant})$$

The length of the vector $\mathbf{R}$ is

$$|\mathbf{R}| = \sqrt{x^2 + y^2 + z^2} \tag{2}$$

Direction angles and direction cosines are direct generalizations of their counterparts in two dimensions. Thus, the direction cosines are (Fig. 12.12)

$$\cos \alpha = \frac{x}{\sqrt{x^2 + y^2 + z^2}} \ , \ \cos \beta = \frac{y}{\sqrt{x^2 + y^2 + z^2}} \ ,$$

$$\text{and} \cos \gamma = \frac{z}{\sqrt{x^2 + y^2 + z^2}} \tag{3}$$

The numbers $m \cos \alpha, m \cos \beta, m \cos \gamma$ (m a constant) are called **direction numbers**.

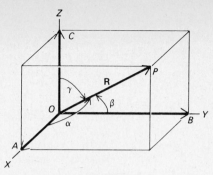

FIG. 12.12

The length of the vector from $A(a_1, a_2, a_3)$ to $B(b_1, b_2, b_3)$ is

$$|\overrightarrow{AB}| = \sqrt{(b_1 - a_1)^2 + (b_2 - a_2)^2 + (b_3 - a_3)^2}$$

and the *direction cosines* are

$$\cos \alpha = \frac{b_1 - a_1}{|\overrightarrow{AB}|}, \quad \cos \beta = \frac{b_2 - a_2}{|\overrightarrow{AB}|}, \quad \text{and} \quad \cos \gamma = \frac{b_3 - a_3}{|\overrightarrow{AB}|} \qquad (4)$$

The scalar products formed with the unit vectors $\mathbf{i}$, $\mathbf{j}$, and $\mathbf{k}$ are as follows:

$$
\begin{array}{lll}
\mathbf{i} \cdot \mathbf{i} = 1 & \mathbf{i} \cdot \mathbf{j} = 0 & \mathbf{i} \cdot \mathbf{k} = 0 \\
\mathbf{j} \cdot \mathbf{i} = 0 & \mathbf{j} \cdot \mathbf{j} = 1 & \mathbf{j} \cdot \mathbf{k} = 0 \\
\mathbf{k} \cdot \mathbf{i} = 0 & \mathbf{k} \cdot \mathbf{j} = 0 & \mathbf{k} \cdot \mathbf{k} = 1
\end{array}
$$

For any two vectors $\mathbf{A} = a_1\mathbf{i} + a_2\mathbf{j} + a_3\mathbf{k}$ and $\mathbf{B} = b_1\mathbf{i} + b_2\mathbf{j} + b_3\mathbf{k}$, their scalar product is

$$\mathbf{A} \cdot \mathbf{B} = a_1 b_1 + a_2 b_2 + a_3 b_3$$

The scalar product is distributive:

$$\mathbf{A} \cdot (\mathbf{B} + \mathbf{C}) = \mathbf{A} \cdot \mathbf{B} + \mathbf{A} \cdot \mathbf{C}$$

and

$$c(\mathbf{A} \cdot \mathbf{B}) = (c\mathbf{A}) \cdot \mathbf{B} = \mathbf{A} \cdot (c\mathbf{B})$$

12.6 The Vector Product

Consider two vectors $\mathbf{A} = a_1\mathbf{i} + a_2\mathbf{j} + a_3\mathbf{k}$ and $\mathbf{B} = b_1\mathbf{i} + b_2\mathbf{j} + b_3\mathbf{k}$ with angle θ between them, and let $\mathbf{e}$ be a unit vector perpendicular to both $\mathbf{A}$ and $\mathbf{B}$ in the direction in which a right-handed screw advances when turned from $\mathbf{A}$ toward $\mathbf{B}$ through the angle θ (Fig. 12.13). The **vector product** (also called the *cross product*) is

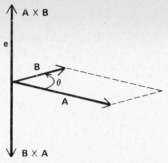

FIG. 12.13

$$A \times B = e |A| |B| \sin \theta \qquad (1)$$

For nonzero vectors A and B, $A \times B = O$ if and only if A and B are parallel or antiparallel, since then $\sin \theta = 0$. The various vector products of the unit vectors are as follows:

$$
\begin{array}{lll}
i \times i = o & i \times j = k & i \times k = -j \\
j \times i = -k & j \times j = o & j \times k = i \\
k \times i = j & k \times j = -i & k \times k = o
\end{array}
$$

For vectors A, B, and C, and scalar c, the following rules hold:

$$
\begin{aligned}
&A \times B = -(B \times A) \text{ (noncommutative)} \\
&c(A \times B) = cA \times B = A \times cB \text{ (distributive)} \\
&A \times (B + C) = (A \times B) + (A \times C) \text{ (distributive)} \\
&(A \times B) \times C \neq A \times (B \times C) \text{ (nonassociative)} \\
&A \cdot B \times C = A \times B \cdot C \text{ (triple scalar product)} \\
&\qquad\qquad = B \cdot C \times A \\
&\qquad\qquad = C \cdot A \times B
\end{aligned}
$$

The vector product may be computed in terms of components by using the following schematic device which is similar to a determinant, but is not a real number. However, the computation is performed as if it were a determinant.

$$A \times B = (a_1 i + a_2 j + a_3 k) \times (b_1 i + b_2 j + b_3 k)$$

$$= \begin{vmatrix} i & j & k \\ a_1 & a_2 & a_3 \\ b_1 & b_2 & b_3 \end{vmatrix} = (a_2 b_3 - a_3 b_2)i + (a_3 b_1 - a_1 b_3)j + (a_1 b_2 - a_2 b_1)k$$

12.7 A Geometric Interpretation of the Vector Product

The number which represents the length of the vector $\mathbf{A} \times \mathbf{B}$ is numerically the same as the number which represents the area of the parallelogram with sides $\mathbf{A}$ and $\mathbf{B}$. Let $\mathbf{A}$, $\mathbf{B}$, and $\mathbf{C}$ be right-handed, noncoplanar vectors and let V be the volume of the parallelepiped with coterminal sides $\mathbf{A}$, $\mathbf{B}$, and $\mathbf{C}$, then

$$V = \mathbf{A} \cdot \mathbf{B} \times \mathbf{C} = \begin{vmatrix} a_1 & a_2 & a_3 \\ b_1 & b_2 & b_3 \\ c_1 & c_2 & c_3 \end{vmatrix} \tag{1}$$

where $\mathbf{A} = a_1\mathbf{i} + a_2\mathbf{j} + a_3\mathbf{k}$, $\mathbf{B} = b_1\mathbf{i} + b_2\mathbf{j} + b_3\mathbf{k}$, and $\mathbf{C} = c_1\mathbf{i} + c_2\mathbf{j} + c_3\mathbf{k}$.

12.8 Differentiation of Vectors in Three Space

Let $P(x, y, z)$ be a point on a curve with parametric equations $x = f(t)$, $y = g(t)$, and $z = h(t)$. Then the vector $\overrightarrow{OP} = \mathbf{R}$ can be written in the form

$$\mathbf{R}(t) = f(t)\mathbf{i} + g(t)\mathbf{j} + h(t)\mathbf{k} \tag{2}$$

Then
$$\mathbf{R}'(t) = f'(t)\mathbf{i} + g'(t)\mathbf{j} + h'(t)\mathbf{k} \tag{3}$$

where (3) is the derivative of $\mathbf{R}$ with respect to t, and if $\mathbf{R}'(t) \neq 0$, then $\mathbf{R}'(t)$ is a vector tangent to the curve and pointing in the direction of increasing t. If $\mathbf{T}$ is a unit tangent vector, then

$$\mathbf{T} = \frac{\dfrac{df}{dt}\mathbf{i} + \dfrac{dg}{dt}\mathbf{j} + \dfrac{dh}{dt}\mathbf{k}}{\sqrt{\left(\dfrac{df}{dt}\right)^2 + \left(\dfrac{dg}{dt}\right)^2 + \left(\dfrac{dh}{dt}\right)^2}} \tag{4}$$

For differentiable vector functions $\mathbf{U}(t)$ and $\mathbf{V}(t)$, and a scalar function $f(t)$, we have

$$\frac{d}{dt}(\mathbf{U} + \mathbf{V}) = \frac{d\mathbf{U}}{dt} + \frac{d\mathbf{V}}{dt}$$

$$\frac{d}{dt}f\mathbf{V} = f\frac{d\mathbf{V}}{dt} + \frac{df}{dt}\mathbf{V}$$

$$\frac{d}{dt}\mathbf{U} \cdot \mathbf{V} = \mathbf{U} \cdot \frac{d\mathbf{V}}{dt} + \frac{d\mathbf{U}}{dt} \cdot \mathbf{V}$$

$$\frac{d}{dt}\mathbf{U} \times \mathbf{V} = \mathbf{U} \times \frac{d\mathbf{V}}{dt} + \frac{d\mathbf{U}}{dt} \times \mathbf{V}$$

12.9 Equations of Line and Plane

If $\mathbf{R}$ is the position vector of $P(x,y,z)$ on a line through $A(x_0,y_0,z_0)$ and $B(x_1,y_1,z_1)$ then

$$\mathbf{R} = \overrightarrow{OA} + t\overrightarrow{AB} \qquad (1)$$

for every value of the scalar t, and conversely. Thus, if $\mathbf{R} = x\mathbf{i} + y\mathbf{j} + z\mathbf{k}$, then

$$x\mathbf{i} + y\mathbf{j} + z\mathbf{k} = x_0\mathbf{i} + y_0\mathbf{j} + z_0\mathbf{k} + t(x_1 - x_0)\mathbf{i} + t(y_1 - y_0)\mathbf{j} + t(z_1 - z_0)\mathbf{k}$$

This yields the following parametric equations of the line:

$$x = x_0 + t(x_1 - x_0)$$
$$y = y_0 + t(y_1 - y_0)$$
$$z = z_0 + t(z_1 - z_0)$$

If we set $a = x_1 - x_0$, $b = y_1 - y_0$, and $c = z_1 - z_0$, and if no one of a, b, or c is zero, the above equations can be written in the symmetric form:

$$\frac{x - x_0}{a} = \frac{y - y_0}{b} = \frac{z - z_0}{c} \qquad (2)$$

Note that a, b, and c are *direction numbers* of the line.

An equation of a plane normal to the line above and passing through (x_0, y_0, z_0) is

$$a(x - x_0) + b(y - y_0) + c(z - z_0) = 0 \qquad (3)$$

Note that a, b, and c are direction numbers of a line *perpendicular* to this plane.

Let a plane be determined by the three points $P_1(x_1, y_1, z_1)$, $P_2(x_2, y_2, z_2)$, and $P_3(x_3, y_3, z_3)$. The vectors

$$\overrightarrow{P_1 P_2} = (x_2 - x_1)\mathbf{i} + (y_2 - y_1)\mathbf{j} + (z_2 - z_1)\mathbf{k}$$

and

$$\overrightarrow{P_1 P_3} = (x_3 - x_1)\mathbf{i} + (y_3 - y_1)\mathbf{j} + (z_3 - z_1)\mathbf{k}$$

lie in the plane. The point $P(x, y, z)$ will lie in the plane if and only if $\overrightarrow{P_1 P}$ is *perpendicular* to $\mathbf{N}$, where $\mathbf{N}$ is the **normal vector** to the plane; that is, P is in the plane if and only if $\overrightarrow{P_1 P} \cdot \mathbf{N} = 0$, or

$$\overrightarrow{P_1P} \cdot \overrightarrow{P_1P_2} \times \overrightarrow{P_1P_3} = \begin{vmatrix} x - x_1 & y - y_1 & z - z_1 \\ x_2 - x_1 & y_2 - y_1 & z_2 - z_1 \\ x_3 - x_1 & y_3 - y_1 & z_3 - z_1 \end{vmatrix} = 0$$

(See Equation (1), Section 12.7.)

which reduces to an equation of the form

$$Ax + By + Cz = D$$

This is the equation of the plane through P_1, P_2, and P_3. Moreover, A, B, and C are direction numbers of a line *normal* to the plane.

PROBLEMS WITH SOLUTIONS

1. Find the magnitude and the direction cosines of the vector
   (a) from $O(0, 0, 0)$ to $P(3, 4, -5)$ and (b) from $A(3, 2, -1)$ to
   $B(4, -1, 6)$.
   Solution.
   (a) $|\overrightarrow{OP}| = \sqrt{9 + 16 + 25} = 5\sqrt{2}$

   $\cos \alpha = \dfrac{3}{5\sqrt{2}}$, $\cos \beta = \dfrac{4}{5\sqrt{2}}$, and $\cos \gamma = \dfrac{-1}{\sqrt{2}}$

   (b) $|\overrightarrow{AB}| = \sqrt{(4 - 3)^2 + (-1 - 2)^2 + (6 + 1)^2} = \sqrt{59}$

   $\cos \alpha = \dfrac{4 - 3}{\sqrt{59}}$, $\cos \beta = \dfrac{-1 - 2}{\sqrt{59}}$, and $\cos \gamma = \dfrac{6 + 1}{\sqrt{59}}$

2. Find symmetric equations of the line through $A(1, 5, -8)$ and
   $B(-2, 3, 4)$, and find the coordinates of the point where the line
   pierces the xy-plane.
   Solution. Direction numbers are $a = 3$, $b = 2$, and $c = -12$; thus,
   the symmetric equations are

   $$\frac{x - 1}{3} = \frac{y + 2}{2} = \frac{z + 8}{-12}$$

   The z-coordinate of the piercing point is, of course, 0; hence,
   $(y + 2)/2 = -8/12$, or $y = -10/3$, and $x = -9$. The coordinates of
   the piercing point are $(-9, -10/3, 0)$.

3. Find the cosine of angle CAB of the triangle $A(0, -1, 0)$, $B(1, 2, 3)$,
   and $C(1, 1, -2)$.
   Solution. Angle CAB is the angle between the vectors $\overrightarrow{AC}$ and $\overrightarrow{AB}$
   (from $\overrightarrow{AC}$ to $\overrightarrow{AB}$).

$$\vec{AC} = i + 2j - 2k$$

$$\vec{AB} = i + 3j + 3k$$

$$\cos CAB = \frac{\vec{AC} \cdot \vec{AB}}{|\vec{AC}| \, |\vec{AB}|}$$

$$= \frac{(1)(1) + (2)(3) + (-2)(-3)}{\sqrt{1 + 4 + 4} \sqrt{1 + 9 + 9}} = \frac{1}{3\sqrt{19}}$$

4. If a line through the origin makes equal angles with the axes, what is the angle?
    Solution. The direction cosines are identical; therefore, since the sum of squares is 1, each is $\frac{1}{3}\sqrt{3}$. Thus, $\theta = \cos^{-1} \frac{1}{3}\sqrt{3}$.

5. Find a unit vector perpendicular to both $A = 3i - 2j + k$ and $B = i - j - 2k$.

    Solution. $A \times B = \begin{vmatrix} i & j & k \\ 3 & -2 & 1 \\ 1 & -1 & -2 \end{vmatrix} = 5i + 7j - k$

    The unit vector is, therefore, $(5i + 7j - k)/5\sqrt{3}$.

6. Find the area of the triangle, two sides of which are $A = 3i - 2j + k$ and $B = 3j + 4k$.
    Solution. $A \times B = e \, |A| \, |B| \sin \theta$. The area of the triangle is

    $$\tfrac{1}{2} |A \times B| = \tfrac{1}{2} |A| \, |B| \sin \theta$$

    $$A \times B = \begin{vmatrix} i & j & k \\ 3 & -2 & 1 \\ 0 & 3 & 4 \end{vmatrix} = -11i - 12j + 9k$$

    Area $= \tfrac{1}{2} |A \times B| = \tfrac{1}{2} \sqrt{121 + 144 + 81} = \tfrac{1}{2} \sqrt{346}$

7. Find the distance d from $P(1, -2, 3)$ to the line through $A(2, 1, -2)$ and $B(1, -1, -3)$ (Fig. 12.14).
    Solution. From the figure, it should be clear that

    $$d = |\vec{AP}| \sin \theta$$

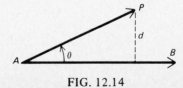

FIG. 12.14

Since $\overrightarrow{AP} \times \overrightarrow{AB} = \mathbf{e}\,|\overrightarrow{AP}|\,|\overrightarrow{AB}|\sin\theta$, it follows that $d = \dfrac{|\overrightarrow{AP} \times \overrightarrow{AB}|}{|\overrightarrow{AB}|}$.

In the special case considered, we have

$$\overrightarrow{AP} = (1-2)\mathbf{i} + (-2-1)\mathbf{j} + (3+2)\mathbf{k}$$

$$\overrightarrow{AB} = (1-2)\mathbf{i} + (-1-1)\mathbf{j} + (-3+2)\mathbf{k}$$

$$\overrightarrow{AP} \times \overrightarrow{AB} = \begin{vmatrix} \mathbf{i} & \mathbf{j} & \mathbf{k} \\ -1 & -3 & 5 \\ -1 & -2 & -1 \end{vmatrix} = 13\mathbf{i} - 6\mathbf{j} - \mathbf{k}$$

Therefore, $d = \dfrac{|13\mathbf{i} - 6\mathbf{j} - \mathbf{k}|}{|-\mathbf{i} - 2\mathbf{j} - \mathbf{k}|} = \dfrac{4\sqrt{11}}{\sqrt{6}}$

8. Find the distance d from $P(1,-2,3)$ to the plane through the three points: $A(2,1,-2)$, $B(1,-1,-3)$, and $C(1,0,1)$ (Fig. 12.15).

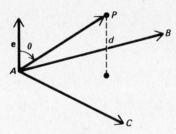

FIG. 12.15

Solution. The vectors $\overrightarrow{AB} = -\mathbf{i} - 2\mathbf{j} - \mathbf{k}$ and $\overrightarrow{AC} = -\mathbf{i} - \mathbf{j} + 3\mathbf{k}$ determine the plane, and

$$\mathbf{e} = \frac{\overrightarrow{AB} \times \overrightarrow{AC}}{|\overrightarrow{AB} \times \overrightarrow{AC}|}$$

Also, $d = |\overrightarrow{AP}|\cos\theta$ (which indicates that a scalar product is called for).

Therefore, $d = \overrightarrow{AP} \cdot \mathbf{e} = \dfrac{\overrightarrow{AP} \cdot \overrightarrow{AB} \times \overrightarrow{AC}}{|\overrightarrow{AB} \times \overrightarrow{AC}|}$

In the special case considered, we have

$$\overrightarrow{AB} \times \overrightarrow{AC} = \begin{vmatrix} \mathbf{i} & \mathbf{j} & \mathbf{k} \\ -1 & -2 & -1 \\ -1 & -1 & 3 \end{vmatrix} = -7\mathbf{i} + 4\mathbf{j} - \mathbf{k}$$

$$\overrightarrow{AP} = (1-2)\mathbf{i} + (-2-1)\mathbf{j} + (3+2)\mathbf{k} = -\mathbf{i} - 3\mathbf{j} + 5\mathbf{k}$$

$$\overrightarrow{AP} \cdot \overrightarrow{AB} \times \overrightarrow{AC} = (-1)(-7) + (-3)(4) + (5)(-1) = -10$$

Hence for the distance d we get

$$\frac{-10}{\sqrt{49 + 16 + 1}} = \frac{-10}{\sqrt{66}}$$

The minus sign indicates that we actually set the problem up with a left-handed system of coordinates, but $|d| = 10/\sqrt{66}$.

9. Find the distance between the line AB (Fig. 12.16), where $A = (1, 0, 1)$ and $B = (0, -1, 2)$, and the line CD, where $C = (-1, 1, 2)$ and $D = (-2, 1, 3)$.

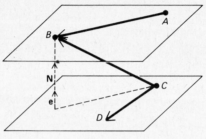

FIG. 12.16

Solution. $\overrightarrow{AB} = -\mathbf{i} - \mathbf{j} + \mathbf{k}$ and $\overrightarrow{CD} = -\mathbf{i} + \mathbf{k}$.

The vector $\mathbf{N} = \overrightarrow{AB} \times \overrightarrow{CD}$ is *normal* to each line, and there are therefore parallel planes P_1 and P_2 such that P_1 contains $\overrightarrow{AB}$ and P_2 contains $\overrightarrow{CD}$. The distance between these two planes is the distance between the two given lines, and the projection of $\overrightarrow{CB}$ onto $\mathbf{N}$ will give us the distance d. Let $\mathbf{N}/|\mathbf{N}| = \mathbf{e}$; then

$$d = |\overrightarrow{CB} \cdot \mathbf{e}| = \left| \overrightarrow{CB} \cdot \frac{\overrightarrow{AB} \times \overrightarrow{CD}}{|\overrightarrow{AB} \times \overrightarrow{CD}|} \right|$$

In the special case considered, we have

$$\mathbf{N} = \overrightarrow{AB} \times \overrightarrow{CD} = \begin{vmatrix} \mathbf{i} & \mathbf{j} & \mathbf{k} \\ -1 & -1 & 1 \\ -1 & 0 & 1 \end{vmatrix} = -\mathbf{i} - \mathbf{k}$$

$$\mathbf{e} = \frac{\mathbf{N}}{|\mathbf{N}|} = \frac{-1}{\sqrt{2}} \mathbf{i} - \frac{1}{\sqrt{2}} \mathbf{k}$$

$$\overrightarrow{CB} = \mathbf{i} - 2\mathbf{j}$$

Hence, $d = (1) \left(-\frac{1}{\sqrt{2}} \right) + (-2) \left(-\frac{1}{\sqrt{2}} \right) = \frac{1}{\sqrt{2}}$

10. Find an equation of the plane through $P_1(1, 2, 8)$, $P_2(2, -1, 6)$, and $P_3(-3, 0, 2)$.
Solution.

$$\begin{vmatrix} x - 1 & y - 2 & z - 8 \\ 2 - 1 & -1 - 2 & 6 - 8 \\ -3 - 1 & 0 - 2 & 2 - 8 \end{vmatrix} = 0$$

Hence, the desired equation is $x + y - z = -5$.

11. Given the curve $\mathbf{R} = (2 + t)\mathbf{i} + 4e^{-t}\mathbf{j} + (3 \sin t)\mathbf{k}$ and the point $P(2, 4, 0)$ on this curve, find (a) equations of the line tangent to $\mathbf{R}$ at P and (b) the equation of the plane normal to $\mathbf{R}$ at P.
Solution.
(a) The coordinates of P correspond to $t = 0$ and the tangent line is given by the vector $\mathbf{R}' = \mathbf{i} - 4e^{-t}\mathbf{j} + (3 \cos t)\mathbf{k}$. From this, we find the direction numbers 1 and $-4e^0 = -4$, and $3 \cos 0 = 3$; therefore, the equations of the tangent line are

$$\frac{x - 2}{1} = \frac{y - 4}{-4} = \frac{y}{3}$$

(b) The equation of the normal plane is

$$(x - 2) - 4(y - 4) + 3z = 0$$
or
$$x - 4y + 3z = -14$$

CHAPTER 13
INFINITE SERIES

13.1 Infinite Sequences

An **infinite sequence** is defined to be *a set of numbers in one-to-one correspondence with the set of positive integers*; that is to say, it is a function whose domain is the set of positive integers and whose range is any countable subset of the real numbers. An example of an infinite sequence is the following.

$$\frac{1}{2}, \frac{1}{2^2}, \frac{1}{2^3}, \frac{1}{2^4}, \ldots, \frac{1}{2^n}, \ldots$$

The first term is $\frac{1}{2}$, the second term is $\frac{1}{2^2}, \ldots$, and the nth term is $\frac{1}{2^n}$. We may express the function by $f(n) = \frac{1}{2^n}$, where n is a positive integer. Sometimes we write u_n for $f(n)$ as in the sequence $u_1, u_2, \ldots, u_n, \ldots$.

We call $f(n)$ or u_n the **general term**, and a sequence is given when and only when the general term is known or when a rule is given by which any term can be determined. The general term or rule cannot be deduced from a finite number of particular terms. For example, $\frac{1}{1}, \frac{1}{2}, \frac{1}{3}, \frac{1}{4}, \ldots$ does not define a sequence since no *unique* rule (general term) can be deduced from the given terms. Although $1/n$ would be one possible general term, there are many others that describe this sequence. If $u_n = \log(n + 1)$, then the sequence is $\log 2, \log 3, \ldots, \log(n + 1), \ldots$. The general term (rule or function) is $f(n) = \log(n + 1)$. In most work involving sequences, it is desirable to have the nth term expressed as a function of n.

13.2 Infinite Series.

From a sequence

$$u_1, u_2, \ldots, u_n, \ldots$$

we can form the partial sums S_i, where

$$S_1 = u_1$$

$$S_2 = u_1 + u_2$$

$$S_n = u_1 + u_2 + \ldots + u_n$$

These are all defined and have meaning. The question arises, does

$$\lim_{n \to \infty} S_n = u_1 + u_2 + \cdots + u_n + \cdots$$

have meaning? If so, we call it a **convergent infinite series**. If not we call it a **divergent infinite series**. Thus, an infinite series (convergent or divergent) is defined as *the limiting value of the nth partial sum*, and we write

$$S = \lim_{n \to \infty} S_n = u_1 + u_2 + \cdots + u_n + \cdots$$

If S exists and is finite, then the series *converges*. If S fails to exist or is infinite, then the series *diverges*.

13.3 Geometric Series.

A **geometric progression** is a sequence in which the nth term is obtained by multiplying the $(n-1)$st term by a constant ratio r. Thus, beginning with a first term a, a finite geometric progression of n terms would read

$$a, ar, ar^2, \cdots, ar^{n-1} \tag{1}$$

The sum $a + ar + \cdots + ar^{n-1}$ of such a progression is given by

$$S_n = a \frac{1 - r^n}{1 - r} \tag{2}$$

In the event (1) is an infinite progression, or sequence, S_n will represent the sum of the first n terms, and the sum of the infinite geometric series $S = a + ar + \cdots + ar^{n-1} + \cdots$ will be

$$S = \lim_{n \to \infty} S_n = \lim_{n \to \infty} a \frac{1 - r^n}{1 - r} \tag{3}$$

Now if $|r| < 1, r^n \longrightarrow 0$ as $n \longrightarrow \infty$, and

$$S = \frac{a}{1 - r} \tag{4}$$

for an infinite geometric series of ratio r with $|r| < 1$; hence, the series converges. Since $|r|^n \longrightarrow \infty$ as $n \longrightarrow \infty$ if $|r| > 1$, the series will diverge in this case. It evidently diverges for $r = 1$.

We summarize: The geometric series

$$S = a + ar + ar^2 + \cdots + ar^{n-1} + \cdots \tag{5}$$

converges when $|r| < 1$ and diverges when $|r| \geqslant 1$.

Example 1. Test the series $1 + \dfrac{1}{2} + \dfrac{1}{4} + \dfrac{1}{8} + \cdots + \dfrac{1}{2^n} + \cdots$ for convergence.

Solution. Here $\dfrac{1}{2^n}$ is actually the $(n+1)$ st term. With $r = \dfrac{1}{2}$, the series converges to the value $S = \dfrac{1}{1 - \frac{1}{2}} = 2$. The sum of the series is said to be 2.

Example 2. Test the series whose nth term is $(-1)^n \dfrac{1}{3^n}$ for convergence.

Solution. The series is a geometric series with $r = -\frac{1}{3}$; therefore, it converges; the series is

$$S = -\frac{1}{3} + \frac{1}{9} - \frac{1}{27} + \cdots + (-1)^n \frac{1}{3^n} + \cdots$$

and

$$S = \frac{-\dfrac{1}{3}}{1 + \dfrac{1}{3}} = -\frac{1}{4}$$

A series can converge in only one way: $\lim\limits_{n \to \infty} S_n$ must exist and be finite. But a series can diverge in two ways: (a) $\lim\limits_{n \to \infty} S_n = \infty$ or (b) $\lim\limits_{n \to \infty} S_n$ fails to exist.

Example 3. The series $1 + 1 + 1 + \cdots$ diverges since

$$\lim_{n \to \infty} S_n = \lim_{n \to \infty} n = \infty$$

This is type (a) divergence.

Example. 4. Test the series $1 - 1 + 1 - 1 + \cdots$ for convergence.

Solution. By inference, the nth term is ± 1, according to whether n is odd or even. Since $S_n = 1$ for n odd and $S_n = 0$ for n even, $S_n \longrightarrow$ no limit as $n \longrightarrow \infty$. The series is divergent and is said to *oscillate*. This is type (b) divergence.

13.4 Tests for Convergence

We list in the form of theorems without proof the following tests for convergence and divergence. Examples will make their applications clear.

For Any Infinite Series

THEOREM I. *In order that a series converge, it is necessary that the general term approach zero as n approaches infinity.*

That is, $\lim_{n \to \infty} u_n = 0$ is a necessary condition for convergence.

This follows immediately from the definition of convergence: unless $\lim_{n \to \infty} u_n = 0$, then $\lim_{n \to \infty} S_n$ cannot exist and be finite. The condition is not sufficient; the series may diverge even though $\lim_{n \to \infty} u_n = 0$. (See Example 1 below.)

For Series of Positive Constant Terms

1. *Comparison Tests*

THEOREM II. *If the terms of a positive series $u_1 + u_2 + \cdots + u_n + \cdots$ are not greater than the corresponding terms in a known convergent series, then the series converges.*

THEOREM III. *If the terms of a positive series $u_1 + u_2 + \cdots + u_n + \cdots$ are not less than the corresponding terms in a known divergent series, then the series diverges.*

Example 1. Test the series $1 + \dfrac{1}{2} + \dfrac{1}{3} + \cdots + \dfrac{1}{n} + \cdots$ for convergence.

Solution. Compare the given series

$$S = 1 + \tfrac{1}{2} + (\tfrac{1}{3} + \tfrac{1}{4}) + (\tfrac{1}{5} + \tfrac{1}{6} + \tfrac{1}{7} + \tfrac{1}{8}) + \cdots$$

with the series

$$T = 1 + \tfrac{1}{2} + (\tfrac{1}{4} + \tfrac{1}{4}) + (\tfrac{1}{8} + \tfrac{1}{8} + \tfrac{1}{8} + \tfrac{1}{8}) + \cdots$$

It is evident that, from the first term, the terms in the given series S are not less than those in the test series T. That is to say, u_i in the given series is greater than or equal to the corresponding ith term in the T series; but the T series evidently diverges, since

$$T = 1 + \frac{1}{2} + \frac{1}{2} + \frac{1}{2} + \cdots$$

Therefore, the given series, known as the **harmonic series**, is divergent.

Example 2. Test the *p*-series, $1 + \frac{1}{2^p} + \frac{1}{3^p} + \cdots + \frac{1}{n^p} + \cdots$, for convergence.

Solution. This series is divergent when $p = 1$, since it then reduces to the harmonic series. For $p > 1$ write the series in the form

$$S = 1 + \left(\frac{1}{2^p} + \frac{1}{3^p} \right) + \left(\frac{1}{4^p} + \frac{1}{5^p} + \frac{1}{6^p} + \frac{1}{7^p} \right) + \cdots$$

Compare with

$$T = 1 + \left(\frac{1}{2^p} + \frac{1}{2^p} \right) + \left(\frac{1}{4^p} + \frac{1}{4^p} + \frac{1}{4^p} + \frac{1}{4^p} \right) + \cdots$$

Each term in T is equal to or greater than the corresponding term in S; but T converges, since

$$T = 1 + \frac{2}{2^p} + \frac{4}{4^p} + \cdots$$

$$= 1 + \frac{1}{2^{p-1}} + \frac{1}{(2^{p-1})^2} + \cdots$$

This is a geometric series with ratio $\frac{1}{2^{p-1}}$ which is less than 1. Therefore, S converges. For $p < 1$, the series S is termwise greater than the harmonic series except for the first term. The S series, therefore, diverges. Hence, the *p*-series is convergent for $p > 1$ and divergent for $p \leqslant 1$.

2. Ratio Test

THEOREM IV. *Form the ratio of the $(n+1)$st term to the nth term, namely, $\frac{u_{n+1}}{u_n}$, and take the limit as $n \longrightarrow \infty$. Let $\lim\limits_{n \to \infty} \frac{u_{n+1}}{u_n} = r$. Then the series of positive terms $u_1 + u_2 + \cdots + u^n + \cdots$*

        (a) *Converges if $r < 1$.*
        (b) *Diverges if $r > 1$.*
        (c) *Test fails if $r = 1$.*

Example 3. Test the series

$$\frac{1}{3} + \frac{1 \cdot 2}{3 \cdot 5} + \frac{1 \cdot 2 \cdot 3}{3 \cdot 5 \cdot 7} + \cdots + \frac{1 \cdot 2 \cdot 3 \cdots n}{3 \cdot 5 \cdot 7 \cdots (2n+1)} + \cdots$$

for convergence.

Solution. $\dfrac{u_{n+1}}{u_n} = \dfrac{\dfrac{1 \cdot 2 \cdot 3 \cdots n(n+1)}{3 \cdot 5 \cdot 7 \cdots (2n+1)(2n+3)}}{\dfrac{1 \cdot 2 \cdot 3 \cdots n}{3 \cdot 5 \cdot 7 \cdots (2n+1)}} = \dfrac{n+1}{2n+3}$

$$\lim_{n \to \infty} \frac{u_{n+1}}{u_n} = \lim_{n \to \infty} \frac{n+1}{2n+3} = \frac{1}{2}$$

Hence, the series converges.

Example 4. Test the series $\dfrac{1}{3} + \dfrac{1}{5} + \dfrac{1}{7} + \cdots + \dfrac{1}{2n+1} + \cdots$ for convergence.

Solution.

$$\lim_{n \to \infty} \frac{u_{n+1}}{u_n} = \lim_{n \to \infty} \frac{\dfrac{1}{2n+3}}{\dfrac{1}{2n+1}}$$

$$= \lim_{n \to \infty} \frac{2n+1}{2n+3} = 1$$

The test fails; this method does not tell us whether the series converges or diverges.

THEOREM V. *If, for a given series,* $\lim\limits_{n \to \infty} \dfrac{u_{n+1}}{u_n} = 1$ *and if* $\dfrac{u_{n+1}}{u_n}$ *is reducible to the form* $\dfrac{u_{n+1}}{u_n} = \dfrac{n^k + an^{k-1} + \cdots}{n^k + bn^{k-1} + \cdots}$, *the series converges if* $b - a > 1$ *and diverges if* $b - a \leqslant 1$.

Example 5. Test the series $\dfrac{1}{3} + \dfrac{1}{5} + \dfrac{1}{7} + \cdots + \dfrac{1}{2n+1} + \cdots$ in Example 4 for convergence.

Solution. We saw that

$$\lim_{n \to \infty} \frac{u_{n+1}}{u_n} = \lim_{n \to \infty} \frac{2n+1}{2n+3} = 1$$

and the ratio test failed.

Here, $\dfrac{u_{n+1}}{u_n} = \dfrac{n + \frac{1}{2}}{n + \frac{3}{2}}$ and $b - a = \dfrac{3}{2} - \dfrac{1}{2} = 1$

Therefore, by Theorem V, the series diverges.

3. Cauchy's Integral Test

THEOREM VI. *Let the general term of a series of positive terms*
$u_1 + u_2 + \cdots$ *be* $u_n = f(n)$. *If* $f(x)$ *is a nonincreasing function of the*
continuous variable x *for* $x \geqslant a$, *then the series converges if* $\int_a^\infty f(x)\, dx$
exists and diverges if this integral fails to exist.

Example 6. Test the p-series $1 + \dfrac{1}{2^p} + \dfrac{1}{3^p} + \cdots + \dfrac{1}{n^p} + \cdots$ for con-

vergence.

Solution.

$$u_n = f(n) = \frac{1}{n^p}$$

$$\int_a^\infty f(x)\, dx = \int_1^\infty \frac{1}{x^p}\, dx$$

$$= \lim_{n \to \infty} \int_1^n \frac{1}{x^p}\, dx$$

$$= \lim_{n \to \infty} \frac{1}{p - 1} \left(1 - \frac{1}{n^{p-1}}\right)$$

If $p > 1$, this limit exists and equals $\dfrac{1}{p - 1}$; therefore, the series con-

verges. If $p < 1$, this limit does not exist and the series diverges. If $p = 1$,

$$\lim_{n \to \infty} \int_1^n \frac{1}{x}\, dx = \lim_{n \to \infty} \log n = \infty$$

The series again diverges.

For Series with Constant Positive and Negative Terms

THEOREM VII. *A series containing positive and negative terms will*
converge, if the corresponding series of the absolute values of the terms

converges. That is to say, $u_1 + u_2 + \cdots + u_n + \cdots$ *converges if* $|u_1| + |u_2| + \cdots + |u_n| + \cdots$ *converges.*

If the number of minus terms is finite, the theorem is obvious, since the dropping of a finite number of terms in any series will not affect the convergence.

Example 7. Test the series

$$1 - \frac{1}{2} + \frac{1}{2^2} - \frac{1}{2^3} - \frac{1}{2^4} + \frac{1}{2^5} - \frac{1}{2^6} - \frac{1}{2^7} - \frac{1}{2^8} + \frac{1}{2^9} - \cdots$$

for convergence.

Solution. The series of absolute values $1 + \frac{1}{2} + \cdots + \frac{1}{2^n} + \cdots$ is a convergent series (geometric series with ratio $r = \frac{1}{2}$). Therefore, the original series converges.

DEFINITION I. *A series is said to* converge absolutely, *if the corresponding series of absolute values converges.*

DEFINITION II. *A series is said to* converge conditionally, *if it converges, but the corresponding series of absolute values diverges.*

DEFINITION III. *An* alternating series *is one in which the signs of the terms alternate.*

THEOREM VIII. *An alternating series converges, if* (a) $\lim\limits_{n \to \infty} u_n = 0$ *and if* (b) *from some point on,* $|u_i| > |u_{i+1}| > |u_{i+2}| > \cdots$.

Example 8. Test the series

$$\frac{1}{\sqrt{2}} - \frac{1}{\sqrt{3}} + \frac{1}{\sqrt{4}} - \cdots + (-1)^{n-1} \frac{1}{\sqrt{n+1}} + \cdots$$

for convergence.

Solution. This is an alternating series in which

(a) $$\lim_{n \to \infty} |u_n| = \lim_{n \to \infty} \left| \frac{1}{\sqrt{n+1}} \right| = 0, i \geqslant 1$$

(b) $$|u_i| > |u_{i+1}|$$

Therefore, the series converges. Moreover, the series

$$\frac{1}{\sqrt{2}} + \frac{1}{\sqrt{3}} + \frac{1}{\sqrt{4}} + \cdots + \frac{1}{\sqrt{n}} + \cdots$$

is a p-series with $p = \frac{1}{2}$. This series diverges and therefore the original series converges conditionally.

Example 9. Test the series $\frac{3}{4} - \left(\frac{3}{4}\right)^2 + \left(\frac{3}{4}\right)^3 - \cdots + (-1)^{n-1}\left(\frac{3}{4}\right)^n + \cdots$ for convergence.

Solution. Since the series of absolute values $\frac{3}{4} + \left(\frac{3}{4}\right)^2 + \cdots + \left(\frac{3}{4}\right)^n + \cdots$ is a geometric series with ratio $r = \frac{3}{4}$, it converges; therefore, the original series converges absolutely.

PROBLEMS WITH SOLUTIONS

In problems 1 through 6, test the series for convergence.

1. $\dfrac{1}{1 \cdot 4} + \dfrac{1}{4 \cdot 7} + \cdots + \dfrac{1}{(3n - 2)(3n + 1)} + \cdots$

   Solution. Write $u_n = \dfrac{1}{(3n - 2)(3n + 1)}$

   in the form $u_n = \dfrac{1}{3}\left[\dfrac{1}{3n - 2} - \dfrac{1}{(3n + 1)}\right]$

   Then $S_n = \dfrac{1}{3}\left(1 - \dfrac{1}{4} + \dfrac{1}{4} - \dfrac{1}{7} + \dfrac{1}{7} - \dfrac{1}{10} + \cdots + \dfrac{1}{3n - 2} - \dfrac{1}{3n - 1}\right)$

   $= \dfrac{1}{3}\left(1 - \dfrac{1}{3n + 1}\right)$

   Hence, $\lim\limits_{n \to \infty} S_n = \lim\limits_{n \to \infty} \dfrac{1}{3}\left(1 - \dfrac{1}{3n + 1}\right) = \dfrac{1}{3}$

   The series converges, and we say the sum of the series is 1/3.

2. $\dfrac{1}{1 \cdot 2} + \dfrac{1}{2 \cdot 2^3} + \cdots + \dfrac{1}{(n) 2^{2n-1}} + \cdots.$

   Solution. First we note that the geometric series with nth term $\dfrac{1}{2^{2n-1}}$ converges. Since, by comparison,

   $$\dfrac{1}{(n) 2^{n-1}} < \dfrac{1}{2^{2n-1}}$$

   the given series converges.

3. $1 + \dfrac{2}{3} + \dfrac{3}{5} + \cdots + \dfrac{1+n}{1+2n} + \cdots$.

Solution. We suspect that $\dfrac{1+n}{1+2n} > \dfrac{1}{n}$, or what is the same thing,

$$\frac{1+n}{1+2n} - \frac{1}{n} > 0$$

That is, $\dfrac{n^2 - n - 1}{(1+2n)\,n} > 0$

and this is seen to be true for $n \geqslant 2$

Since $\dfrac{1}{n}$ is the nth term of the harmonic series, which diverges, the given series diverges.

4. $\dfrac{1}{2 \log^2 2} + \dfrac{1}{3 \log^2 3} + \cdots + \dfrac{1}{(n+1) \log^2 (n+1)} + \cdots$.

Solution. By the integral test,

$$\int_1^\infty \frac{dx}{(x+1) \log^2 (x+1)}$$

$$= \lim_{n \to \infty} \left(\frac{-1}{\log (x+1)} \right)\Bigg|_1^n = \frac{1}{\log 2}$$

This shows that the series converges.

5. $\dfrac{a}{1!} + \dfrac{a^2}{2!} + \cdots + \dfrac{a^n}{n!} + \cdots$ (a = constant).

Solution. By the ratio test,

$$\frac{a^{n+1}/(n+1)!}{a^n/n!} = \frac{a^{n+1}\,n!}{(n+1)!\,a^n}$$

$$= \frac{a}{n+1}$$

This shows that the series converges for any constant a.

6. $u_n = \dfrac{3^n\,n!}{n^n}$.

Solution. The ratio test yields

$$\frac{u_{n+1}}{u_n} = \frac{\dfrac{3^{n+1} \cdot (n+1)!}{(n+1)^{n+1}}}{\dfrac{3^n \cdot n!}{n^n}} = \frac{3n^n}{(n+1)^n} = \frac{3}{\left(1 + \dfrac{1}{n}\right)^n}$$

$$\lim_{n \to \infty} \frac{3}{\left(1 + \dfrac{1}{n}\right)^n} = \frac{3}{e} > 1$$

Thus, the series diverges.

PROBLEMS WITH ANSWERS

In Problems 7 through 18, test the series for convergence.

7. $1 + \dfrac{1}{2} + \dfrac{1}{4} + \dfrac{1}{6} + \cdots + \dfrac{1}{2n} + \cdots$.

                                                      Ans. divergent

8. $1 - \dfrac{1}{2} + \dfrac{1}{4} - \dfrac{1}{6} + \cdots$.

                                   Ans. conditionally convergent

9. $\dfrac{5}{3} + \dfrac{5}{5^2} + \dfrac{5}{7^3} + \cdots + \dfrac{5}{(2n+1)^n} \cdots$.

                                           Ans. convergent

10. $1\left(\dfrac{2}{3}\right) + 2\left(\dfrac{2}{3}\right)^2 + 3\left(\dfrac{2}{3}\right)^3 + \cdots + n\left(\dfrac{2}{3}\right)^n + \cdots$

                                           Ans. convergent

11. $\dfrac{1}{10} + \dfrac{2!}{10^2} + \dfrac{3!}{10^3} + \cdots + \dfrac{n!}{10^n} + \cdots$.

                                           Ans. divergent

12. $\dfrac{2}{1} + \dfrac{2^2}{2^2} + \dfrac{2^3}{3^2} + \cdots + \dfrac{2^n}{n^2} + \cdots$.

                                           Ans. divergent

13. The series for which $u_n = \dfrac{1}{n^4}$.

                                           Ans. convergent

14. $\dfrac{1}{\log 2} + \dfrac{1}{\log 3} + \cdots + \dfrac{1}{\log n} + \cdots$.

                                           Ans. divergent

15. $\dfrac{1}{2} - \dfrac{2}{3} + \dfrac{3}{4} - \dfrac{4}{5} + \cdots + (-1)^{n+1} \dfrac{n}{n+1} + \cdots$.

                                           Ans. divergent

16. $1 - \dfrac{1}{3^3} + \dfrac{1}{3^6} - \cdots + (-1)^{n+1} \dfrac{1}{3^{3n-3}} + \cdots$.

                             Ans. absolutely convergent

17. $u_n = \dfrac{2^n \cdot n!}{n^n}$.

Ans. convergent

18. $u_n = \dfrac{e^n \cdot n!}{n^n}$.

Ans. divergent

13.5 Power Series

A series of the form

$$a_0 + a_1 x + a_2 x^2 + \cdots + a_n x^n + \cdots \tag{1}$$

involving positive integral powers of a variable x and constant coefficients a_i is called a **power series in** x. It is an immediate generalization of the polynomial $a_0 + a_1 x + \cdots + a_n x^n$. A series of the form

$$a_0 + a_1(x - a) + a_2(x - a)^2 + \cdots + a_n(x - a)^n + \cdots \tag{2}$$

is called a *power series in* $x - a$.

If in the power series (1) a particular value is assigned to x, the series is reduced to a series of constants which may or may not converge. It is apparent that (1) converges for the value $x = 0$, but (1) might not converge for any other value of x. Or, again, a power series might converge for all values of x. If (1) converges for $x = b$, then it will converge for all values of x such that $|x| < b$. If (1) diverges for $x = b$, then it will diverge when $|x| > b$. Let us apply the ratio test to the power series (1).

$$\lim_{n \to \infty} \left| \frac{u_{n+1}}{u_n} \right| = \lim_{n \to \infty} \left| \frac{a_{n+1} x^{n+1}}{a_n x^n} \right| = \lim_{n \to \infty} |x| \left| \frac{a_{n+1}}{a_n} \right| \tag{3}$$

If this limit is equal to a number $L < 1$, the series *converges*. This says that (1) will converge if

$$\lim_{n \to \infty} |x| \cdot \left| \frac{a_{n+1}}{a_n} \right| < 1, \text{ or if } |x| < \lim_{n \to \infty} \left| \frac{a_n}{a_{n+1}} \right|$$

THEOREM IX. *The power series* $a_0 + a_1 x + \cdots + a_n x^n + \cdots$ *converges when* $|x| < \lim\limits_{n \to \infty} \left| \dfrac{a_n}{a_{n+1}} \right|$ *and diverges when* $|x| > \lim\limits_{n \to \infty} \left| \dfrac{a_n}{a_{n+1}} \right|$ *; the series may or may not converge when* $|x| = \lim\limits_{n \to \infty} \left| \dfrac{a_n}{a_{n+1}} \right|$.

The totality of points x at which the series converges makes up the **interval of convergence**, and the endpoints of the interval are

$$\pm \lim_{n \to \infty} \left| \frac{a_n}{a_{n+1}} \right|.$$

Example 1. Find the interval of convergence of the series

$$x - \frac{x^3}{3} + \frac{x^5}{5} - \cdots + (-1)^{n-1} \frac{x^{2n-1}}{2n-1} + \cdots .$$

Solution. $\lim_{n \to \infty} \left| \frac{a_n}{a_{n+1}} \right| = \lim_{n \to \infty} \left| \frac{2n+1}{2n-1} \right| = 1$

Therefore, the series converges for $|x| < 1$ and the interval of convergence is $-1 < x < 1$. At $x = -1$, the series is alternating and is convergent. At $x = 1$, the series is again a convergent alternating series. Therefore, the series converges when $-1 \leqslant x \leqslant 1$.

Example 2. Find the interval of convergence for the series

$$1 + 2!x + 3!x^2 + \cdots + (n+1)!x^n + \cdots .$$

Solution. $\lim_{n \to \infty} \left| \frac{a_n}{a_{n+1}} \right| = \lim_{n \to \infty} \left| \frac{n!}{(n+1)!} \right|$

$$= \lim_{n \to \infty} \left| \frac{1}{n+1} \right| = 0$$

Therefore, the series converges only for $x = 0$.

Example 3. Find the interval of convergence of the series

$$1 + \frac{2x}{2!} + \frac{2^2 x^2}{3!} + \frac{2^3 x^3}{4!} + \cdots + \frac{2^n x^n}{(n+1)!} + \cdots$$

Solution. $\lim_{n \to \infty} \left| \frac{a_n}{a_{n+1}} \right| = \lim_{n \to \infty} \left| \frac{2^n}{(n+1)!} \cdot \frac{(n+2)!}{2^{n+1}} \right|$

$$= \lim_{n \to \infty} \left| \frac{n+2}{2} \right| = \infty$$

Therefore, the series converges for all values of x, and the interval of convergence is infinite.

PROBLEMS WITH SOLUTIONS

In Problems 1 through 3, find the interval of convergence.
1. $1 + 2x^2 + 4x^4 + \cdots + 2^{n-1} x^{2(n-1)} + \cdots .$

Solution. $\lim_{n \to \infty} \left| \frac{u_{n+1}}{u_n} \right| = \lim_{n \to \infty} \left| \frac{2^n x^{2n}}{2^{n-1} x^{2(n-1)}} \right| = 2|x|^2$

The series diverges for $x = \pm \frac{1}{2} \sqrt{2}$, and the interval of convergence is $-\frac{1}{2} \sqrt{2} < x < \frac{1}{2} \sqrt{2}$.

2. $\dfrac{x}{1^2} + \dfrac{x^2}{2^2} + \dfrac{x^3}{3^2} + \cdots + \dfrac{x^n}{n^2} + \cdots.$

Solution. $\lim\limits_{n \to \infty} \dfrac{u_{n+1}}{u_n} = \lim\limits_{n \to \infty} \dfrac{n^2}{(n+1)^2} |x| = |x|$

The series converges for $x = \pm 1$, and the interval of convergence is $-1 \leqslant x \leqslant 1$.

3. $\dfrac{(x+3)}{1 \cdot 2} + \dfrac{(x+3)^2}{3 \cdot 4} + \cdots + \dfrac{(x+3)^n}{(2n-1)(2n)} + \cdots.$

Solution. $\lim\limits_{n \to \infty} \dfrac{u_{n+1}}{u_n} = |x + 3|$

The series converges for $x = -4$ and -2; the interval of convergence is $-4 \leqslant x \leqslant 2$.

PROBLEMS WITH ANSWERS

In Problems 4 through 8, find the interval of convergence.

4. $1 + 3x + (3x)^2 + (3x)^3 + \cdots + (3x)^n + \cdots.$ *Ans.* $-\frac{1}{3} < x < \frac{1}{3}$

5. $x - \dfrac{3x^3}{1!} + \dfrac{5x^5}{2!} - \dfrac{7x^7}{3!} + \cdots + (-1)^{n-1} \dfrac{(2n-1)x^{2n-1}}{(n-1)!} + \cdots.$

Ans. $-\infty < x < \infty$

6. $1 + x + \dfrac{x^2}{\sqrt{2}} + \dfrac{x^3}{\sqrt[3]{2}} + \cdots + \dfrac{x^n}{\sqrt[n]{2}} + \cdots.$ *Ans.* $-1 < x < 1$

7. $1 + x + \dfrac{x^2}{2} + \dfrac{x^3}{3} + \cdots + \dfrac{x^n}{n} + \cdots.$ *Ans.* $-1 \leqslant x < 1$

8. $(x-2) - \dfrac{(x-2)^2}{2} + \dfrac{(x-2)^3}{3} - \cdots + (-1)^{n-1} \dfrac{(x-2)^n}{n} + \cdots.$

Ans. $1 < x \leqslant 3$

13.6 Some Properties of Series

It is often desirable to make use of one or more of the following general properties of infinite series.

PROPERTY I. *Regrouping of the terms of a convergent series by the insertion of parentheses will neither affect the convergence nor the value of the series.*

Regrouping by inserting parentheses may, however, change a conditionally convergent series into just a plain convergent series. In general, parentheses may not be removed unless the series converges absolutely.

PROPERTY II. *Rearrangement of the terms of an absolutely convergent series in any manner whatsoever will leave the sum of the series unaltered.*

PROPERTY III. *In an absolutely convergent series, the series of plus signs converges and also the series of minus signs converges.*
If the series of plus signs converges to P and if the series of minus signs converges to $-M$, then the whole series converges to $P - M$.

PROPERTY IV. *In a conditionally convergent series, the series of plus signs diverges and also the series of minus signs diverges.*

PROPERTY V. *The terms of a conditionally convergent series may be rearranged so as to make the value of the series any desired quantity.*

PROPERTY VI. *The sum of a convergent alternating series will not differ numerically from the sum of the first n terms by more than the $(n + 1)$ st term.*
The error made in using the sum of the first few terms as an approximation to the sum of the whole series will not exceed the magnitude of the first term omitted.

PROPERTY VII. *If the corresponding terms of two convergent series $U = u_1 + u_2 + \cdots u_n + \cdots$ and $V = v_1 + v_2 + \cdots + v_n + \cdots$ are added or subtracted, the sum of the new series thus formed will be $U \pm V = (u_1 \pm v_1) + (u_2 \pm v_2) + \cdots (u_n + v_n) + \cdots$.*

PROPERTY VIII. *The interval of convergence of the power series $a_0 + a_1 x + \cdots + a_n x^n + \cdots$ is $(-R, +R,)$, where R is given by $R = \lim\limits_{n \to \infty} \left| \dfrac{a_n}{a_{n+1}} \right|$.*

The endpoints $-R$ and $+R$ may or may not be included. The series converges absolutely for all values of $|x| < R$.

PROPERTY IX. *Let $f(x) = a_0 + a_1 x + \cdots + a_n x^n + \cdots$ and $g(x) = b_0 + b_1 x + \cdots + b_n x^n + \cdots$; then*

(a) *Sum:* $f(x) \pm g(x) = (a_0 \pm b_0) + (a_1 \pm b_1) x + \cdots + (a_n \pm b_n) x^n + \cdots$

for every value of x for which each series converges. That is, the sum or difference of two power series converges within the smaller interval of convergence.

(b) *Product:* $f(x) \cdot g(x) = a_0 b_0 + (a_0 b_1 + a_1 b_0) x$
$$+ (a_0 b_2 + a_1 b_1 + a_2 b_0) x^2 + \cdots$$

for every value of x for which each series converges absolutely.

That is, the product of two power series converges within the smaller interval of absolute convergence. Note that since a power series always converges absolutely except, perhaps, at an endpoint, the condition of absolute convergence is not much more restrictive; the endpoints of the smaller series of convergence are the only points that may affect the validity of the product property.

(c) *Quotient:* $\dfrac{f(x)}{g(x)} = \dfrac{a_0 + a_1 x + \cdots + a_n x^n + \cdots}{b_0 + b_1 x + \cdots + b_n x^n + \cdots}$, $b_0 \neq 0$

$$= q_1 + q_2 x + \cdots + q_n x^n + \cdots$$

for every value of x for which $a_0 + a_1 x + \cdots a_n x^n + \cdots$ *converges and for which at the same time* $|b_1 x| + |b_2 x^2| + \cdots < |b_0|$.

PROPERTY X. *Within the interval of convergence, a power series may be differentiated termwise in order to obtain the derivative of the function which the series represents; thus,*

$$\frac{df}{dx} = a_1 + 2a_2 x + \cdots + na_n x^{n-1} + \cdots$$

This is valid within the interval of convergence, but may or may not be valid at the endpoints.

PROPERTY XI. *Between any two limits within the interval of convergence, a power series may be integrated termwise in order to obtain the definite integral of the function which the series represents; that is,*

$$\int_a^b f(x)\,dx = \int_a^b a_0\,dx + \int_a^b a_1 x\,dx + \cdots + \int_a^b a_n x^n\,dx + \cdots$$

This is valid for all a and b such that $-R < a < b < R$. It may or may not be possible to extend the limits of integration to include the endpoints.

13.7 Maclaurin's Series

When a known function $f(x)$ is written in the form of an infinite series, the function is said to be *expanded* in an infinite series and the infinite series is said to represent the function *in the interval of convergence.*

Example 1. Since $1 + x + x^2 + \cdots + x^n + \cdots$ is a geometric series, it will converge to $\dfrac{1}{1-x}$ for $|x| < 1$. Hence,

$$\frac{1}{1-x} = 1 + x + x^2 + \cdots + x^n + \cdots, -1 \leqslant x < 1$$

and the function $\frac{1}{1-x}$ is expanded in an infinite series. The series representation of the function is valid only when $-1 \leqslant x < 1$, but the function exists and is continuous everywhere except at $x = 1$.

A given function $f(x)$ may be expanded in a power series as follows:

$$f(x) = a_0 + a_1 x + a_2 x^2 + a_3 x^3 + \cdots + a_n x^n + \cdots \qquad (1)$$

where the coefficients a_i are to be determined. Then, by Property X, Section 13.6, it is permissible to write

$$f'(x) = a_1 + 2a_2 x + 3a_3 x^2 + \cdots + na_n x^{n-1} + \cdots \qquad (2)$$

$$f''(x) = 2a_2 + 2 \cdot 3a_3 x^2 + \cdots + n(n-1)a_n x^{n-2} + \cdots \qquad (3)$$

$$f^{(n)}(x) = n!a_n + (n+1)!a_{n+1} x + \cdots \qquad (4)$$

From these equations, the coefficients a_i can be determined. In (1), set $x = 0$, then

$$a_0 = f(0) \qquad (5)$$

In $(2), (3), \cdots$, and, (4), set $x = 0$, and compute the a's to be

$$a_1 = f'(0), \ a_2 = \frac{f''(0)}{2!}, \cdots \qquad (6)$$

$$a_n = \frac{1}{n!} f^{(n)}(0) \qquad (7)$$

We thus obtain the expansion of a known function $f(x)$ in **Maclaurin's series**, which is

$$f(x) = f(0) + \frac{f'(0)}{1!} x + \frac{f''(0)}{2!} x^2 + \cdots + \frac{f^{(n)}(0)}{n!} x^n + \cdots \qquad (8)$$

The expansion is said to be about the origin, or in the *neighborhood* of the origin.

Example 2. Expand e^x in a Maclaurin series, and determine the interval of convergence.

Solution. Here, for all $n, f^{(n)}(x) = e^x$. Therefore, the expansion is

$$e^x = 1 + x + \frac{x^2}{2!} + \frac{x^3}{3!} + \cdots + \frac{x^n}{n!} + \cdots$$

Since $\lim\limits_{n\to\infty}\left|\dfrac{a_n}{a_{n+1}}\right| = \infty$, the expansion is valid for all x.

Example 3. Expand $\sin x$ in a Maclaurin series, and determine the interval of convergence.

Solution. $\qquad f(x) = \sin x, f'(x) = \cos x$

$$f''(x) = -\sin x, f'''(x) = -\cos x, \cdots$$

The $\sin 0 = 0$ and $\cos 0 = 1$; therefore, the expansion is

$$\sin x = x - \frac{x^3}{3!} + \frac{x^5}{5!} - \frac{x^7}{7!} + \cdots$$

The series converges for all values of x, since

$$\lim_{n\to\infty}\left|\frac{a_n}{a_{n+1}}\right| = \infty$$

PROBLEMS WITH SOLUTIONS

In Problems 1 through 3, expand the given function in a Maclaurin series, and find the interval of convergence.

1. $f(x) = e^{ax}$.
   Solution. $f(0) = 1, f'(0) = a, f''(0) = a^2, \cdots, f^{(n)}(0) = a^n, \cdots$.

   Therefore, $e^{ax} = 1 + \dfrac{ax}{1!} + \dfrac{a^2 x^2}{2!} + \cdots + \dfrac{a^n x^n}{n!} + \cdots$

   Since $\lim\limits_{n\to\infty}\left|\dfrac{a_n}{a_{n+1}}\right| = \infty$, the series converges for each real number x.

2. $f(x) = \sin ax$.
   Solution. $f(0) = 0, f'(0) = a, f''(0) = 0, f'''(0) = -a^3, \cdots$.
   The Maclaurin series is

   $$\sin ax = \frac{ax}{1!} - \frac{a^3 x^3}{3!} + \frac{a^5 x^5}{5!} - \cdots + (-1)^{n-1}\frac{a^{2n-1} x^{2n-1}}{(2n-1)!} + \cdots$$

   Since $\lim\limits_{n\to\infty}\left|\dfrac{a_n}{a_{n+1}}\right| = \infty$, the series converges for each real number x.

3. $f(x) = \log(1-x)$.
   Solution. $f(0) = 0, f'(0) = -1, f''(0) = -1,$

   $$f'''(0) = -2!, f^{(4)}(0) = -3!, \cdots, f^{(n)}(0) = -n!$$

   $$\log(1-x) = -x - \frac{x^2}{2} - \frac{x^3}{3} - \cdots \frac{x^n}{n} - \cdots$$

Since $\lim_{n\to\infty} \dfrac{a_n}{a_{n+1}} = 1$ and since $\log(1-1)$ does not exist, the interval of convergence is $1 \leqslant x < 1$.

PROBLEMS WITH ANSWERS

In Problems 4 through 9, verify the Maclaurin expansions of the given functions.

4. $\cos x = 1 - \dfrac{x^2}{2!} + \dfrac{x^4}{4!} - \cdots + (-1)^{n-1} \dfrac{x^{2n-2}}{(2n-2)!} + \cdots, \ -\infty < x < \infty.$

5. $\tan x = x + \dfrac{x^3}{3} + \dfrac{2x^5}{15} + \cdots, \ -\dfrac{\pi}{2} < x < \dfrac{\pi}{2}.$

(Do not try to get the general term as it is quite difficult to obtain.)

6. $\log(1+x) = x - \dfrac{x^2}{2} + \dfrac{x^3}{3} - \cdots + (-1)^{n-1} \dfrac{x^n}{n} + \cdots, \ -1 < x \leqslant 1.$

7. $\log\left(\dfrac{1+x}{1-x}\right) = 2\left(x + \dfrac{x^3}{3} + \dfrac{x^5}{5} + \cdots + \dfrac{x^{2n-1}}{2n-1} + \cdots\right), \ -1 < x < 1.$

8. $\tan^{-1} x = x - \dfrac{x^3}{3} + \dfrac{x^5}{5} - \cdots + (-1)^{n-1} \dfrac{x^{2n-1}}{2n-1} + \cdots, \ -1 \leqslant x \leqslant 1.$

9. $(1+x)^m = 1 + \dfrac{mx}{1!} + \dfrac{m(m-1)}{2!}x^2 + \cdots$

$$+ \dfrac{m(m-1)\cdots(m-n+2)}{(n-1)!}x^{n-1} + \cdots.$$

Note that this is the binomial series which reduces to a polynomial when m is zero or a positive integer. When m is not a positive integer, the series converges for $-1 < x < 1$. The series will also converge at the left endpoint $x = -1$ if $m > 0$; it will converge at the right endpoint $x = 1$ if $m > -1$.

13.8 Taylor's Series

A function f may be expanded about a point a instead of about the origin; that is, $f(x)$ can be represented by a series of the form

$$f(x) = a_0 + a_1(x-a) + a_2(x-a)^2 + \cdots + a_n(x-a)^n + \cdots \tag{1}$$

The coefficients a_i are computed by repeated differentiations of this relation and subsequent evaluations at the point $x = a$; thus, we have

$$f'(x) = a_1 + 2a_2(x - a) + 3a_3(x - a)^2 + \cdots + na_n(x - a)^{n-1} + \cdots \quad (2)$$

$$f''(x) = 2a_2 + 2 \cdot 3a_3(x - a) + \cdots + n(n - 1)a_n(x - a)^{n-2} + \cdots \quad (3)$$

$$f^{(n)}(x) = n!a_n + (n + 1)!(x - a) + \cdots \quad (4)$$

From these, we compute

$$a_i = \frac{f^{(i)}(a)}{i!}, \quad i = 0, 1, 2, \cdots \quad (5)$$

and the series expansion of $f(x)$ in the neighborhood of the point a becomes

$$f(x) = f(a) + \frac{f'(a)}{1!}(x - a) + \frac{f''(a)}{2!}(x - a)^2 + \cdots + \frac{f^{(n)}(a)}{n!}(x - a)^n + \cdots$$
$$(6)$$

This is known as **Taylor's series**. When $a = 0$, it reduces as a special case to Maclaurin's series. If we set $x = a + h$, another form of Taylor's series is obtained, namely,

$$f(a + h) = f(a) + \frac{f'(a)}{1!}h + \frac{f''(a)}{2!}h^2 + \cdots + \frac{f^{(n)}(a)}{n!}h^n + \cdots \quad (7)$$

In (6), let $r_n(x)$ denote the remainder of the series after n terms, or in symbols,

$$f(x) = f(a) + \frac{f'(a)}{1!}(x - a) + \cdots + \frac{f^{(n-1)}(a)}{(n - 1)!}(x - a)^{n-1} + r_n(x)$$

$$= S_n(x) + r_n(x) \quad (8)$$

It can be shown that $r_n(x) = \dfrac{f^{(n)}(x_1)}{n!}(x - a)^n$, where $a < x_1 < x$. When (8) is written in the form

$$f(x) = f(a) + f'(a)(x - a) + \cdots + \frac{f^{(n-1)}(a)}{(n - 1)!}(x - a)^{n-1} \quad (9)$$

$$+ \frac{f^{(n)}(x_1)}{n!}(x - a)^n, \, a < x_1 < x$$

it is known as **Taylor's formula with remainder**. For (7), Taylor's formula with the remainder takes on the form

$$f(a + h) = f(a) + f'(a)h + \cdots + \frac{f^{(n-1)}(a)}{(n-1)!} h^{n-1} + \frac{f^{(n)}(a + \theta h)}{n!} h^n,$$

$$0 < \theta < 1 \quad (10)$$

THEOREM X. *The Taylor's series expansion of a function $f(x)$ will be a valid representation of the function for those values of x and only for those values of x for which* $\lim\limits_{n \to \infty} r_n(x) = 0$.

If a Taylor's series is an alternating series, the absolute value of the whole remainder, namely, $|r_n(x)|$, will not exceed the absolute value of the first term in the remainder. That is to say, in an alternating series

$$|f(x) - S_n(x)| \leqslant \left| \frac{f^{(n)}(a)}{n!} (x - a)^n \right|$$

Omitting the obvious remarks about absolute values, this says that *the error committed by using the sum of the first n terms as an approximation to $f(x)$ will not exceed the first term omitted.* In general,

$$r_n(x) = \frac{f^{(n)}(x_1)}{n!} (x - a)^n$$

where x_1 is some point that lies between a and x, but it is impossible to calculate $r_n(x)$ with no detailed information about this point. However, if x_1' is arbitrarily chosen so as to make $f^{(n)}(x_1')$ as large as possible, then

$$|r_n(x)| \leqslant \left| \frac{f^{(n)}(x_1')}{n!} (x - a)^n \right| \quad (11)$$

The error made, therefore, in breaking off at the nth term will not exceed

$$\left| \frac{f^{(n)}(x_1')}{n!} (x - a)^n \right|$$

Example 1. Expand $\sin x$ in a Taylor's series about the point $\frac{\pi}{6}$.

Solution. $f(x) = \sin x$, $f'(x) = \cos x$, etc. These evaluated at $\frac{\pi}{6}$ become

$$f\left(\frac{\pi}{6}\right) = \frac{1}{2} , \; f'\left(\frac{\pi}{6}\right) = \frac{\sqrt{3}}{2} , \text{ etc.}$$

Therefore,

$$\sin x = \frac{1}{2} + \frac{\sqrt{3}}{2}\left(x - \frac{\pi}{6}\right) - \frac{1}{2}\frac{\left(x - \frac{\pi}{6}\right)^2}{2!} - \frac{\sqrt{3}}{2}\frac{\left(x - \frac{\pi}{6}\right)^3}{3!} + \cdots,$$

$$-\infty < x < \infty$$

Example 2. Expand e^x in powers of $(x - a)$.

Solution.
Since $f(x) = e^x$, $f^{(i)}(x) = e^x$, and $f^{(i)}(a) = e^a$ the expansion becomes

$$e^x = e^a\left[1 + (x - a) + \frac{(x - a)^2}{2!} + \cdots + \frac{(x - a)^n}{n!} + \cdots\right] \quad -\infty < x < \infty$$

PROBLEMS WITH ANSWERS

In Problems 1 through 3, verify the Taylor expansions.

1. $\sin x = \sin a + (x - a)\cos a - \dfrac{(x - a)^2}{2!}\sin a - \dfrac{(x - a)^3}{3!}\cos a + \cdots,$

$$-\infty < x < \infty.$$

2. $\cos x = \cos a - (x - a)\sin a - \dfrac{(x - a)^2}{2!}\cos a + \dfrac{(x - a)^3}{3!}\sin a + \cdots,$

$$-\infty < x < \infty.$$

3. $\log x = \log a + \dfrac{1}{a}(x - a) - \dfrac{1}{2a^2}(x - a)^2 + \cdots$

$$+ \frac{(-1)^n}{(n - 1)a^{n-1}}(x - a)^{n-1} + \cdots, a > 0, 0 < x \leqslant 2a.$$

13.9 Application of Series to Computation

(1) *Increments.* We have seen before (Section 8.11) that $\Delta y \cong f'(x)\Delta x$. A clearer picture of this is gotten by rewriting (7) in Section 13.8, setting $a = x$ and $h = \Delta x$. This gives

$$f(x + \Delta x) = f(x) + \frac{f'(x)}{1!}\Delta x + \frac{f''(x)}{2!}(\Delta x)^2 + \cdots + \frac{f^{(n)}(x)}{n!}(\Delta x)^n + \cdots$$

$$(1)$$

Now transpose $f(x)$ and write $\Delta y = f(x + \Delta x) - f(x)$, then (1) reads

$$\Delta y = f'(x)\Delta x + \frac{f''(x)}{2!}(\Delta x)^2 + \cdots + \frac{f^{(n)}(x)}{n!}(\Delta x)^n + \cdots \qquad (2)$$

From this, as a first approximation for small increments, we get

$$dy \cong \Delta y \cong f'(x)\,\Delta x \tag{3}$$

As a second approximation

$$dy \cong \Delta y \cong f'(x)\,\Delta x + \frac{f''(x)}{2!}(\Delta x)^2 \tag{4}$$

A still better approximation can be obtained by taking more terms in the expansion (2).

(2) *Computation by Series.* Long before this point in his mathematical development, the student may have wondered how the tables of the trigonometric, logarithmic, and exponential functions are obtained. The answer is that they were largely made from computations by series.

Example 1. Compute the approximate value of *e* by series.

Solution. The Maclaurin expansion for e^x is

$$e^x = 1 + x + \frac{x^2}{2!} + \frac{x^3}{3!} + \cdots + \frac{x^n}{n!} + \cdots$$

If $x = 1$, this becomes

$$e = 1 + 1 + \frac{1}{2!} + \frac{1}{3!} + \frac{1}{4!} + \cdots + \frac{1}{n!} + \cdots$$

$$= 1 + 1 + 0.5 + 0.166667 + 0.041667$$
$$+ 0.008333 + 0.001389 + 0.000198 + \cdots$$

Hence, $e \cong 2.718254$ which is correct to four decimals.

Example 2. Compute $\sin 10° = \sin \dfrac{\pi}{18}$ correct to five decimals.

Solution. $\sin x = x - \dfrac{x^3}{3!} + \dfrac{x^5}{5!} - \dfrac{x^7}{7!} + \cdots$

$$\sin 10° = \frac{\pi}{18} - \frac{1}{3!}\left(\frac{\pi}{18}\right)^3 + \frac{1}{5!}\left(\frac{\pi}{18}\right)^5 - \cdots$$

Now this is an alternating series and the error committed in using only a few terms will not exceed the numerical value of the first term omitted (see Property VI, Section 13.6).

$$\sin 10° = 0.174532 - 0.000886 + 0.000001 - \cdots$$

The third term does not affect the fifth decimal place. Therefore, taking only the first two terms in the expansion, we compute correct to five places $\sin 10° = 0.17364$.

Note that the sum of the first two terms is 0.173646, and a better approximation to $\sin 10°$ is 0.17365, the value listed in a five-place table of sines.

Example 3. Find $\cos 44°$ correct to five decimals.

Solution. Maclaurin's series for the cosine is

$$\cos x = 1 - \frac{x^2}{2!} + \frac{x^4}{4!} - \cdots$$

But for $x = 44° = 11\pi/45$, this series will not converge very rapidly. We should use Taylor's series instead, and expand about the point $\frac{\pi}{4}$. Thus,

$$\cos x = \frac{\sqrt{2}}{2} \left[1 - \left(x - \frac{\pi}{4} \right) - \frac{1}{2!} \left(x - \frac{\pi}{4} \right)^2 + \frac{1}{3!} \left(x - \frac{\pi}{4} \right)^3 + \cdots \right]$$

from which it follows that

$$\cos 44° = \frac{\sqrt{2}}{2} \left[1 + \frac{\pi}{180} - \frac{1}{2!} \left(\frac{\pi}{180} \right)^2 - \frac{1}{3!} \left(\frac{\pi}{180} \right)^3 + \cdots \right]$$

which converges rapidly; thus,

$$\cos 44° = 0.707106 \left[1 + 0.017453 - \tfrac{1}{2}(0.017453)^2 - \tfrac{1}{6}(0.017453)^3 + \cdots \right]$$

$$= 0.707106 + 0.012341 - 0.000107 - 0.0000006 + \cdots$$

$$= 0.71934.$$

Note that the error committed by using only three terms does not exceed the fourth term, since the largest value of $f'''(x)$ in the interval is $\frac{\sqrt{2}}{2}$ (see (11), Section 13.8).

PROBLEMS WITH SOLUTIONS

1. Compute the value of $e^{1/10}$ correct to five decimals, using the Maclaurin series.

   Solution. When $x = \frac{1}{10}$, the Maclaurin expansion of e^x becomes

   $$e^{1/10} = 1 + 0.1 + 0.005 + 0.000167 + 0.0000004 \cdots$$

Since $r_4 \left(\dfrac{1}{10} \right) \leqslant \left| \dfrac{e^{1/10}}{240,000} \right|$, and since $e^{1/10} < 1.2$, we can stop at the fourth term and get $e^{1/10} = 1.105171 = 1.10517$ correct to five decimals.

2. Compute the value of sin 30° correct to five decimals, using the Maclaurin series.

Solution. $\sin x = \dfrac{x}{1!} - \dfrac{x^3}{3!} + \dfrac{x^5}{5!} \cdots$

When $x = 30^\circ = \dfrac{1}{6}\pi$ radian = 0.523599 radian, this becomes

$$\sin (0.523599) = 0.523599 - 0.023925 + 0.000328 \cdots.$$

Since this is an alternating series, the error will not exceed the absolute value of the first term not used, which in this case is 0.000002. Sin 30° (using the above three terms) = 0.500002 = 0.50000 correct to five decimals.

3. Compute the value of cos 60° correct to five decimals with a Taylor series about $x = \dfrac{1}{2}\pi$ in radians.

Solution. $\text{Cos } x = \dfrac{-\left(x - \dfrac{1}{2}\pi\right)}{1!} + \dfrac{\left(x - \dfrac{1}{2}\pi\right)^3}{3!} - \dfrac{\left(x - \dfrac{1}{2}\pi\right)^5}{5!} \cdots$

When $x = 60^\circ = \dfrac{1}{3}\pi$ radians, this becomes the same expansion and answer as those of Problem 3.

4. Compute the value of sin 15° correct to five decimals with a Maclaurin series.

Solution. When $x = 15^\circ = \dfrac{1}{12}\pi$ radian, the Maclaurin expansion of sin x becomes

$$\sin \left(\dfrac{1}{12}\pi \right) = 0.261799 - 0.002991 + 0.000010 \cdots$$

The next term is −0.000000017, which can be forgotten. The first three terms then give sin 15° = 0.258818 = 0.25882 correct to five decimals.

5. Compute the value of $\cos 75°$ correct to five decimals with a Taylor series about $\frac{1}{2}\pi$.

Solution. When $x = 75° = \frac{5}{12}\pi$, the Taylor series about $\frac{1}{2}\pi$ gives the same expansion and answer of Problem 5.

PROBLEMS WITH ANSWERS

6. Expand e^{-x} in a Maclaurin series and compute e^{-2} correct to five decimals.

Ans.   0.13533

7. By using a Maclaurin expansion, show that $\log 2 = 0.6931$ correct to four places

8. Show that for angles less than about $1°46'$, $\sin\theta = \theta$ with accuracy to five decimals.

9. Use the Maclaurin series expansion of $\tan^{-1} x$ and the relation $\frac{\pi}{4} = \tan^{-1}\frac{1}{7} + 2\tan^{-1}\frac{1}{3}$ to compute $\pi = 3.1416$ correct to three places.

10. Expand $\sin x$ in a Taylor series, and compute $\sin 61° = 0.8746$ correct to four places.

11. Expand $\log x$ in a Taylor series, and use four terms to compute $\log 1.2$, and estimate the error.
    Ans. $\log 1.2 = 0.182667$ with an error not in excess of 0.0004; the correct value of $\log 1.2$ to four decimals is $\log 1.2 = 0.1823$, so that our answer, based on four terms, is correct to three places.

CHAPTER 14
HYPERBOLIC FUNCTIONS

14.1 Relation between Exponential and Trigonometric Functions

The following expansions were obtained in Chapter 13.

$$e^x = 1 + x + \frac{x^2}{2!} + \frac{x^3}{3!} + \cdots + \frac{x^n}{n!} + \cdots \tag{1}$$

$$\sin x = x - \frac{x^3}{3!} + \frac{x^5}{5!} - \cdots + (-1)^{n-1} \frac{x^{2n-1}}{(2n-1)!} + \cdots \tag{2}$$

$$\cos x = 1 - \frac{x^2}{2!} + \frac{x^4}{4!} - \cdots + (-1)^{n-1} \frac{x^{2n-2}}{(2n-2)!} + \cdots \tag{3}$$

It can be shown that these series converge for all values of x, real or complex. Indeed when $x = \alpha + i\beta$, these series will serve as definitions of $e^{\alpha+i\beta}$, $\sin(\alpha + i\beta)$, $\cos(\alpha + i\beta)$, respectively. For $x = i\theta$, a pure imaginary number, (1) becomes

$$e^{i\theta} = 1 + i\theta - \frac{\theta^2}{2!} - i\frac{\theta^3}{3!} + \frac{\theta^4}{4!} + \cdots \tag{4}$$

since $i = \sqrt{-1}$, $i^2 = -1$, $i^3 = -i$, $i^4 = 1$, etc. Multiplying (2) by i and writing θ for x yields

$$i \sin \theta = i\theta - \frac{i\theta^3}{3!} + \frac{i\theta^5}{5!} - \cdots \tag{5}$$

For $x = \theta$ (3) becomes

$$\cos \theta = 1 - \frac{\theta^2}{2!} + \frac{\theta^4}{4!} - \cdots \tag{6}$$

By adding (5) and (6) we get

$$e^{i\theta} = \cos \theta + i \sin \theta \quad \text{(compare with (4))} \tag{7}$$

This is a remarkable relation, and it is generally known as Euler's identity. It exhibits a very simple connection between $\sin \theta$, $\cos \theta$, and $e^{i\theta}$. Evidently $e^{i(-\theta)} = \cos(-\theta) + i \sin(-\theta)$, or

$$e^{-i\theta} = \cos \theta - i \sin \theta \tag{8}$$

Solving (7) and (8) simultaneously for $\sin\theta$ and $\cos\theta$, we get

$$\sin\theta = \frac{e^{i\theta} - e^{-i\theta}}{2i} \tag{9}$$

$$\cos\theta = \frac{e^{i\theta} + e^{-i\theta}}{2} \tag{10}$$

These relations are very important in advanced mathematics, and (9) and (10) could be used as definitions of $\sin\theta$ and $\cos\theta$.

14.2 Hyperbolic Functions

In many branches of applied mathematics, there are functions very similar to the right-hand members of (9) and (10) that are of definite importance. These are $\dfrac{e^{\theta} - e^{-\theta}}{2}$ and $\dfrac{e^{\theta} + e^{-\theta}}{2}$, where the exponents are now real. Although these are just simple combinations of the exponential functions e^{θ} and $e^{-\theta}$, they are used so extensively that tables have been prepared for them and names given to them. For reasons that will shortly be made clear, they are called the **hyperbolic sine** and the **hyperbolic cosine**, respectively, of the variable θ. These are written **sinh** θ and **cosh** θ; that is, by definition,

$$\sinh\theta = \frac{e^{\theta} - e^{-\theta}}{2} \tag{1}$$

$$\cosh\theta = \frac{e^{\theta} + e^{-\theta}}{2} \tag{2}$$

In order to make clear the reference to a hyperbola in these definitions, we first reconsider the trigonometric definition of $\sin\theta$ (Fig. 14.1). Let $P(x, y)$ be a point on the circle $x^2 + y^2 = a^2$. Now the area

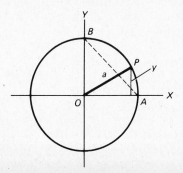

Fig. 14.1

of the sector OAP is equal to $\frac{1}{2}a^2\theta$ and the area of triangle OAB is $\frac{1}{2}a^2$. Thus, the angle θ can be thought of as the ratio of the area of the sector to the area of the triangle, that is,

$$\theta = \frac{\text{Sector } OAP}{\triangle OAB} \tag{3}$$

Moreover, it turns out that

$$\sin \theta = \frac{\triangle OAP}{\triangle OAB} = \frac{y}{a} \tag{4}$$

which is in agreement with the usual definition of $\sin \theta$, where θ is the angle AOP.

In an analogous manner, if the rectangular hyperbola $x^2 - y^2 = a^2$ is used (Fig. 14.2), it turns out that if

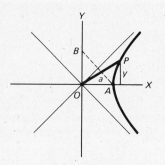

Fig. 14.2

$$\theta = \frac{\text{Sector } OAP}{\triangle OAB} \tag{5}$$

then

$$\sinh \theta = \frac{\triangle OAP}{\triangle OAB} = \frac{y}{a} \tag{6}$$

where, now, P is a point on the hyperbola. For

$$\theta = \frac{\frac{1}{2}xy - \int_a^x \sqrt{x^2 - a^2}\, dx}{\frac{1}{2}a^2}$$

or $\qquad a^2\theta = xy - 2\int_a^x \sqrt{x^2 - a^2}\, dx$

$$= xy - \left[x\sqrt{x^2 - a^2} - a^2\, \log\,(x + \sqrt{x^2 - a^2}) \right]_a^x$$

which, after reducing,

$$= a^2\, \log\, \frac{x + y}{a}$$

Therefore, $\qquad\qquad\qquad e^\theta = \frac{x + y}{a} \qquad\qquad\qquad (7)$

Similarly, $\qquad\qquad\qquad e^{-\theta} = \frac{a}{x + y} \qquad\qquad\qquad (8)$

From (7) and (8), it follows that

$$\frac{e^\theta - e^{-\theta}}{2} = \frac{1}{2}\left(\frac{x + y}{a} - \frac{a}{x + y} \right)$$

$$= \frac{y}{a} = \sinh\theta$$

Note that θ is *not* the angle AOP in this case as it is in the case of the circular function $\sin\theta$.

Other hyperbolic functions are defined as follows:

$$\tanh\theta = \frac{\sinh\theta}{\cosh\theta} = \frac{e^\theta - e^{-\theta}}{e^\theta + e^{-\theta}}$$

$$\coth\theta = \frac{1}{\tanh\theta}, \quad \text{sech}\,\theta = \frac{1}{\cosh\theta}, \quad \text{and}\ \ \text{csch}\,\theta = \frac{1}{\sinh\theta}$$

A system of hyperbolic functions can be developed comparable to that of the trigonometric functions.

Example 1. Show that $\cosh^2\theta - \sinh^2\theta = 1$.

Solution. $\qquad\qquad \cosh^2\theta = \frac{(e^\theta + e^{-\theta})^2}{4}$

$$= \frac{e^{2\theta} + 2 + e^{-2\theta}}{4}$$

$$\sinh^2 \theta = \frac{e^{2\theta} - 2 + e^{-2\theta}}{4}$$

By subtraction, the result follows.

Example 2. Sketch the graph of $y = \sinh x$.

Solution.
$$y = \sinh x = \frac{e^x - e^{-x}}{2}$$

Using the exponential definition of $\sinh x$, we develop the following table of values.

x	0	1	2	3	∞
$\sinh x$	0	1.2	3.6	10.0	∞

Since $\sinh(-x) = -\sinh x$, the graph (Fig. 14.3) is symmetric with respect to the origin.

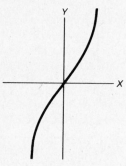

Fig. 14.3

Whereas the direct hyperbolic functions are given in terms of exponentials, the inverse hyperbolic functions involve logarithms. For example, let

$$y = \sinh^{-1} x$$

Then
$$x = \sinh y = \frac{e^y - e^{-y}}{2}$$

which, on multiplying both sides by e^y, becomes a quadratic in e^y, namely,

$$e^{2y} - 2xe^y - 1 = 0$$

Solving this we get

$$e^y = x \pm \sqrt{x^2 + 1}$$

where the minus sign must be discarded since e^y is always positive. Finally, therefore, we have

$$y = \sinh^{-1} x = \log(x + \sqrt{x^2 + 1}) \qquad (9)$$

which holds for all values of x.

PROBLEMS WITH ANSWERS

1. Show that $\operatorname{sech}^2 \theta = 1 - \tanh^2 \theta$.

2. Show that $\operatorname{csch}^2 \theta = \coth^2 \theta - 1$.

3. Show that $\sinh(x + y) = \sinh x \cosh y + \cosh x \sinh y$.

4. Sketch the graph of $y = \cosh x$.

5. Sketch the graph of $y = \tanh x$.

6. Show that $\sinh ix = i \sin x$.

7. Show that $\cosh ix = \cos x$.

8. Show that $\cosh^{-1} x = \log(x \pm \sqrt{x^2 - 1})$, $x \geqslant 1$.

9. Show that $\tanh^{-1} x = \dfrac{1}{2} \log \dfrac{1 + x}{1 - x}$, $-1 < x < 1$.

14.3   Differentiation and Integration of Hyperbolic Functions

By making use of the exponential definitions of the hyperbolic functions, the following rules of differentiation are easily verified.

$$\frac{d}{dx} \sinh u = \frac{d}{dx} \left(\frac{e^u - e^{-u}}{2} \right) \qquad (1)$$

$$= \frac{e^u + e^{-u}}{2} \frac{du}{dx}$$

$$= \cosh u \frac{du}{dx}$$

$$\frac{d}{dx} \cosh u = \sinh u \frac{du}{dx} \tag{2}$$

$$\frac{d}{dx} \tanh u = \operatorname{sech}^2 u \frac{du}{dx} \tag{3}$$

$$\frac{d}{dx} \coth u = -\operatorname{csch}^2 u \frac{du}{dx} \tag{4}$$

$$\frac{d}{dx} \operatorname{sech} u = -\operatorname{sech} u \tanh u \frac{du}{dx} \tag{5}$$

$$\frac{d}{dx} \operatorname{csch} u = -\operatorname{csch} u \coth u \frac{du}{dx} \tag{6}$$

Further, from the relations given previously, formula (9) and Problems 8 and 9, Section 14.2, for the inverse hyperbolic functions, it follows that

$$\frac{d}{dx} \sinh^{-1} u = \frac{1}{\sqrt{u^2 + 1}} \frac{du}{dx}, \text{ for all } u \tag{7}$$

$$\frac{d}{dx} \cosh^{-1} u = \frac{\pm 1}{\sqrt{u^2 - 1}} \frac{du}{dx}, \quad u > 1 \tag{8}$$

$$\frac{d}{dx} \tanh^{-1} u = \frac{1}{1 - u^2} \frac{du}{dx}, \quad -1 < u < 1 \tag{9}$$

To find the integrals of the elementary hyperbolic functions, we may make use of the exponential forms of the functions and, of course, the usual methods of integration.

$$\int \sinh u \, du = \tfrac{1}{2} \int (e^u - e^{-u}) \, du \tag{10}$$

$$= \tfrac{1}{2} (e^u + e^{-u}) + C$$

$$= \cosh u + C$$

$$\int \cosh u \, du = \sinh u + C \tag{11}$$

$$\int \tanh u \, du = \log \cosh u + C \tag{12}$$

$$\int \coth u \, du = \log \sinh u + C \tag{13}$$

Example 1. Find $\int \operatorname{sech} u \, du$.

Solution. Write

$$\operatorname{sech} u = \frac{1}{\cosh u}$$

$$= \frac{\cosh u}{\cosh^2 u}$$

$$= \frac{\cosh u}{1 + \sinh^2 u}$$

Thus,

$$\int \operatorname{sech} u \, du = \int \frac{\cosh u \, du}{1 + \sinh^2 u}$$

$$= \int \frac{d(\sinh u)}{1 + \sinh^2 u}$$

$$= \tan^{-1} (\sinh u) + C$$

Example 2. Find $\int x \cosh x \, dx$.

Solution. Integrating by parts, we get

$$\int x \cosh x \, dx = x \sinh x - \int \sinh x \, dx$$

$$= x \sinh x - \cosh x + C$$

Many of the integrals that we have already met can be expressed in terms of hyperbolic functions. For example, by a slight generalization of formulas (7), (8), and (9), above, it follows that

$$\int \frac{du}{\sqrt{u^2 + a^2}} = \sinh^{-1} \frac{u}{a} + C, \quad \text{for all } u \tag{14}$$

$$= \log (u + \sqrt{u^2 + a^2}) + C$$

$$\int \frac{du}{\sqrt{u^2 - a^2}} = \cosh^{-1} \frac{u}{a} + C, \quad u \geqslant a \tag{15}$$

$$= \log (u + \sqrt{u^2 - a^2}) + C$$

$$\int \frac{du}{a^2 - u^2} = \frac{1}{a} \tanh^{-1} \frac{u}{a} + C, \quad -a < u < a \tag{16}$$

$$= \frac{1}{2a} \log \frac{a + u}{a - u} + C$$

PROBLEMS WITH ANSWERS

1. Write the equation of the catenary $y = \dfrac{a}{2}(e^{x/a} + e^{-x/a})$ in terms of a hyperbolic function.

   Ans. $y = a \cosh \dfrac{x}{a}$

2. Show that $\dfrac{d}{dx} \coth^{-1} u = \dfrac{1}{1 - u^2} \dfrac{du}{dx}$, $u^2 > 1$.

3. Show that $\displaystyle\int \operatorname{csch} u \, du = \log \tanh \dfrac{u}{2} + C$.

4. Show that $\displaystyle\int \sqrt{u^2 + a^2} \, du = \dfrac{u}{2} \sqrt{u^2 + a^2} + \dfrac{a^2}{2} \sinh^{-1} \dfrac{u}{a} + C$.

CHAPTER 15
ELEMENTARY DIFFERENTIAL EQUATIONS

15.1 Separable Equations

The problem of integration met with in earlier chapters can be stated as follows: *Where f is a known function, solve the equation* $\dfrac{dy}{dx} = f(x)$ *for* $y(x)$. This is the simplest example of a differential equation, which is any equation involving a derivative; $D_x y = f(x)$ has the solution $y = \int f(x)\,dx + C$. In this chapter, we consider several types of differential equations and their solutions.

An equation of the form

$$\frac{dy}{dx} = f(x)\,g(y) \tag{1}$$

is called *separable* since it can be written in the form $\dfrac{dy}{g(y)} = f(x)\,dx$, which can be integrated immediately to give its solution,

$$\int \frac{dy}{g(y)} = \int f(x)\,dx + C \tag{2}$$

Example. Solve the differential equation $dy/dx = xy$.

Solution. This equation can be rewritten in the form $dy/y = x\,dx$; whence,

$$\int \frac{dy}{y} = \int x\,dx + C$$

or

$$\log y = \frac{x^2}{2} + C$$

is the solution. We could call $C = \log C'$, in which case the solution can be put in the form

$$\log \frac{y}{C'} = \frac{x^2}{2}$$

$$\frac{y}{C'} = e^{x^2/2}$$

$$y = C'e^{x^2/2}$$

15.2   Homogeneous Equations

If the identity $f(x,y) \equiv f(kx,ky)$ holds for x, y, and every number k, then f is said to be a *homogeneous* function of degree zero. For such a function f the differential equation

$$D_x y = f(x, y) \tag{1}$$

is called a **homogeneous differential equation**. The transformation $y = vx$ reduces (1) to a separable differential equation. This follows since if $y = vx$, then

$$D_x y = v + x D_x v = f(x, vx) = f(1, v)$$

The last equality is a result of the homogeneity of f, where x plays the role of k, and the equation separates into

$$D_x v = \left(\frac{1}{x}\right)(f(1, v) - v)$$

The solution is given by

$$\int \frac{dv}{f(1, v) - v} = \int \frac{dx}{x} + C \tag{2}$$

Example. Solve the equation

$$x \frac{dy}{dx} = y + \sqrt{x^2 + y^2}$$

Solution. We rewrite this equation in the form

$$\frac{dy}{dx} = \frac{y}{x} + \frac{\sqrt{x^2 + y^2}}{x}$$

Since $\dfrac{y}{x} + \dfrac{\sqrt{x^2 + y^2}}{x} \equiv \dfrac{ky}{kx} + \dfrac{\sqrt{k^2 x^2 + k^2 y^2}}{kx}$, the equation is homogeneous, and we set $y = vx$; we get

$$v + x\frac{dv}{dx} = \frac{vx}{x} + \frac{\sqrt{x^2 + v^2 x^2}}{x}$$

$$= v + \sqrt{1 + v^2}$$

$$x \frac{dv}{dx} = \sqrt{1 + v^2}$$

$$\int \frac{dv}{\sqrt{1 + v^2}} = \int \frac{dx}{x}$$

$$\log (v + \sqrt{1 + v^2}) = \log x + C$$

$$\log \left(\frac{y}{x} + \sqrt{1 + \frac{y^2}{x^2}} \right) = \log x + C$$

$$\log (y + \sqrt{x^2 + y^2}) = C$$

$$y + \sqrt{x^2 + y^2} = e^C = k$$

$$\sqrt{x^2 + y^2} = k - y$$

$$x^2 + y^2 = k^2 - 2ky + y^2$$

Finally, the solution takes the simple form, $x^2 = k^2 - 2ky$.

15.3 Linear Equations

Linear differential equations of the **first order** are of the form

$$\frac{dy}{dx} + Py = Q \tag{1}$$

where P and Q are functions of x. Here, integration can be performed by making the transformation

$$y = uz$$

where u and z are functions of x to be determined. Then

$$\frac{dy}{dx} = u \frac{dz}{dx} + z \frac{du}{dx} .$$

Substituting this in (1) yields

$$u \frac{dz}{dx} + z \frac{du}{dx} + Puz = Q$$

Hence,

$$u \frac{dz}{dx} + z \left(\frac{du}{dx} + Pu \right) = Q \tag{2}$$

Now we compute u such that

$$\frac{du}{dx} + Pu = 0$$

This is now separable, and any solution of it will suffice. Thus,

$$\frac{du}{u} = -P\,dx$$

or
$$\log u = -\int P\,dx$$

$$u = e^{-\int P dx}$$

When this is substituted into (2), we get

$$(e^{-\int P\,dx})\frac{dz}{dx} = Q$$

or
$$dz = Q\,e^{\int P dx}\,dx$$

and
$$z = \int Q\,e^{\int P dx}\,dx + C$$

Therefore, $$y = uz = e^{-\int P\,dx}\left(\int Q e^{\int P\,dx}dx + C\right) \qquad (3)$$

which is the general solution of (1).

Example. Find the general solution of $\dfrac{dy}{dx} - \dfrac{3}{x}y = x^3$.

Solution.
$$y = e^{\int 3\,dx/x}\left(\int x^3 e^{-\int 3 dx/x}\,dx + C\right)$$

$$= e^{\log x^3}\left(\int x^3 e^{-\log x^3}\,dx + C\right)$$

$$= x^3\left(\int \frac{x^3}{x^3}\,dx + C\right)$$

$$= x^4 + Cx^3$$

15.4   Equations Reducible to Linear Form

An equation of this type is of the form

$$\frac{dy}{dx} + Py = Qy^n \qquad (1)$$

The substitution $z = y^{-n+1}$ reduces it to

$$\frac{dz}{dx} + (1 - n)Pz = (1 - n)Q$$

This equation can be solved for z by (3) in the preceding section, when **appropriately** modified.

Example. Solve $\dfrac{dy}{dx} + \dfrac{y}{x} = y^2 \log x$.

Solution.

$$y^{-1} = e^{\int dx/x} \left(- \int \log x \cdot e^{-\int dx/x}\, dx + C \right)$$

$$= x \left(- \int \frac{\log x}{x}\, dx + C \right)$$

$$= x \left(- \tfrac{1}{2} \log^2 x + C \right)$$

$$\frac{2}{y} = -x \log^2 x + C'x, \quad C' = 2C$$

15.5 Exact Equations

An equation of this type is of the form

$$M(x, y)\, dx + N(x, y)\, dy = 0 \tag{1}$$

where $M = \dfrac{u}{x}$ and $N = \dfrac{u}{y}$, from which it follows that $\dfrac{M}{y} = \dfrac{N}{x}$. In this case, the integration is accomplished as follows.

$$u = \int M\, \partial x + f(y) \tag{2}$$

The x indicates that the integration is with respect to x; hence, the constant of integration is some function of y. Similarly, it must be the case that

$$u = \int N\, \partial y + g(x) \tag{3}$$

The two u's must be the same so that f and g can be readily determined. Then the general solution of (1) is $u = C$. For if $u = C$, then

$$du = \frac{u}{x}\, dx + \frac{u}{y}\, dy = 0$$

This is equation (1).

Example. Find the general solution of

$$\frac{2xy + 1}{y}\, dx + \frac{y - x}{y^2}\, dy = 0$$

Solution. If this is exact, then $\dfrac{\partial M}{\partial y} = \dfrac{\partial N}{\partial x}$, and we have

$$\frac{\partial}{\partial y}\left(\frac{2xy + 1}{y}\right) = -\frac{1}{y^2}$$

$$\frac{\partial}{\partial x}\left(\frac{y - x}{y^2}\right) = -\frac{1}{y^2}$$

The equation is therefore exact and

$$u = \int \frac{2xy + 1}{y}\, \partial x + f(y)$$

$$= \frac{x^2 y + x}{y} + f(y)$$

$$= x^2 + \frac{x}{y} + f(y)$$

and also

$$u = \int \frac{y - x}{y^2}\, \partial y + g(x)$$

$$= \int\left(\frac{1}{y} - \frac{x}{y^2}\right) \partial y + g(x)$$

$$= \log y + \frac{x}{y} + g(x)$$

If we take $f(y) = \log y$ and $g(x) = x^2$, then

$$u = x^2 + \frac{x}{y} + \log y$$

The general solution is $x^2 + \dfrac{x}{y} + \log y = C$.

15.6 Second-Order, Linear, Homogeneous Equations with Constant Coefficients

An equation of this type is of the form

$$a\frac{d^2 y}{dx^2} + b\frac{dy}{dx} + cy = 0, \quad a \neq 0 \tag{1}$$

Since the derivative of e^{mx} is me^{mx}, we guess that for some m, e^{mx} is a solution; hence,

$$am^2 e^{mx} + bme^{mx} + ce^{mx} = 0, \quad \text{or}$$

$$am^2 + bm + c = 0 \tag{2}$$

This is called the **auxiliary equation,** and its roots are

$$m_1 = \frac{-b + \sqrt{b^2 - 4ac}}{2a} \quad \text{and} \quad m_2 = \frac{-b - \sqrt{b^2 - 4ac}}{2a}$$

The general solution of (1) takes the form

$$y = C_1 e^{m_1 x} + C_2 e^{m_2 x} \tag{3}$$

where $b^2 - 4ac > 0$, $m_1 \neq m_2$, and C_1 and C_2 are arbitrary constants.

$$y = C_1 e^{mx} + C_2 x e^{mx}$$

where $b^2 - 4ac = 0$, $m_1 = m_2 = m$, and C_1 and C_2 are arbitrary constants.

$$y = e^{\alpha x} (A \cos \beta x + B \sin \beta x)$$

where $b^2 - 4ac < 0$, $m_1 = \alpha + i\beta$, $m_2 = \alpha - i\beta$ (or $m = \alpha \pm i\beta$), and A and B are arbitrary constants.

Example 1. Find the general solution of $3 \dfrac{d^2 y}{dx^2} + \dfrac{dy}{dx} - 2y = 0$.

Solution. The auxiliary equation is $3m^2 + m - 2 = 0$, $m_1 = \frac{2}{3}$, $m_2 = 2$, and the general solution is $y = C_1 e^{2x/3} + C_2 e^{2x}$.

Example 2. Find the general solution of $\dfrac{d^2 y}{dx^2} - 6 \dfrac{dy}{dx} + 9y = 0$.

Solution. The auxiliary equation is $m^2 - 6m + 9 = 0$, $m_1 = m_2 = 3$, and the general solution is $y = C_1 e^{3x} + C_2 x e^{3x}$.

Example 3. Find the general solution of $\dfrac{d^2 y}{dx^2} + \dfrac{dy}{dx} + y = 0$.

Solution. The auxiliary equation is $m^2 + m + 1 = 0$, $m = \alpha \pm i\beta = -\dfrac{1}{2} \pm i \dfrac{\sqrt{3}}{2}$, and the general solution is

$$y = e^{-x/2} \left(A \cos \frac{\sqrt{3}}{2} x + B \sin \frac{\sqrt{3}}{2} x \right)$$

PROBLEMS WITH SOLUTIONS

In Problems 1 through 13, find the general solution of each equation.

1. $\dfrac{dy}{dx} = \dfrac{10y}{3x(y - 5)}$.

Solution. $\dfrac{y-5}{y}\,dy = \dfrac{10}{3}\,\dfrac{dx}{x}$

$$y - 5\,\log y = \frac{10}{3}\,\log x + C$$

2. $\dfrac{dy}{dx} - y = \sin x.$

Solution. $y = e^{\int dx}\left(\int \sin x \cdot e^{-\int dx}\,dx + C\right)$

$\qquad\quad = e^x\left(\int e^{-x}\,\sin x\,dx + C\right)$

$\qquad\quad = e^x\left[\dfrac{e^{-x}\,(-\sin x - \cos x)}{2} + C\right]$

$\qquad\quad = Ce^x - \tfrac{1}{2}(\sin x + \cos x)$

3. $(xy^2 + x^2 y)\,dx + x^3\,dy = 0.$

Solution. This is a homogeneous equation; hence, the transformation $y = xv$ produces the separable equation

$$\frac{dx}{x} + \frac{dv}{v^2 + 2v} = 0$$

The solution is $\log\left(x\,\dfrac{v}{v+2}\right) = C.$ The general solution of the original equation is

$$\frac{xy}{y + 2x} = e^C = C'$$

4. An isotope A of iodine decays at a rate which is proportional to the amount present at any time. If only half of the original amount A_0 remains after 8 days, find the amount present at any time t.

Solution. $\qquad\qquad\qquad\qquad \dfrac{dA}{dt} = -kA$

or $\qquad\qquad\qquad\qquad\quad A = Ce^{-kt}$

The boundary conditions are

$\qquad\quad A = A_0, \quad$ when $t = 0, \quad$ therefore $C = A_0$ and

$\qquad\quad A = \tfrac{1}{2}A_0, \quad$ when $t = 8$

Hence,
$$\frac{A_0}{2} = A_0 e^{-8k}$$

or
$$k = \frac{1}{8} \log 2$$

Finally, $A = A_0 e^{(-t \log 2)/8}$.

5. The circuit illustrated in Fig. 15.1 has the equation

$$L \frac{di}{dt} + Ri = E$$

where L = inductance in henries, i = current in ampheres, t = time in seconds, R = resistance in ohms, and E = constant emf. Find the current as a function of time if no current is in the circuit at $t = 0$.

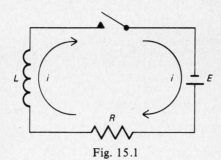

Fig. 15.1

Solution.
$$L \frac{di}{dt} = E - Ri$$

$$\frac{L \, di}{E - Ri} = dt$$

Integrate from 0 to t:

$$\frac{-L}{R} [\log (E - Ri) - \log E] = t$$

$$\log \left(\frac{E - Ri}{E} \right) = - \frac{Rt}{L}$$

$$\frac{E - Ri}{E} = e^{-Rt/L}$$

and
$$i = \frac{E}{R}(1 + e^{-Rt/L})$$

6. The motion of a simple pendulum is described by the equation

$$ML \frac{d^2\theta}{dt^2} + Mg \sin \theta = 0$$

Solution. For very small θ, $\sin \theta \cong \theta$ so that the motion is described approximately by

$$\frac{d^2\theta}{dt^2} + \frac{g}{L} \theta = 0 \qquad (1)$$

The auxiliary equation is $m^2 + \dfrac{g}{L} = 0$; and the general solution of (1) is

$$\theta = A \cos \mu t + B \sin \mu t \qquad (2)$$

where $\mu^2 = g/L$. Equation (2) can be rewritten in the form

$$\theta = \sqrt{A^2 + B^2} \left[\frac{A}{\sqrt{A^2 + B^2}} \cos \mu t + \frac{B}{\sqrt{A^2 + B^2}} \sin \mu t \right]$$

$$= \sqrt{A^2 + B^2} \, (\sin \alpha \cos \mu t + \cos \alpha \sin \mu t)$$

where $\sin \alpha = A/\sqrt{A^2 + B^2}$ and $\cos \alpha = B/\sqrt{A^2 + B^2}$. Thus, $\theta = C \sin (\mu t + \alpha)$, where C and α are arbitrary constants. This shows that the motion is periodic (a sine wave); hence, it is called *simple harmonic motion*.

7. $(x + y) \, dx + (2y + x) \, dy = 0$.

Solution. This is both homogeneous and exact. It can be written in the form

$$x \, dx + 2y \, dy + (y \, dx + x \, dy) = 0$$

Thus, $\qquad\qquad x \, dx + 2y \, dy + d(xy) = 0$

It follows that the general solution is

$$\frac{x^2}{2} + y^2 + xy = C$$

8. $(1 + y^2) \, dx - (xy + y + y^3) \, dy = 0$.

Solution. This equation is linear when written in the form

$$\frac{dx}{dy} - \frac{y}{1 + y^2} x = y$$

The general solution is

$$x = \exp\left(-\int \frac{y}{1+y^2}\,dy\right)\left[\int y \exp\left(\int \frac{y}{1+y^2}\,dy\right)dy + C\right]$$

$$= \exp\left(-\frac{1}{2}\log(1+y^2)\right)\left(\int y \exp\left[\frac{1}{2}\log(1+y^2)\,dy + C\right]\right)$$

$$= \frac{1}{\sqrt{1+y^2}}\left[\frac{1}{3}(1+y^2)^{3/2} + C\right]$$

$$= \tfrac{1}{3}(1+y^2) + C(1+y^2)^{-1/2}$$

9. $\dfrac{dy}{dx} - xy = -y^3 e^{-x^2}$.

Solution. This is reducible to a linear equation; therefore, the general solution is

$$y^{-2} = e^{-2\int x\,dx}\,(e^{2\int x\,dx}\,2e^{-x^2}\,dx + C)$$

$$= e^{-x^2}\left(\int 2\,dx + C\right)$$

$$= e^{-x^2}(2x + C)$$

10. $(x\,e^{y/x} + y)\,dx - x\,dy = 0$.

Solution. This equation is homogeneous; hence, we make the transformation $y = vx$, $dy = v\,dx + x\,dv$; then we have

$$x(e^v + v)\,dx - xv\,dx - x^2\,dv = 0$$

or

$$\frac{dx}{x} - \frac{dv}{e^v} = 0$$

$$\log x + e^{-v} = C$$

The general solution is

$$\log x + e^{-y/x} = C$$

11. $3\dfrac{d^2 y}{dx^2} + 5\dfrac{dy}{dx} - 2y = 0$.

Solution. The auxiliary equation is

$$(3m - 1)(m + 2) = 0$$

The general solution is

$$y = C_1 e^{x/3} + C_2 e^{-2x}$$

12. $\dfrac{d^2 y}{dx^2} - 10 \dfrac{dy}{dx} + 25 = 0.$

Solution. The auxiliary equation is $(m - 5)^2 = 0$, and the general solution is

$$y = C_1 e^{5x} + C_2 x e^{5x}$$

13. $\dfrac{d^2 y}{dx^2} + 4y = 0.$

Solution. The auxiliary equation is $m^2 + 4 = 0$, and $m_1 = 2i$ and $m_2 = -2i$. The general solution is, therefore,

$$y = C_1 \cos 2x + C_2 \sin 2x$$

PROBLEMS WITH ANSWERS

In the Problems 14 through 26, find the general solution of each equation.

14. $\dfrac{dy}{dx} + \dfrac{1}{x} y = x^2 .$ *Ans.* $y = \dfrac{1}{4} x^3 + \dfrac{C}{x}$

15. $e^y dx + (x e^y - 2y) dy = 0.$ *Ans.* $x e^y - y^2 = C$

16. $\dfrac{d^2 y}{dx^2} + 4 \dfrac{dy}{dx} + 4y = 0.$ *Ans.* $y = C_1 e^{-2x} + C_2 x e^{-2x}$

17. $\dfrac{dy}{dx} = \dfrac{-4}{y + 5}.$ *Ans.* $y^2 + 10y + 8x = C$

18. $(2x - 10y^3) \dfrac{dy}{dx} + y = 0.$ *Ans.* $x = 2y^3 + Cy^{-2}$

19. $\dfrac{d^2 y}{dx^2} + \dfrac{dy}{dx} + 7y = 0.$ *Ans.* $y = e^{-x/2} \left(A \cos \dfrac{\sqrt{27}}{2} + B \sin \dfrac{\sqrt{27}}{2} \right)$

20. $\dfrac{dy}{dx} = \dfrac{y - x}{y + x}.$ *Ans.* $\log (x^2 + y^2) + 2 \operatorname{Tan}^{-1} \dfrac{y}{x} = C$

21. $\sin x \cos y \, dx + \sin y \, dy = 0.$ *Ans.* $\cos x \cos y = C$

22. $(x + 2y)x \, dx + xy \, dy = 0.$ *Ans.* $\log (x + y) + \dfrac{x}{x + y} = C$

23. $\dfrac{dy}{dx} + \dfrac{y}{x} = \sin x.$ *Ans.* $y = \dfrac{\sin x}{x} - \cos x + \dfrac{C}{x}$

24. $\dfrac{d^2 y}{dx^2} - 6\dfrac{dy}{dx} + 25y = 0.$ *Ans.* $y = e^{3x} (A \cos 4x + B \sin 4x)$

25. $x \, dy - y \, dx = x\sqrt{x^2 - y^2} \, dx.$ *Ans.* $\operatorname{Sin}^{-1} \dfrac{y}{x} = y + C$

26. $\dfrac{(x + y) \, dx - (x - y) \, dy}{x^2 + y^2} = 0.$ *Ans.* $\log (x^2 + y^2) + \operatorname{Tan}^{-1} \dfrac{x}{y} = C$

APPENDIX A
SAMPLE EXAMINATIONS
DIFFERENTIAL CALCULUS

EXAMINATION I

1. Find $\dfrac{dy}{dx}$ in each case:

   a. $x = e^{\tan y}$.

   b. $y = \sqrt{\dfrac{x^2 + 2}{1 - 3x}}$.

   c. $y = \sec(5x^2 - 7)$.

   d. $x = \theta + \sin\theta,\ y = 1 - \cos\theta$.

   e. $y = 2^x + \operatorname{Sin}^{-1}(1 - 2x)$.

2. A cistern is in the shape of an inverted cone (vertex down) with its diameter equal to its height, each being 10 ft. How fast is the water pouring in when it is 3 ft. deep and rising at the rate of 4 in./min.? (Leave the answer in terms of π.)

3. Find the maximum and minimum values of

$$y = \frac{1 + x + x^2}{1 - x + x^2}$$

4. A particle is projected vertically upward with an initial velocity of 640 ft./sec. The height y after t sec. is given by $y = 640t - 16t^2$.

   a. How high will the particle rise?

   b. What will the particle's velocity be when it strikes the ground?

   c. How long will the particle remain in the air?

5. Find the angle at which the curve $x^3 - 3xy^2 - 2y + 4 = 0$ cuts the Y-axis.

6. The height h of a tower is deduced from observing the angular elevation θ at a fixed distance b from the foot of the tower. Find:

   a. The error due to a small error in the observed value of θ

   b. The relative error

   c. The error in calculated height, if $b = 100$ ft., $\theta = 30°$, and the error in the angle θ is $1'$ ($= 0.0003$ rad.)

7. Sketch the curve $\rho = \dfrac{1}{1 - \cos \theta}$, and find its slope at $\theta = \dfrac{\pi}{3}$.

8. Find the value of the curvature of $x^{1/2} + y^{1/2} = 2$ at the point $(1, 1)$.

9. Compute:

a. $\lim\limits_{x \to \infty} \dfrac{e^x}{x^n}$, n a positive integer

b. $\lim\limits_{x \to 0} \dfrac{\log x}{\log \sin x}$

10. A particle moves along the parabola $y^2 = 4x$. At a particular point (x, y), the velocity component in the y direction is 1. Find the velocity component in the x direction and the tangential velocity.

11. Solve the differential equation $\dfrac{d^2 y}{dx^2} + 16y = 0$.

12. Find the angle θ between the two vectors $\mathbf{A} = 9\mathbf{i} + \mathbf{j} - 6\mathbf{k}$ and $\mathbf{B} = 4\mathbf{i} - 6\mathbf{j} + 5\mathbf{k}$.

EXAMINATION II

1. Find $\dfrac{dy}{dx}$ in each case:

a. $y = (1 - x^2)^3$
b. $y = \log(1 - x)$
c. $y = \sin e^{-x}$
d. $y = \csc^3 2x$
e. $y = (1 - x)^2 (x^3 + 1)$
f. $xy^2 + \log(x + y) = 1$

2. Find the minimum value of $y = x^2 + \dfrac{1}{x^2}$.

3. Find (a) the slope of the curve $x = a(\theta - \sin \theta)$, $y = a(1 - \cos \theta)$ at $\theta = 60°$ and (b) the slope of $\rho = a(1 - \cos \theta)$ at $\theta = \dfrac{\pi}{6}$.

4. The diameter of a circle is found by measurement to be 5.2 in. with a maximum error of 0.05 in. Find the percentage error made in computing the area.

5. A rectangular storage space adjacent to a factory is to contain 7200 sq. ft. and is to be fenced off. If no fence is needed along the factory wall, what must be the dimensions requiring the least amount of fencing?

6. Evaluate:

a. $\lim\limits_{x \to \infty} \dfrac{\log x^n}{x^2}$

b. $\lim\limits_{x \to \frac{\pi}{2}} \left(x - \dfrac{\pi}{2} \right) \tan x$

7. Find the radius of curvature of $y = \sin x$ at $x = \dfrac{\pi}{3}$.

8. A point P moves in accord with the equations $x = a \cos t$, $y = b \sin t$.

    a. Find the least positive value of t for which the speed of P is a maximum.

    b. At that instant, find the magnitude and direction of the acceleration vector.

9. Find the equation of the line normal to the curve $y = x^3 + 6x^2 - 2x + 8$ at the point of inflection.

10. Solve the differential equation $xy' - y \log y = x^2 y$.

11. Find a vector that is perpendicular to the plane determined by the line $x - 2 = y + 1 = 2z$ and the point $(2, -6, 1)$.

INTEGRAL CALCULUS

EXAMINATION III

1. Solve the following integrals:

    a. $\displaystyle\int x \tan x^2 \, dx$

    b. $\displaystyle\int \dfrac{2x + 1}{(x - 1)(x + 2)} \, dx$

    c. $\displaystyle\int_0^{1/2} 3xe^{2x} \, dx$

2. Find the area under one arch of the curve $x = a\theta$, $y = a(1 - \cos \theta)$.

3. A trough 10 ft. long, whose cross section is a triangle 4 ft. deep and 6 ft. across the top, is full of water. Find the total force on one end.

4. Find the value of $\bar{x}$ for the area bounded by the circles $\rho = 4 \cos \theta$ and $\rho = 8 \cos \theta$.

5. Find the moment of inertia about the X-axis of the solid formed by revolving the portion of the curve $2y - 4 + x^2 = 0$, lying in the first quadrant, about the Y-axis.

6. Expand $\cos x$ in a Taylor series in powers of $\left(x - \dfrac{\pi}{4}\right)$.

7. Using Simpson's rule and four strips, find the approximate area bounded by $y = \dfrac{1}{100} x^4$, $x = 1$, $y = 0$, and $x = 5$.

8. Find (a) the equation of the plane tangent to $x^2 + y^2 + z^2 = 14$ at $(1, 2, 3)$, and (b) the directional derivative of $z = xy^2$ at $(1, 2, 4)$ in the direction $\alpha = \text{Tan}^{-1} 2$.

9. Find (a) the sum of the series $1 - \dfrac{1}{2!} + \dfrac{1}{3!} - \dfrac{1}{4!} + \cdots$ correct to the first three decimal places and (b) the interval of convergence of the series $x - \dfrac{x^2}{2} + \dfrac{x^3}{3} - \dfrac{x^4}{4} + \cdots$, and test the endpoints.

10. Use cylindrical coordinates and triple integration to find the volume which lies inside $x^2 + y^2 = z$ and outside $x^2 + y^2 = 4(z - 1)$.

11. Solve the differential equation $(2xy^{-3} - 3x^2)\,dx + 3(1 - x^2)y^{-4}\,dy = 0$.

12. Find equations of the line tangent to $x = t^3 - 2, y = t^4, z = t^2$ at the point $(6, 16, 4)$.

EXAMINATION IV

1. Evaluate:

a. $\displaystyle\int 3x\sqrt{1 - 2x^2}\,dx$

b. $\displaystyle\int \sin^2(2x - 5)\,dx$

c. $\displaystyle\int_0^{2\sqrt{2}} (8x^2 - x^4)\,dx$

2. Find the area, regardless of sign, enclosed between $y = \sin x$ and $y = \cos x$ between consecutive points of intersection.

3. Find the area of the surface of revolution generated by revolving the arc of $y = x^2$ about the Y-axis from $(0, 0)$ to $(2, 4)$.

4. A vertical cylindrical cistern of radius 8 ft. and depth 10 ft. is full of an oil weighing w lbs./cu. ft. Calculate the work necessary to pump the oil to a height of 15 ft. above the top of the cistern.

5. Find the length of that part of the curve $9y^2 = 4x^3$ joining $(0, 0)$ and $(3, 2\sqrt{3})$.

6. Find the value of $\bar{y}$ for the area of the segment of the ellipse $x^2 + 4y^2 = 4$ in the first quadrant cut off by the line $x + 2y - 2 = 0$.

7. The resistance R (ohms) of a circuit was found by using the formula $R = E/i$, where E = voltage (volts) and i = current (amperes). If there is an error of 0.1 amp. in reading $i = 10$ and 0.1 volt in reading $E = 100$, what is the approximate maximum error in the computed value of R?

8. The volume of a cylinder is increasing at the rate of 2π cu. ft./min. How is the radius changing when $r = 4$ ft. and $h = 20$ ft., if the height is increasing at the rate of 6 ft./min.?

9. Expand $\tan x$ in the neighborhood of $x = \dfrac{\pi}{4}$, finding the first four terms.

10. a. Set up the triple integral for the volume of the ellipsoid $\dfrac{x^2}{a^2} + \dfrac{y^2}{b^2} + \dfrac{z^2}{c^2} = 1$.

b. A volume is generated by revolving the area of the ellipse $\dfrac{(x-h)^2}{a^2} + \dfrac{(y-k)^2}{b^2} = 1$ about the line $x = h - a$. Use Pappus' theorem to find the volume, given the area of the ellipse $A = \pi ab$.

11. Solve the differential equation $(x^2 + 1)y' - xy = 1$.

12. Find an equation of the plane tangent to the surface $(x-2)^2 + (y-1)^2 + z^2 = 14$ at the point $(3, 3, 3)$.

ANSWERS TO EXAMINATIONS

Examination I

1. a. $\dfrac{1}{x} \cos^2 y$ d. $\dfrac{\sin \theta}{1 + \cos \theta}$

    b. $\dfrac{1}{2} \sqrt{\dfrac{1 - 3x}{x^2 + 2}} \left(\dfrac{6 + 2x - 3x^2}{(1 - 3x)^2} \right)$ e. $2^x \log 2 - \dfrac{1}{\sqrt{x - x^2}}$

    c. $10x \sec(5x^2 - 7) \tan(5x^2 - 7)$

2. $\dfrac{3}{4}\pi$ cu. ft./min.

3. max. 3, min. $\dfrac{1}{3}$

4. a. 6400 ft., b. -640 ft./sec., c. 40 sec.

5. $\text{Cot}^{-1} 6$

6. a. $dh = b \sec^2 \theta \, d\theta$, b. $\dfrac{dh}{h} = \sec \theta \csc \theta \, d\theta$, c. 0.04 ft.

7. $\dfrac{1}{3}\sqrt{3}$

8. $\dfrac{1}{4}\sqrt{2}$

9. a. ∞; b. 1

10. $\dfrac{y}{2}, \sqrt{x + 1}$

11. $y = C_1 \cos 4x + C_2 \sin 4x$

12. $\theta = \pi/2$

Examination II

1. a. $-6x(1 - x^2)^2$ b. $\dfrac{1}{x - 1}$

c. $-e^{-x} \cos e^{-x}$

e. $3x^2(1-x)^2 - 2(1-x)(x^3+1)$

d. $-6 \csc^3 2x \cot 2x$

f. $-\dfrac{1+y^2(x+y)}{1+2xy(x+y)}$

2. 2
3. (a) $\sqrt{3}$, (b) 1
4. 1.92%.
5. 60 ft. $\times$ 120 ft. (long side along the factory)
6. a. 0, b. -1
7. $\dfrac{\sqrt{3}}{12}(5)^{3/2}$
8. a. $t = \dfrac{\pi}{2}$, b. $|a| = b, \phi = \dfrac{\pi}{2}$
9. $x - 14y + 394 = 0$
10. $\log y = x^2 + Cx$
11. $19\mathbf{i} + 5\mathbf{j} - 10\mathbf{k}$

Examination III

1. a. $-\frac{1}{2} \log \cos x^2 + C$
   b. $\log(x-1)(x+2) + C$
   c. $\frac{3}{4}$
2. $2\pi a^2$ sq. units
3. $16w$
4. $\frac{14}{3}$
5. $\frac{16}{3}\pi$
6. $\cos x = \dfrac{\sqrt{2}}{2}\left[1 - \dfrac{\left(x - \dfrac{\pi}{4}\right)}{1!} - \dfrac{\left(x - \dfrac{\pi}{4}\right)^2}{2!} + \dfrac{\left(x - \dfrac{\pi}{4}\right)^3}{3!} + \cdots\right]$
7. 6.253 sq. units
8. (a) $x + 2y + 3z - 14 = 0$, (b) $\frac{12}{5}\sqrt{5}$
9. (a) 0.632, (b) $-1 < x \leqslant 1$
10. $\frac{2}{3}\pi$ cu. units
11. $(x^2 - 1)y^{-3} - x^3 = C$
12. $\dfrac{x-6}{3} = \dfrac{y-16}{4} = \dfrac{z-4}{2}$

Examination IV

1. a. $-\frac{1}{2}(1 - 2x^2)^{3/2} + C$

    b. $\frac{1}{2} x - \frac{1}{8} \sin 2 (2x - 5) + C$

    c. $\frac{256}{15} \sqrt{2}$

2. $2\sqrt{2}$ sq. units

3. $\frac{\pi}{6} (17\sqrt{17} - 1)$ sq. units

4. $12,800 \pi w$ ft.-lbs.

5. $\frac{14}{3}$

6. $\bar{y} = \dfrac{2}{3(\pi - 2)}$

7. 0.11 ohm

8. Decreasing at the rate of $\frac{47}{80}$ ft./min.

9. $\tan x = 1 + 2 \left(x - \dfrac{\pi}{4} \right) + 2 \left(x - \dfrac{\pi}{4} \right)^2 + \dfrac{8}{3} \left(x - \dfrac{\pi}{4} \right)^3 + \cdots$

10. a. $V = 8 \displaystyle\int_0^a \int_0^{b\sqrt{1 - \frac{x^2}{a^2}}} \int_0^{c\sqrt{1 - \frac{x^2}{a^2} - \frac{y^2}{b^2}}} dz\, dy\, dx$

    b. $2\pi^2 a^2 b$ cu. units

11. $(x - y)^2 = C(x^2 + 1)$

12. $x - 2y + 3z = -6$

APPENDIX B
TABLE OF INTEGRALS

Note: Constants of integration have been omitted.

1. $\int u^n \, du = \dfrac{u^{n+1}}{n+1}$, $n \neq -1$

2. $\int \dfrac{du}{u} = \log u$

3. $\int \dfrac{du}{u(a+bu)} = -\dfrac{1}{a} \log \left(\dfrac{a+bu}{u} \right)$

4. $\int \dfrac{du}{u(a+bu)^2} = \dfrac{1}{a(a+bu)} - \dfrac{1}{a^2} \log \left(\dfrac{a+bu}{u} \right)$

5. $\int \dfrac{du}{u^2(a+bu)} = -\dfrac{1}{au} + \dfrac{b}{a^2} \log \left(\dfrac{a+bu}{u} \right)$

6. $\int u \sqrt{a+bu} \, du = -\dfrac{2(2a-3bu)(a+bu)^{3/2}}{15b^2}$

7. $\int u^2 \sqrt{a+bu} \, du = \dfrac{2(8a^2 - 12abu + 15b^2u^2)(a+bu)^{3/2}}{105b^3}$

8. $\int u^m \sqrt{a+bu} \, du = \dfrac{2u^m(a+bu)^{3/2}}{b(2m+3)} - \dfrac{2am}{b(2m+3)} \int u^{m-1} \sqrt{a+bu} \, du$

9. $\int \dfrac{u \, du}{\sqrt{a+bu}} = -\dfrac{2(2a-bu)\sqrt{a+bu}}{3b^2}$

10. $\int \dfrac{u^2 \, du}{\sqrt{a+bu}} = \dfrac{2(8a^2 - 4abu + 3b^2u^2)\sqrt{a+bu}}{15b^3}$

11. $\int \dfrac{u^m \, du}{\sqrt{a+bu}} = \dfrac{2u^m \sqrt{a+bu}}{b(2m+1)} - \dfrac{2am}{b(2m+1)} \int \dfrac{u^{m-1} \, du}{\sqrt{a+bu}}$

12. $\displaystyle\int \frac{du}{u\sqrt{a + bu}} = \frac{2}{\sqrt{-a}} \tan^{-1} \sqrt{\frac{a + bu}{-a}}$, for $a < 0$

13. $\displaystyle\int \frac{du}{u\sqrt{a + bu}} = \frac{1}{\sqrt{a}} \log \frac{\sqrt{a + bu} - \sqrt{a}}{\sqrt{a + bu} + \sqrt{a}}$, for $a > 0$

14. $\displaystyle\int \frac{du}{u^m\sqrt{a + bu}} = - \frac{\sqrt{a + bu}}{a(m - 1)u^{m-1}} - \frac{b(2m - 3)}{2a(m - 1)}\int \frac{du}{u^{m-1}\sqrt{a + bu}}$

15. $\displaystyle\int \frac{\sqrt{a + bu}\, du}{u} = 2\sqrt{a + bu} + a \int \frac{du}{u\sqrt{a + bu}}$

16. $\displaystyle\int \frac{\sqrt{a + bu}\, du}{u^m} = - \frac{(a + bu)^{3/2}}{a(m - 1)u^{m-1}} - \frac{b(2m - 5)}{2a(m - 1)}\int \frac{\sqrt{a + bu}\, du}{u^{m-1}}$

17. $\displaystyle\int \sqrt{2au - u^2}\, du = \frac{u - a}{2}\sqrt{2au - u^2} + \frac{a^2}{2}\cos^{-1}\frac{a - u}{a}$

18. $\displaystyle\int u\sqrt{2au - u^2}\, du = - \frac{3a^2 + au - 2u^2}{6}\sqrt{2au - u^2}$

$$+ \frac{a^3}{2}\cos^{-1}\frac{a - u}{a}$$

19. $\displaystyle\int u^m\sqrt{2au - u^2}\, du = - \frac{u^{m-1}(2au - u^2)^{3/2}}{m + 2}$

$$+ \frac{a(2m + 1)}{m + 2}\int u^{m-1}\sqrt{2au - u^2}\, du$$

20. $\displaystyle\int \frac{\sqrt{2au - u^2}\, du}{u} = \sqrt{2au - u^2} + a\cos^{-1}\frac{a - u}{a}$

21. $\displaystyle\int \frac{\sqrt{2au - u^2}\, du}{u^m} = - \frac{(2au - u^2)^{3/2}}{a(2m - 3)u^m} + \frac{m - 3}{a(2m - 3)}\int \frac{\sqrt{2au - u^2}\, du}{u^{m-1}}$

22. $\displaystyle\int \frac{du}{\sqrt{2au - u^2}} = \cos^{-1}\frac{a - u}{a}$

23. $\displaystyle\int \frac{du}{\sqrt{2au + u^2}} = \log(u + a + \sqrt{2au + u^2})$

24. $\displaystyle\int \frac{u\,du}{\sqrt{2au - u^2}} = -\sqrt{2au - u^2} + a\,\cos^{-1}\frac{a - u}{a}$

25. $\displaystyle\int \frac{du}{u\sqrt{2au - u^2}} = -\frac{\sqrt{2au - u^2}}{au}$

26. $\displaystyle\int \sqrt{\frac{a + u}{b + u}}\,du = \sqrt{(a + u)(b + u)} + (a - b)\log\left(\sqrt{a + u} + \sqrt{b + u}\right)$

27. $\displaystyle\int \sqrt{\frac{a - u}{b + u}}\,du = \sqrt{(a - u)(b + u)} + (a + b)\sin^{-1}\sqrt{\frac{u + b}{a + b}}$

28. $\displaystyle\int \sqrt{\frac{a + u}{b - u}}\,du = -\sqrt{(a + u)(b - u)} - (a + b)\sin^{-1}\sqrt{\frac{b - u}{a + b}}$

29. $\displaystyle\int \frac{du}{\sqrt{(u - a)(b - u)}} = 2\sin^{-1}\sqrt{\frac{u - a}{b - a}}$

30. $\displaystyle\int \frac{du}{a^2 + u^2} = \frac{1}{a}\tan^{-1}\frac{u}{a}$

31. $\displaystyle\int \frac{du}{a^2 - u^2} = \frac{1}{2a}\log\frac{a + u}{a - u},\ u^2 < a^2$

32. $\displaystyle\int \frac{du}{u^2 - a^2} = \frac{1}{2a}\log\frac{u - a}{u + a},\ u^2 > a^2$

33. $\displaystyle\int \frac{du}{(u^2 + a^2)^n} = \frac{1}{2(n - 1)a^2}\left[\frac{u}{(u^2 + a^2)^{n-1}} + (2n - 3)\int\frac{du}{(u^2 + a^2)^{n-1}}\right]$

34. $\displaystyle\int (u^2 \pm a^2)^{1/2}\,du = \frac{u}{2}\sqrt{u^2 \pm a^2} \pm \frac{a^2}{2}\log\left(u + \sqrt{u^2 \pm a^2}\right)$

35. $\displaystyle\int \frac{du}{(u^2 \pm a^2)^{1/2}} = \log\left(u + \sqrt{u^2 \pm a^2}\right)$

36. $\displaystyle\int \frac{u^2\,du}{(u^2 \pm a^2)^{1/2}} = \frac{u}{2}\sqrt{u^2 \pm a^2} - \frac{\pm a^2}{2}\log\left(u + \sqrt{u^2 \pm a^2}\right)$

37. $\displaystyle\int \frac{u^2\,du}{(u^2 \pm a^2)^{3/2}} = -\frac{u}{\sqrt{u^2 \pm a^2}} + \log\left(u + \sqrt{u^2 \pm a^2}\right)$

38. $\int \dfrac{du}{u(u^2 + a^2)^{1/2}} = -\dfrac{1}{a} \log \left(\dfrac{a + \sqrt{u^2 + a^2}}{u} \right)$

39. $\int \dfrac{du}{u(u^2 - a^2)^{1/2}} = \dfrac{1}{a} \sec^{-1} \dfrac{u}{a}$

40. $\int \dfrac{du}{u^2(u^2 \pm a^2)^{1/2}} = -\dfrac{\sqrt{u^2 \pm a^2}}{\pm a^2 u}$

41. $\int \dfrac{(u^2 + a^2)^{1/2} \, du}{u} = \sqrt{u^2 + a^2} - a \log \left(\dfrac{a + \sqrt{u^2 + a^2}}{u} \right)$

42. $\int \dfrac{(u^2 - a^2)^{1/2} \, du}{u} = \sqrt{u^2 - a^2} - a \sec^{-1} \dfrac{u}{a}$

43. $\int \dfrac{(u^2 \pm a^2)^{1/2} \, du}{u^2} = -\dfrac{\sqrt{u^2 \pm a^2}}{u} + \log(u + \sqrt{u^2 \pm a^2})$

44. $\int (a^2 - u^2)^{1/2} \, du = \dfrac{u}{2} \sqrt{a^2 - u^2} + \dfrac{a^2}{2} \sin^{-1} \dfrac{u}{a}$

45. $\int \dfrac{du}{(a^2 - u^2)^{1/2}} = \sin^{-1} \dfrac{u}{a}$

46. $\int \dfrac{du}{(a^2 - u^2)^{3/2}} = \dfrac{u}{a^2 \sqrt{a^2 - u^2}}$

47. $\int \dfrac{u^2 \, du}{(a^2 - u^2)^{1/2}} = -\dfrac{u}{2} \sqrt{a^2 - u^2} + \dfrac{a^2}{2} \sin^{-1} \dfrac{u}{a}$

48. $\int (a^2 - u^2)^{3/2} \, du = \dfrac{u}{4}(a^2 - u^2)^{3/2} + \dfrac{3a^2 u}{8} \sqrt{a^2 - u^2} + \dfrac{3a^4}{8} \sin^{-1} \dfrac{u}{a}$

49. $\int u^2 (a^2 - u^2)^{1/2} \, du = \dfrac{u}{8}(2u^2 - a^2) \sqrt{a^2 - u^2} + \dfrac{a^4}{8} \sin^{-1} \dfrac{u}{a}$

50. $\int e^{au} \, du = \dfrac{e^{au}}{a}$

51. $\displaystyle\int b^{au}\, du = \frac{b^{au}}{a \log b}$

52. $\displaystyle\int u e^{au}\, du = \frac{e^{au}}{a^2}(au - 1)$

53. $\displaystyle\int u^n e^{au}\, du = \frac{u^n e^{au}}{a} - \frac{n}{a}\int u^{n-1} e^{au}\, du$

54. $\displaystyle\int \log u\, du = u \log u - u$

55. $\displaystyle\int u^n \log u\, du = u^{n+1}\left[\frac{\log u}{n+1} - \frac{1}{(n+1)^2}\right]$

56. $\displaystyle\int e^{au} \log u\, du = \frac{e^{au} \log u}{a} - \frac{1}{a}\int \frac{e^{au}}{u}\, du$

57. $\displaystyle\int \frac{du}{u \log u} = \log(\log u)$

58. $\displaystyle\int \sin u\, du = -\cos u$

59. $\displaystyle\int \cos u\, du = \sin u$

60. $\displaystyle\int \tan u\, du = \log \sec u$

61. $\displaystyle\int \cot u\, du = \log \sin u$

62. $\displaystyle\int \sec u\, du = \log(\sec u + \tan u)$

$$= \log \tan\left(\frac{u}{2} + \frac{\pi}{4}\right)$$

63. $\int \csc u \, du = \log (\csc u - \cot u)$

$$= \log \tan \frac{u}{2}$$

64. $\int \sin^2 u \, du = \frac{u}{2} - \frac{1}{4} \sin 2u$

65. $\int \sin^3 u \, du = - \frac{1}{3} \cos u \, (\sin^2 u + 2)$

66. $\int \sin^4 u \, du = \frac{3}{8} u - \frac{\sin 2u}{4} + \frac{\sin 4u}{32}$

67. $\int \sin^n u \, du = - \frac{\sin^{n-1} u \cos u}{n} + \frac{n-1}{n} \int \sin^{n-2} u \, du$

68. $\int \cos^2 u \, du = \frac{u}{2} + \frac{1}{4} \sin 2u$

69. $\int \cos^3 u \, du = \frac{1}{3} \sin u \, (\cos^2 u + 2)$

70. $\int \cos^4 u \, du = \frac{3}{8} u + \frac{\sin 2u}{4} + \frac{\sin 4u}{32}$

71. $\int \cos^n u \, du = \frac{\cos^{n-1} u \sin u}{n} + \frac{n-1}{n} \int \cos^{n-2} u \, du$

72. $\int \tan^n u \, du = \frac{\tan^{n-1} u}{n-1} - \int \tan^{n-2} u \, du$

73. $\int \cot^n u \, du = - \frac{\cot^{n-1} u}{n-1} - \int \cot^{n-2} u \, du$

74. $\int \sec^n u \, du = \frac{\sin u}{(n-1) \cos^{n-1} u} + \frac{n-2}{n-1} \int \sec^{n-2} u \, du$

75. $\int \csc^n u \, du = - \frac{\cos u}{(n-1) \sin^{n-1} u} + \frac{n-2}{n-1} \int \csc^{n-2} u \, du$

76. $\int \cos^m u \sin^n u \, du = \dfrac{\cos^{m-1} u \sin^{n+1} u}{m+n}$

$$+ \frac{m-1}{m+n} \int \cos^{m-2} u \sin^n u \, du$$

77. $\int \cos^m u \sin^n u \, du = -\dfrac{\sin^{n-1} u \cos^{m+1} u}{m+n}$

$$+ \frac{n-1}{m+n} \int \cos^m u \sin^{n-2} u \, du$$

78. $\int \sin mu \sin nu \, du = -\dfrac{\sin(m+n)u}{2(m+n)} + \dfrac{\sin(m-n)u}{2(m-n)}$

79. $\int \cos mu \cos nu \, du = \dfrac{\sin(m+n)u}{2(m+n)} + \dfrac{\sin(m-n)u}{2(m-n)}$

80. $\int \sin mu \cos nu \, du = -\dfrac{\cos(m+n)u}{2(m+n)} - \dfrac{\cos(m-n)u}{2(m-n)}$

81. $\int u \sin u \, du = \sin u - u \cos u$

82. $\int u^2 \sin u \, du = 2u \sin u - (u^2 - 2) \cos u$

83. $\int u^m \sin au \, du = \dfrac{u^{m-1}}{a^2} (m \sin au - au \cos au)$

$$- \frac{m(m-1)}{a^2} \int u^{m-2} \sin au \, du$$

84. $\int u \cos u \, du = \cos u + u \sin u$

85. $\int u^2 \cos u \, du = 2u \cos u + (u^2 - 2) \sin u$

86. $\int u^m \cos au\, du = \dfrac{u^{m-1}}{a^2}(au \sin au + m \cos au)$

$$- \frac{m(m-1)}{a^2}\int u^{m-2} \cos au\, du$$

87. $\int e^{au} \sin nu\, du = \dfrac{e^{au}(a \sin nu - n \cos nu)}{a^2 + n^2}$

88. $\int e^{au} \cos nu\, du = \dfrac{e^{au}(n \sin nu + a \cos nu)}{a^2 + n^2}$

89. $\int \sin^{-1}\dfrac{u}{a}\,du = u \sin^{-1}\dfrac{u}{a} + \sqrt{a^2 - u^2}$

90. $\int \cos^{-1}\dfrac{u}{a}\,du = u \cos^{-1}\dfrac{u}{a} - \sqrt{a^2 - u^2}$

91. $\int \tan^{-1}\dfrac{u}{a}\,du = u \tan^{-1}\dfrac{u}{a} - \dfrac{a}{2} \log(a^2 + u^2)$

92. $\int \cot^{-1}\dfrac{u}{a}\,du = u \cot^{-1}\dfrac{u}{a} + \dfrac{a}{2} \log(a^2 + u^2)$

93. $\int \sec^{-1}\dfrac{u}{a}\,du = u \sec^{-1}\dfrac{u}{a} - a \log(u + \sqrt{u^2 - a^2})$

94. $\int \csc^{-1}\dfrac{u}{a}\,du = u \csc^{-1}\dfrac{u}{a} + a \log(u + \sqrt{u^2 - a^2})$

95. $\int \sinh u\, du = \cosh u$

96. $\int \cosh u\, du = \sinh u$

97. $\int \tanh u\, du = \log \cosh u$

98. $\displaystyle\int \coth u \; du = \log \sinh u$

99. $\displaystyle\int \text{sech} \; u \; du = 2 \tan^{-1} e^u = \tan^{-1} \sinh u$

100. $\displaystyle\int \text{csch} \; u \; du = \log \tanh \frac{u}{2} = \tan^{-1} \cosh u$

Index